HEINEMANN MODULAR MATHEMATICS
for
LONDON AS AND A-LEVEL

Decision Mathematics 1

John Hebborn   Keith Parramore   Joan Stephens

1 Algorithms 1
2 Algorithms on graphs 21
3 Decision making [illegible] 73
4 Critical path analysis 99
5 Flows in networks 133
6 Linear programming 159
7 Matchings 203

Appendix 1: Further work 223

Examination style papers 239

1
2
3
4
5
6
7

Heinemann Educational Publishers,
Halley Court, Jordan Hill, Oxford, OX2 8EJ
a division of Reed Educational & Professional Publishing Ltd

OXFORD FLORENCE PRAGUE MADRID ATHENS
MELBOURNE AUCKLAND KUALA LUMPUR SINGAPORE
TOKYO IBADAN NAIROBI KAMPALA JOHANNESBURG
GABORONE PORTSMOUTH NH (USA) CHICAGO MEXICO CITY
SAO PAULO

Editoral adviser: Susie Jameson

First published 1997

97 98 99 10 9 8 7 6 5 4 3

ISBN 0435 51813 5

Original design by Geoffrey Wadsley: additional design work by Jim Turner

Typeset and illustrated by Tech-Set Limited, Gateshead, Tyne & Wear

Printed in Great Britain by The Bath Press, Bath

About this book

This book is designed to provide you with the best preparation possible for your London Modular Mathematics D1 examination. The series authors are examiners and exam moderators themselves and have a good understanding of the exam board's requirements.

Finding your way around

To help to find your way around when you are studying and revising use the:

- **edge marks** (shown on the front page) – these help you to get to the right chapter quickly;
- **contents list** – this lists the headings that identify key syllabus ideas covered in the book so you can turn straight to them;
- **index** – if you need to find a topic the **bold** number shows where to find the main entry on a topic.

Remembering key ideas

We have provided clear explanations of the key ideas and techniques you need throughout the book. Key ideas you need to remember are listed in a **summary of key points** at the end of each chapter and marked like this in the chapters:

■ **For all events on a critical path $e_i = l_i$.**

Exercises and exam questions

In this book questions are carefully graded so they increase in difficulty and gradually bring you up to exam standard.

- **exam style practice papers** – these are designed to help you prepare for the exam itself;
- **answers** are included at the end of the book – use them to check your work.

Contents

1 Algorithms 1

1.1 What *is* an algorithm? 1

1.2 Some examples of algorithms 2
Using flow charts to represent algorithms 4

1.3 Sorting algorithms 7
Bubble sort 7
Quick sort 10

1.4 Bin packing algorithms 13
Full-bin combinations 14
First fit algorithms 14
First fit decreasing algorithm 15

Summary of key points 20

2 Algorithms on graphs 21

2.1 Modelling using graphs 21

2.2 Definitions of terms used in graph theory 25

2.3 Other ways of representing graphs 31

2.4 Trees 36

2.5 Networks 37

2.6 Mathematical modelling 41

2.7 The minimum spanning tree (or minimum connector) 43

2.8 Prim's algorithm: finding a minimum spanning tree from a graph 46
Prim's algorithm applied to the computer network 47
Applying Prim's algorithm to a matrix 48
Applying the matrix form of Prim's algorithm to the computer network 50
How this method works 51

2.9 Kruskal's algorithm for finding a minimum spanning tree 52
Why is Kruskal's algorithm greedy? 55

2.10 **Finding the shortest path through a network** 59
A complex network problem 60
Algorithmic complexity 62

2.11 **Dijkstra's algorithm** 63

Summary of key points 71

3 **Decision making in graphs** 73

3.1 **The travelling salesman problem (TSP)** 73

3.2 **The difficulty with the TSP** 75

3.3 **An upper bound to the practical problem** 77

3.4 **A lower bound for the classical problem** 82

3.5 **The route inspection problem** 90
Alternative ways of pairing vertices 91
The route inspection algorithm 92

Summary of key points 98

4 **Critical path analysis** 99

4.1 **Precedence tables** 99

4.2 **Activity networks** 103
Drawing an activity network 104
Using dummies 107

4.3 **Analysing the project (the critical path algorithm)** 111
Earliest event time e_i (forward scan) 112
Latest event time l_i (backward scan) 113

4.4 **Critical events and critical activities** 116

4.5 **Time analysis of a network** 117
Meaning of the total float 119

4.6 **Using the float** 120
Scheduling 120
Resource levelling 123

Summary of key points 131

5 **Flows in networks** 133

5.1 **Flow networks** 133

5.2 **Capacities and flows** 134

5.3 **The maximum flow–minimum cut theorem** 137
How is the max flow–min cut theorem useful? 139

5.4 **Finding the maximum flow using augmentation – a labelling algorithm** 141

5.5 **Flows in undirected edges** 147

5.6 **Spotting flow augmenting paths** 150

Summary of key points 157

6 Linear programming 159

6.1 **Formulating linear programming problems** 160

6.2 **Graphical solutions for two-variable problems** 167
Sets of points defined by a linear inequality 167
Sets of points defined by a collection of inequalities 170

6.3 **Feasible solutions of a linear programming problem** 172

6.4 **Finding the optimal solution of a linear programming problem** 172

6.5 **Extreme points and optimality** 181

6.6 **Integer valued solutions** 183

6.7 **The algebraic method for solving linear programming problems** 189
Slack variables 189

6.8 **Basic solutions** 191

6.9 **Simplex method** 192

Summary of key points 202

7 Matchings 203

7.1 **Modelling using a bipartite graph** 203

7.2 **Matchings** 208

7.3 **Improving a matching using an alternating path: the matching improvement algorithm** 212
Changing the status 215

Summary of key points 221

Appendix: Further work 223

Linear programming 223
Linear programming in networks 223

Shortest path 223
Linear programming in transmission networks 225
Allocation and transportation 226

Decision making in graphs 229
Converting a practical problem into its classical equivalent 230
Complexity of the TSP 233
More about upper and lower bounds 234

Examination style paper D1a 239

Examination style paper D1b 245

Answers 251

List of symbols and notation 289

Index 293

Algorithms 1

1.1 What *is* an algorithm?

When you buy a new piece of equipment, such as a video recorder, the package usually includes a set of step-by-step instructions for installing it or setting it up. Such a set of instructions is called an **algorithm**.

In this chapter, we will be considering algorithms of various kinds, beginning with some well known algorithms from the 'everyday world', moving then to some algorithms from 'elementary mathematics' and then to some algorithms that have been developed in the areas of sorting and packing.

Before considering specific algorithms we give a definition of an algorithm and a list of its essential properties.

- **An algorithm is a set of precise instructions which if followed will solve a problem.**

The word *precise* in the definition is important. It means that there must be *no ambiguity* in any instruction and that after a particular instruction has been obeyed there must be *no ambiguity* as to which instruction is to be carried out next.

ambiguous – having two or more possible meanings

Not all sets of instructions constitute an algorithm. A set of instructions would fail to be an algorithm if one of the instructions was ambiguous or if after completing a particular instruction it was not clear which instruction is to be carried out next.

It is also implicit in the definition that the problem will be solved in a finite time so that the set of instructions should include an instruction to stop. This instruction must be reached after carrying out a *finite* number of instructions.

1.2 Some examples of algorithms

There are many things in everyday life which are algorithms but for which we have used other more commonplace names. Here are three examples from knitting, cooking and repairing broken china.

Example 1
The instructions for knitting a man's cardigan include the following set of instructions for knitting the rib.

Using No 3 mm needles cast on 115 sts.
1st row — S1, K1 * P1, K1 repeat from * to the last st, K1.
2nd row — S1 * P1, K1, repeat from * to end.
Repeat 1st and 2nd rows 11 times.
(Abbreviations: K knit, P purl, S1 slip one stitch, st. stitch)

Example 2
The following recipe was given in a magazine:

Celebration Cornflake Crunch (Makes 12–14)

125 g Plain chocolate
250 g Christmas pudding, cooked
60 g Crunchy Nut Cornflakes
Bun tray, lined with paper cases

Melt the chocolate in a bowl over a pan of hot water.
Break the Christmas pudding into pieces and stir into the chocolate.
Fold in the cornflakes and spoon into the paper cases.
Leave to set.

Example 3
On a pack of strong clear adhesive are printed the following instructions for repairing a broken glazed china cup.

Directions for use.
1 Make sure surfaces to be joined are clean, dry and free from grease.
2 Apply a thin layer of adhesive to each surface and leave for at least one minute before pressing the surfaces together.

With a little thought you will be able to recall many examples from your own experience of algorithms.

In your earlier courses on mathematics you will have learned how to add, subtract, multiply and divide numbers. In each of these cases you will have used a method – a set of instructions – which, when carried out precisely, will produce the correct answer. For example, if asked to add 256 and 845 you will probably write:

$$\begin{array}{r} 256+ \\ \underline{845} \\ \underline{1101} \end{array}$$

This does not indicate the order in which you obtained the digits in the answer. A set of instructions would indicate that you work from the right to the left, first adding the units, then the tens and finally the hundreds.

It is not always easy to produce an algorithm. On the other hand it is not necessary to understand the precise logic behind an algorithm to be able to use it.

Consider the usual algorithm for multiplying two numbers, that is:

(multiplicand) × (multiplier)

1. Multiply successively by each figure of the multiplier taken from right to left.
2. Write these intermediate results one beneath the other, shifting each line one place to the left.
3. Finally add all these rows to obtain the answer.

Example 4

Multiply 178×26.

1. $178 \times 6 = 1068$
   $178 \times 2 = 356$
2. 1068
   356
3. 4628

You would of course probably write this simply as:

$$\begin{array}{r} 178\times \\ \underline{26} \\ 3560 \\ \underline{1068} \\ \underline{4628} \end{array}$$

There are, however, other methods of multiplying two numbers. Here is one, sometimes called the Russian peasant's algorithm, although it has also been attributed to the ancient Egyptians. It uses only doubling, halving and addition. Here is a statement of the algorithm:

1. Write multiplier and multiplicand side by side.
2. Make two columns, one under each operand, by repeating the following rule until the number under the multiplier is 1:
   (a) Divide number under multiplier by 2 ignoring fractions.
   (b) Multiply number under multiplicand by 2.
3. Cross out each row in which the number under the multiplier is even.

4. Add up the numbers that remain in the column under the multiplicand.

Example 5

Multiply 178×26.

Multiplier	Multiplicand
~~26~~	~~178~~
13	356
~~6~~	~~712~~
3	1424
1	2848
Total of remaining rows	4628

Using flow charts to represent algorithms

All the algorithms discussed so far have been described in words. Another way of presenting an algorithm is in the form of a flow chart. In a flow chart (or flow diagram) it is customary to use three kinds of blocks:

1. A rectangular block ▭ which contains operations to be carried out. These blocks have only one route leading from them.
2. A diamond shape block ◇ containing questions that can be answered 'yes' or 'no'. These blocks have two alternative routes out, the route out depending on the answer to the question in the block.
3. An oval shape block ⬭ indicating a terminus.

Example 1 may be given by flow chart A.

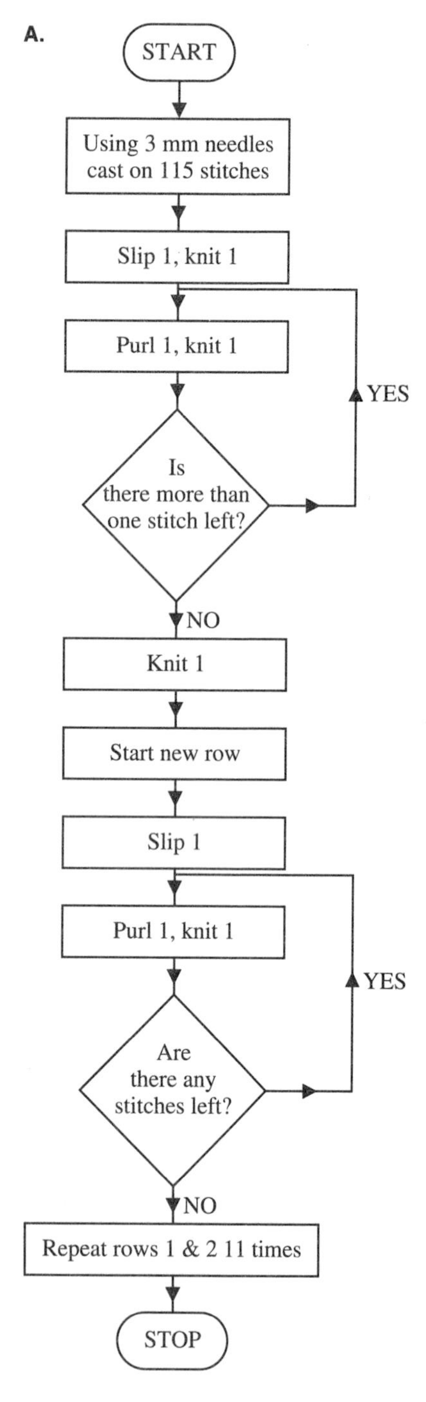

Notice that in this flow diagram there are lines which leave decision blocks and then re-enter the diagram at an earlier point. Lines like these are called **loops**.

You have probably encountered the quadratic equation $ax^2 + bx + c = 0$. The solutions of this equation, if they exist, are obtained from the formula

$$\frac{-b \pm \sqrt{b^2 - 4ac}}{2a}$$

You can solve this equation by using the algorithm given by the following flow chart B.

B.

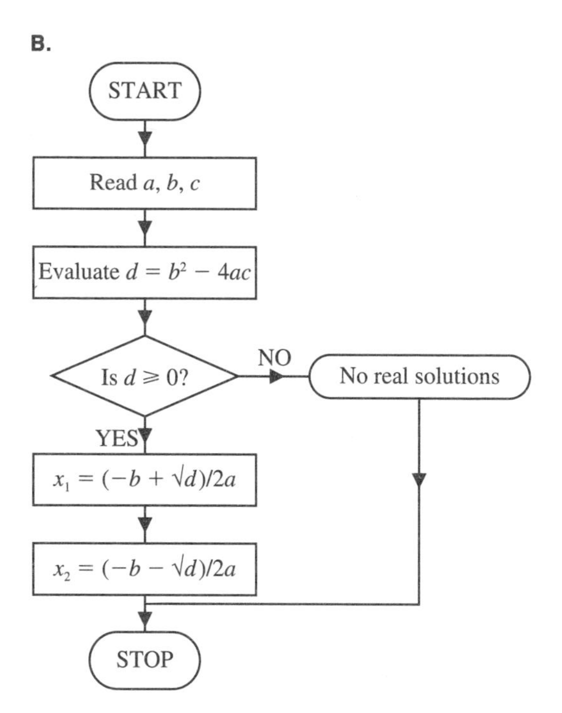

Example 6

Use the algorithm given by flow chart B to find the roots of $x^2 - 21x - 162 = 0$.

Read a, b, c: $a = 1, \quad b = -21, \quad c = -162$

Evaluate d: $d = (-21)^2 - 4(1)(-162) = 441 + 648 = 1089$

d is $\geqslant 0$, $(\sqrt{d} = \sqrt{1089} = 33)$

Evaluate x_1 $\quad x_1 = \dfrac{(-b + \sqrt{d})}{2a} = \dfrac{(21 + 33)}{2} = 27$

Evaluate x_2 $\quad x_2 = \dfrac{(-b - \sqrt{d})}{2a} = \dfrac{(21 - 33)}{2} = -6$

A famous algorithm invented by Euclid – the Euclidean algorithm – for finding the highest common factor of two integers A and B is given by flow chart C on page 6.

Example 7

Use the Euclidean algorithm, given by flow chart C, with $A = 35$ and $B = 112$.

Since $3 \times 35 = 105$ and $4 \times 35 = 140$ in this case $Q = 3$ and $R = 112 - (3 \times 35) = 112 - 105 = 7$

So $R \neq 0$

Now new B is 35 and new A is 7.

Hence $Q = 5$ and $R = 35 - (7 \times 5) = 0$

$R = 0$ and current value of A is 7

So 7 is the highest common factor of 35 and 112.

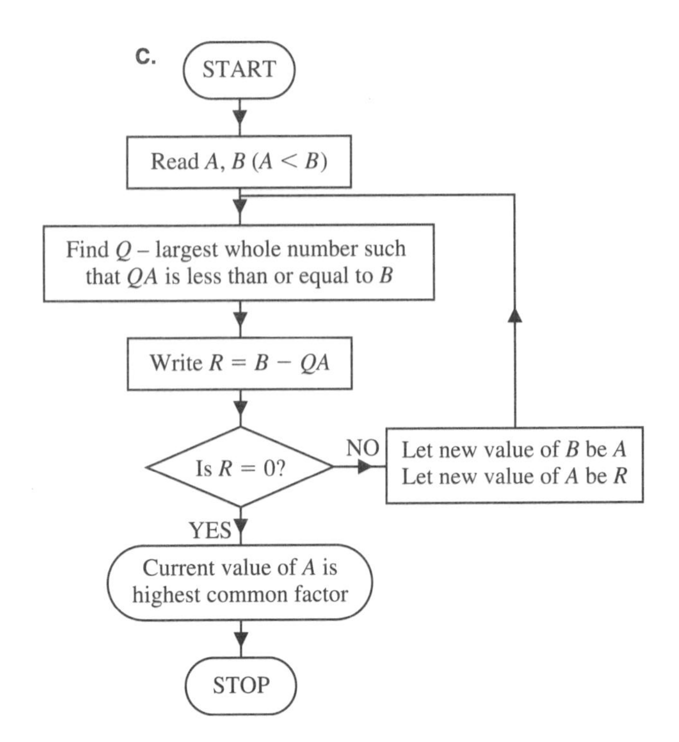

Exercise 1A

1 Use the algorithm given by flow chart B on page 5 to find real solutions, if they exist, of:

(a) $x^2 - x - 2 = 0$

(b) $3x^2 - 14x - 5 = 0$

(c) $2x^2 + 3x + 5 = 0$.

2 Use the Euclidean algorithm given in flow chart C above to find the highest common factor of:

(a) 169 and 39

(b) 972 and 666.

3 Implement the algorithm given by flow chart D and state what the algorithm actually produces.

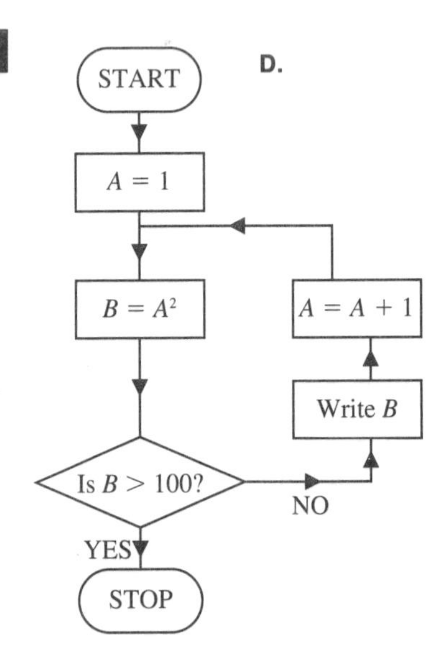

4 Implement the algorithm given by flow chart E and state what the algorithm actually produces.

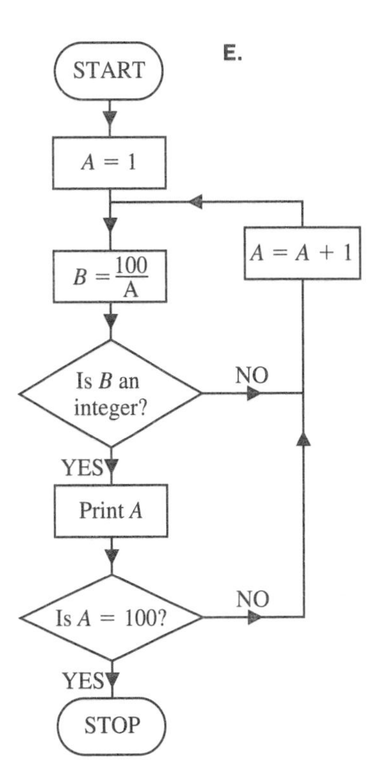

5

$$\begin{array}{ccccc} & & 1 & & \\ & 1 & & 1 & \\ 1 & & 2 & & 1 \end{array}$$

The diagram shows the first three rows of what is known as Pascal's triangle. Construct the next five rows given that

(i) the border consists entirely of 1s,

(ii) each of the other numbers is the sum of the two numbers immediately above it.

6 The first three terms of the Fibonacci sequence are 1, 1, 2. The remaining numbers are obtained by adding the two previous numbers in the sequence. Write down the first 12 numbers of the sequence.

1.3 Sorting algorithms

Suppose a list of students is given in alphabetical order and you want to arrange them in a new order according to the marks they obtained in an examination. This is an example of a **sorting** problem. In this case probably the most useful list would be the one with the marks in descending order.

In Kruskal's algorithm for finding a minimum spanning tree (discussed later in chapter 2) the weights of the edges give you a list of numbers. You have to work through this list from the smallest to the largest and so the first step is to produce a new list in ascending order. This is again a **sorting** problem.

Given a list of people's surnames in a particular club the most useful directory to produce is one in which these surnames are in alphabetical order. This requires sorting to take place.

Bubble sort

This is one of the simplest sorting algorithms. Just as bubbles in a fizzy drink rise to the surface, in a bubble sort the largest or smallest number 'rises' to one end in a list.

- **The bubble sort algorithm makes repeated passes through a list of numbers. On each pass: adjacent numbers in the list are compared, and switched if they are in the wrong order.**

In the first pass of the algorithm the first and second numbers are compared, then the second and third, and so on.

Each comparison refers to the list in the order resulting from all previous comparisons.

Example 8

Consider the list

16, 9, 4, 6, 12, 3, 8, 7

which is to be sorted into *ascending* order.

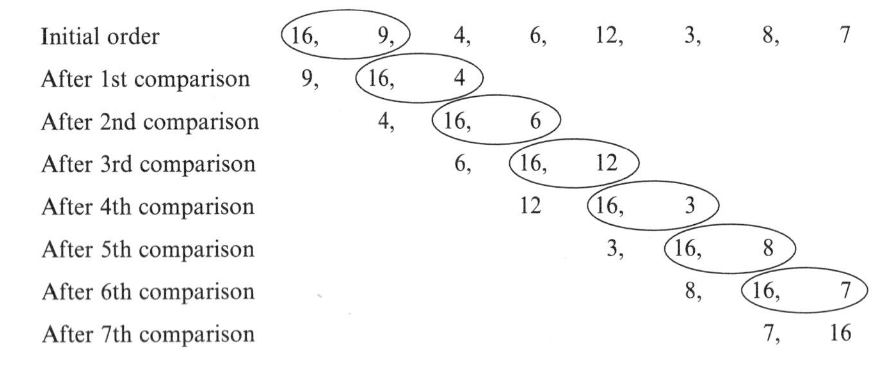

At the end of this **first pass** the largest number, 16, is in its correct position. You can see from the above diagram how the 'bubble' has moved to the right.

Each iteration within the bubble sort algorithm succeeds in placing at least one number in its correct position.

> **Iterate** means repeat. Each pass of the algorithm through the list is called an **iteration**.

Having completed the first pass the list is now

9, 4, 6, 12, 3, 8, 7, 16

The algorithm is now applied to the sublist with the 16 removed. This is called the *second pass.*

The result of the second pass is

4, 6, 9, 3, 8, 7, 12

So now the two largest numbers of the original list, namely 12 and 16, are in their correct positions.

The result of the *third pass* is

4, 6, 3, 8, 7, 9

That for the fourth pass is

4, 3, 6, 7, [8]

and the fifth pass gives

3, 4, 6, [7]

A sixth pass produces no change and so the list is sorted, the final list being

3, 4, 6, 7, 8, 9, 12, 16

We can apply the same method to producing a list in *descending* order.

Example 9
Consider the same list

16, 9, 4, 6, 12, 3, 8, 7

and sort it into decreasing order.

The first pass is

(16, 9,) 4, 6, 12, 3, 8, 7
16, (9, 4)
9, (4, 6)
6 (4, 12)
12, (4, 3)
4, (3, 8)
8, (3, 7)
7, 3

At the end of this pass the smallest number [3] is in its correct place. The other passes give

2nd pass 16, 9, 12, 6, 8, 7, [4]
3rd pass 16, 12, 9, 8, 7, [6]
4th pass 16, 12, 9, 8, 7

No change so the list in descending order is 16, 12, 9, 8, 7, 6, 4, 3.

Example 10

The 5 members of a club have the surnames

Jordan, Smith, Adams, Evans, Kapasi

Use the bubble sort algorithm to sort this list in alphabetical order.

Denote the names by J, S, A, E, K.

The first pass is

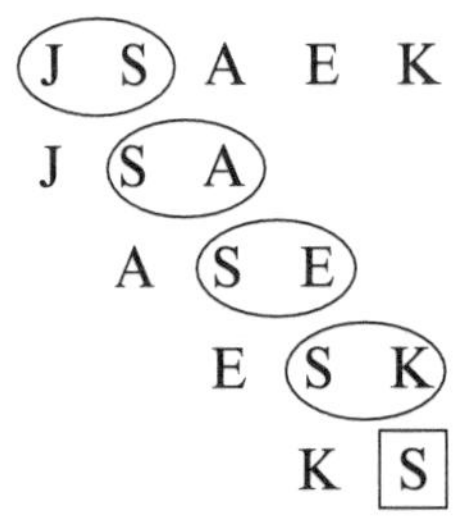

At the end of this pass S (Smith) is in the correct place.

The other passes give

2nd pass A E J K

3rd pass A E J

No change so list in alphabetical order is

A	E	J	K	S
Adams	Evans	Jordan	Kapasi	Smith

Quick sort

A more efficient method of sorting, in that it in general requires fewer comparisons, is the method using the **quick sort algorithm** introduced by Hoare in 1962.

Step 1 The quick sort algorithm proceeds by first selecting a specific number from the list which is used as a kind of pivot. Some authors choose the first number but in this book the number at the mid-point of this list will be chosen. (If there is an even number of numbers in the list you may choose either of two numbers in the middle, it does not matter which.)

As an example, let's apply the quick sort algorithm to sort the list in example 8 in ascending order.

Let L be the list 16, 9, 4, 6, 12, 3, 8, 7.

Choose 6 as the pivot; then you have:

16, 9, 4, (6), 12, 3, 8, 7
↑

Step 2 Now write all the numbers *smaller* than 6 to the *left* of 6, reading the original list from left to right. This creates a sublist L_1.

In a similar way write all the numbers *larger* than 6 to the *right* of 6 reading the original list from left to right. This creates a sublist L_2.

For the above example the application of step 2 gives

4, 3, (6), 16, 9, 12, 8, 7

sublist L_1 ↑ sublist L_2

Step 3 Apply step 1 and step 2 to each separate sublist until each sublist contains only one number.

(a) Consider L_1: 4(3) we choose 3 in step 1 (pivot).
↑

Applying step 2 gives 3 4
L_3

there are no numbers less than 3. Sublist L_3 consists only of 4 and so this list has only one number in it.

(b) Consider now L_2: 16, 9, (12), 8, 7.
↑

We choose 12 in step 1 (pivot).
Applying step 2 gives

9, 8, 7, (12), 16
L_4 L_5

Sublist L_5 has only 1 number.
So we now consider L_4.
Application of step 1 and step 2 gives

7 (8) 9
L_6 ↑ L_7

As sublists L_6 and L_7 consist of just one number each, the sorting is complete. Combining the various steps gives:

$$4, 3, (6), 16, 9, 12, 8, 7$$
L_1: 4, 3 → (3), 4 (L_3: 4)
L_2: 16, 9, 12, 8, 7 → 9, 8, 7 (L_4), (12), 16 (L_5)
L_4: 9, 8, 7 → 7 (L_6), (8), 9 (L_7)

So starting at the left-hand side the final sorted list is:

$$3, 4, 6, 7, 8, 9, 12, 16$$

Exercise 1B

1 The marks obtained in an examination by 5 students were:

Anne (68), Barry (42), Clare (70), David (45), Eileen (50) and Greg (55).

Use the bubble sort algorithm to sort these marks

(a) in ascending order

(b) in descending order.

2 Use the quick sort algorithm to sort the list

$$6, 8, 4, 5, 10, 2, 9$$

in ascending order.

3 The members of a club have surnames

Monro, Jones, Patel, Wilson and Farmer

Use the bubble sort algorithm to sort these into alphabetical order.

4 Use the bubble sort algorithm to sort the following list in ascending order:

$$4, -3, -5, 0, 2, 5, -1$$

5 The times recorded by 7 competitors for a given distance were:

$$9.8, 9.2, 9.6, 9.7, 9.1, 8.9, 9.0$$

(a) Use the bubble sort algorithm to sort these in ascending order.

(b) Use the quick sort algorithm to sort these in ascending order.

(c) Which of the methods is the most efficient in this case?

1.4 Bin packing algorithm

There is a whole class of problems which may be modelled by the 'bin packing problem'. Some of these will be included in the exercises. Consider a set of bins all of the same cross-section and the same height as shown in the figure.

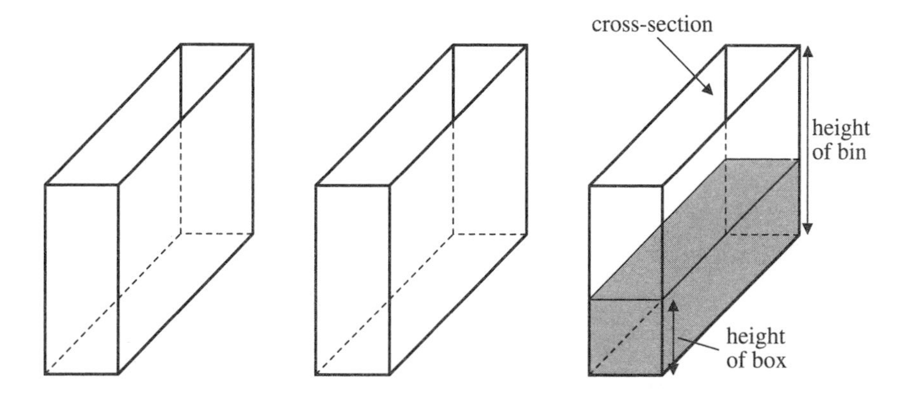

The bin packing problem is how to pack into the bins a number of boxes of the same cross-section as the bins but of varying heights, so as to use as few bins as possible.

Example 11

Suppose the bins are 1.5 m tall and that you have 10 boxes A, B, C, D, ..., J with the heights shown in the table.

Box	A	B	C	D	E	F	G	H	I	J
Height (m)	0.8	0.6	0.7	0.5	0.9	0.4	0.3	0.6	0.5	0.6

To save working with decimals, let's define the capacity of a bin as (height $\times$ 10), and the size of a box as (height $\times$ 10). In this case each bin has capacity 15 and the boxes are of size:

A(8), B(6), C(7), D(5), E(9), F(4), G(3), H(6), I(5), J(6)

The total size of all the boxes is

$$8 + 6 + 7 + 5 + 9 + 4 + 3 + 6 + 5 + 6 = 59$$

If you divide this by 15, the capacity of a bin, you get $\frac{59}{15} = 3\frac{14}{15}$. This indicates that *at least 4 bins* are needed. However, because of the sizes of the boxes, it may not be possible to find a solution (packing) using just 4 bins.

Full-bin combinations

For a problem involving only a few bins and boxes it is possible to look for combinations of boxes which fill a bin. This is not a practical method for larger problems and so other algorithms are considered below.

Example 12

For the data given in example 11 find full-bin combinations.

In this case you can easily see that:

Box A + Box C = 8 + 7 — so fill bin 1
Box B + Box E = 6 + 9 — so fill bin 2

The next largest boxes remaining are H and J, each of size 6.

Box H + Box J + Box G = 6 + 6 + 3 — so fill bin 3

The remaining boxes D, I and F give

Box D + Box I + Box F = 5 + 5 + 4 = 14

and so these may be fitted into bin 4. The situation can be summarised by a diagram:

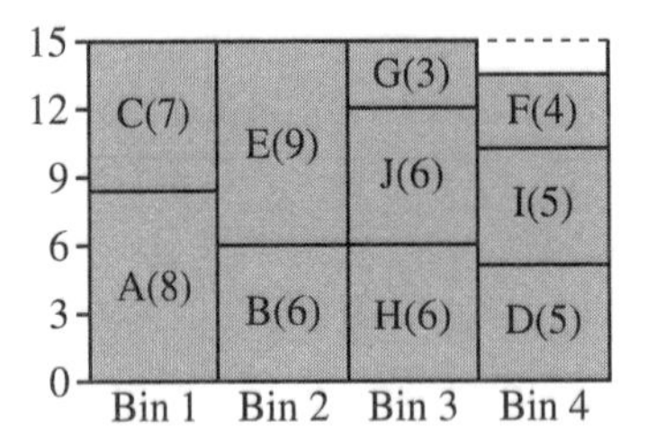

So all the boxes can in fact be fitted into 4 bins.

First fit algorithm

This is an algorithm which provides a method of dealing with more complicated problems. It may be stated as follows:

- ***Taking the boxes in the order listed* place the next box to be packed in the *first* available bin that can take that box.**

Example 13

Apply the first fit algorithm to the data given in example 11.

Applying the algorithm to this situation gives:

Box A(8) into Bin 1 leaving space of 7
Box B(6) into Bin 1 leaving space of 1
Box C(7) into Bin 2 leaving space of 8
Box D(5) into Bin 2 leaving space of 3
Box E(9) into Bin 3 leaving space of 6
Box F(4) into Bin 3 leaving space of 2
Box G(3) into Bin 2 leaving space of 0
Box H(6) into Bin 4 leaving space of 9
Box I(5) into Bin 4 leaving space of 4
Box J(6) into Bin 5 leaving space of 9

The situation can be summarised as:

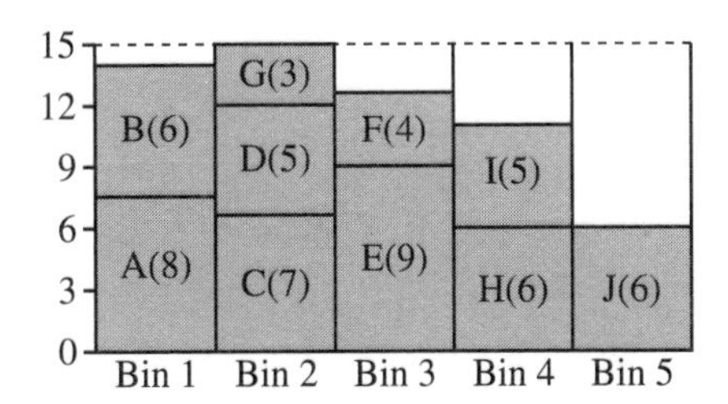

To get a picture of what is happening, try to build up this diagram as you assign boxes to bins.

Notice that the application of this algorithm has produced a solution requiring 5 bins, although the best (optimum) solution requires only 4 bins as we have seen above.

For the bin packing problem there is no known efficient algorithm that will always produce the optimal or best solution. The above is an example of an algorithm that attempts to find a good solution. Such algorithms are called **heuristic algorithms**.

First fit decreasing algorithm

From the above example you can see that if there is a box of a large size towards the end of the list it will probably have to go into a bin on its own, since any spare capacity of the other bins is likely to be distributed in small amounts. This suggests that the first fit algorithm is more likely to give the optimal solution if the boxes are reordered in *descending order of size* before allocation to bins is started. The first fit decreasing algorithm has two steps.

Step 1 Reorder the boxes in *decreasing* order of size using one of the sorting algorithms.

Step 2 Apply the first fit algorithm to the reordered list.

Example 14

Apply the first fit decreasing algorithm to the data given in example 11.

Reordering the boxes in decreasing order of size gives:

Box	E	A	C	B	H	J	D	I	F	G
Size	9	8	7	6	6	6	5	5	4	3

Applying the first fit algorithm to this list gives:

Box E(9) into Bin 1 leaving space of 6
Box A(8) into Bin 2 leaving space of 7
Box C(7) into Bin 2 leaving space of 0
Box B(6) into Bin 1 leaving space of 0
Box H(6) into Bin 3 leaving space of 9
Box J(6) into Bin 3 leaving space of 3
Box D(5) into Bin 4 leaving space of 10
Box I(5) into Bin 4 leaving space of 5
Box F(4) into Bin 4 leaving space of 1
Box G(3) into Bin 3 leaving space of 0

This solution is summarised by:

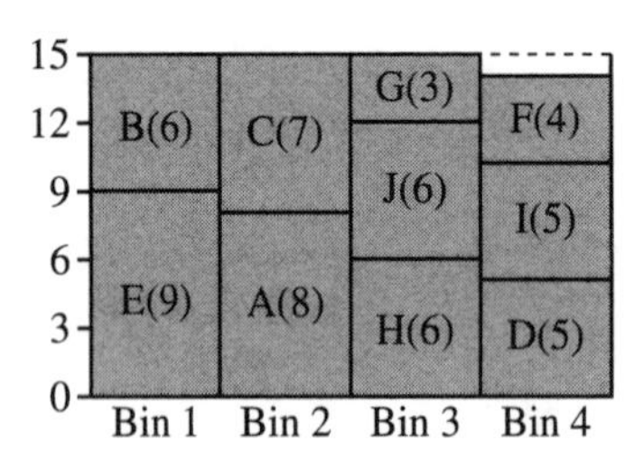

In this case the algorithm does give an optimal solution. Although this algorithm is generally more efficient than the first fit algorithm it is not guaranteed to always give an optimal solution.

Example 15

The time, in hours, taken to produce articles A, B, C,..., K are given in the table.

Article	A	B	C	D	E	F	G	H	I	J	K
Time (hours)	2	2	3	3	4	4	4	5	7	7	7

Determine the numbers of workers required to produce these articles in a 12-hour shift using

(a) the first fit algorithm,
(b) the first fit decreasing algorithm.
(c) Is it possible to obtain a better solution than either of those given by (a) or (b)?

In this example, each 'bin' is a 12-hour shift for one worker.

(a) The first fit algorithm applied to this situation gives:
In bin 1 you can pack A(2), B(2), C(3) and D(3) — leaving space of 2
In bin 2 you can pack E(4), F(4) and G(4) — no space left
In bin 3 you can pack H(5) and I(7) — no space left
In box 4 you can pack J(7) — leaving space of 5
In box 5 you can pack K(7) — leaving space of 5

This solution is summarised by the diagram below:

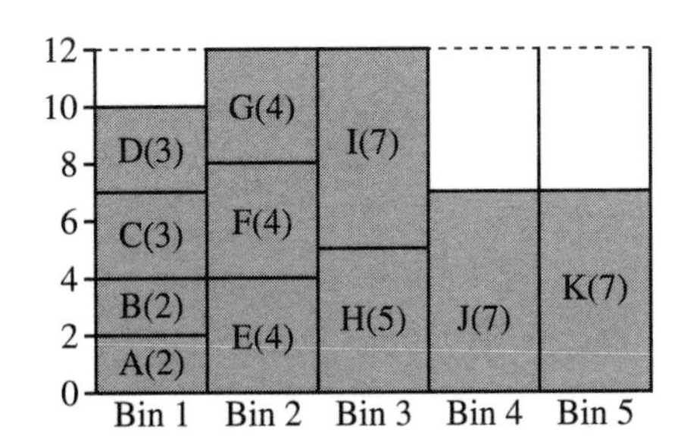

This solution requires 5 bins, that is 5 workers to produce the articles.

(b) Reordering the list in decreasing order of size gives:

K	J	I	H	G	F	E	D	C	B	A
7	7	7	5	4	4	4	3	3	2	2

It is particularly useful in this case to construct the diagram as the assignments to bins are made:

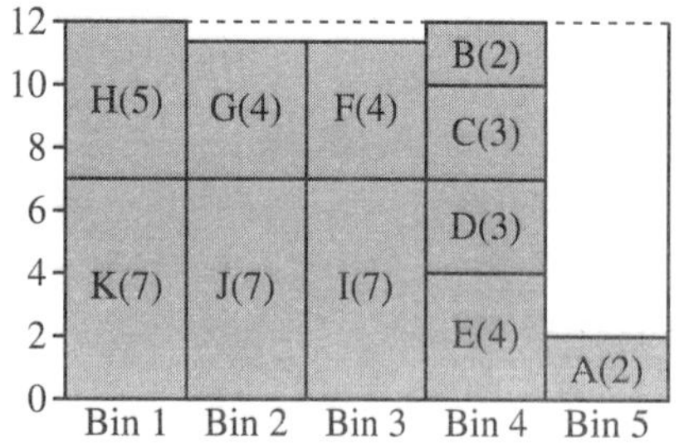

Place K(7) in Bin 1 leaving space of 5
Place J(7) in Bin 2 leaving space of 5
Place I(7) in Bin 3 leaving space of 5
Place H(5) in Bin 1 leaving space of 0
Place G(4) in Bin 2 leaving space of 1
Place F(4) in Bin 3 leaving space of 1
Place E(4) in Bin 4 leaving space of 8
Place D(3) in Bin 4 leaving space of 5
Place C(3) in Bin 4 leaving space of 2
Place B(2) in Bin 4 leaving space of 0
Place A(2) in Bin 5 leaving space of 10

Again this solution requires 5 bins, that is 5 workers.

(c) You can see from the solution in (b) that there is only A(2) in bin 5. In other words, worker 5 would only be working for 2 hours! The question is, can the assignments to the other bins be reorganised so that A(2) may be fitted in?

Both bins 2 and 3 have 1 hour spare.
If you replace G(4) in bin 2 by B(2) and C(3), and F(4) in bin 3 by A(2) and D(3) then both bins 2 and 3 are completely full. You may then place F(4) and G(4) in bin 4 which again is now completely full.

The solution is then:

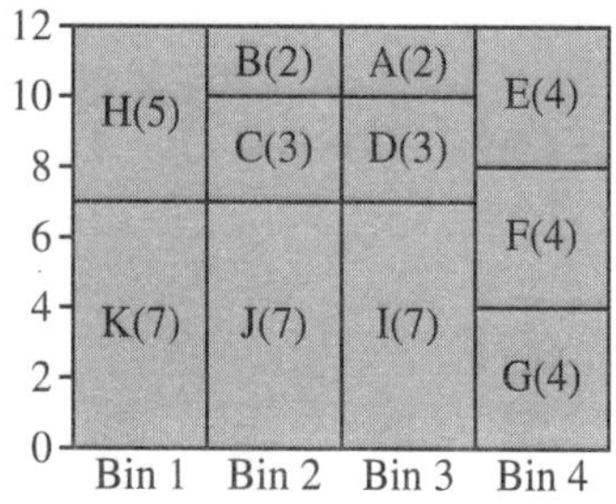

This solution involves only 4 bins. That is, 4 workers can complete the task.

Exercise 1C

1 A project is to be completed in 13 days. The activities involved in the project and their duration in days are given in the table

A	B	C	D	E	F	G	H	I	J
3	8	7	5	8	4	5	4	4	4

To determine how many workers are required

(a) apply the first fit algorithm

(b) apply the first fit decreasing algorithm.

(c) Is it possible to obtain a better solution than either (a) or (b)?

2 A small ferry which sails between two of the islands in the Hebrides has 3 lanes each 20 m long on its car deck. The vehicles waiting to be loaded are:

Petrol tanker	13 m	Small van	3 m	Truck	7 m
Small truck	6 m	Coach	12 m	Car	3 m
Car	4 m	Lorry	11 m		

Determine, by using the first fit decreasing algorithm, if all the vehicles can be taken on one trip.

3 A project consists of 8 activities whose durations are as follows:

A	B	C	D	E	F	G	H
2	4	3	1	5	4	2	3

Use full-bin combinations to determine the minimum number of workers needed to finish the project in 12 hours.

4 A certain kind of pipe is sold in 10 m lengths. For a particular job the following lengths are required.

2 m, 8 m, 4 m, 5 m, 2 m, 5 m, 4 m

By looking for full-bin combinations, or otherwise, find the number of 10 m lengths required for the job.

5 Joan decided that she wanted to record a number of programmes on the video recorder. The lengths of the programmes were

45 min, 1 h, 35 min, 15 min, 40 min, 30 min, 50 min, 55 min and 25 min

Help her to decide how many 2 h tapes she requires, using

(a) the first fit algorithm

(b) the first fit decreasing algorithm

(c) full-bin combinations.

SUMMARY OF KEY POINTS

1 An **algorithm** is a set of precise instructions which if followed will solve a problem.

2 **The bubble sort algorithm**
To sort a list compare adjacent members of the list moving from left to right and switch them if they are in the wrong order. Continue this process until a pass produces no change.

3 **The quick sort algorithm**
Step 1 Select a specific number (the pivot) from the list, say the middle one.
Step 2 Write all numbers smaller than the pivot to the left of the pivot, reading the original list from left to right, and so create sublist L_1.
Write all numbers larger than the pivot to the right of the pivot, reading the original list from left to right, and so create a sublist L_2.
Step 3 Apply step 1 and step 2 to each sublist until all the sublists contain only one number.

4 **First fit algorithm**
Taking the boxes in the order listed place the next box to be packed in the *first* available bin that can take that box.

5 **First fit decreasing algorithm**
Step 1 Reorder the boxes in *decreasing* order of size.
Step 2 Apply the first fit algorithm to the reordered list.

Algorithms on graphs

2

2.1 Modelling using graphs

You will be used to drawing graphs of functions such as $y = x^2$. However, there is also an area of mathematics called **graph theory**, in which the word 'graph' has a different meaning. In graph theory:

- **a graph consists of a finite number of points (usually called vertices or nodes) connected by lines (called edges or arcs)**

Here is an example of a graph:

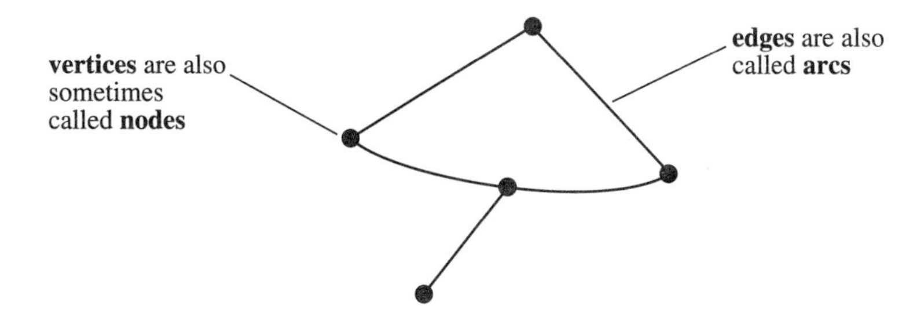

Other examples of graphs are road, train and tube maps. These are graphs of real geographical networks.

This chapter shows you how to use graphs to model many real-life situations and solve problems. It also introduces some algorithms you can use to help solve graph problems.

Here are some examples of situations, and graphs that model them:

Example 1

Mrs H lives in Ashford and wishes to know which towns she can reach directly from Ashford by train. She found the following graph in *Rail routes in Surrey*, which models the rail network that includes Ashford.

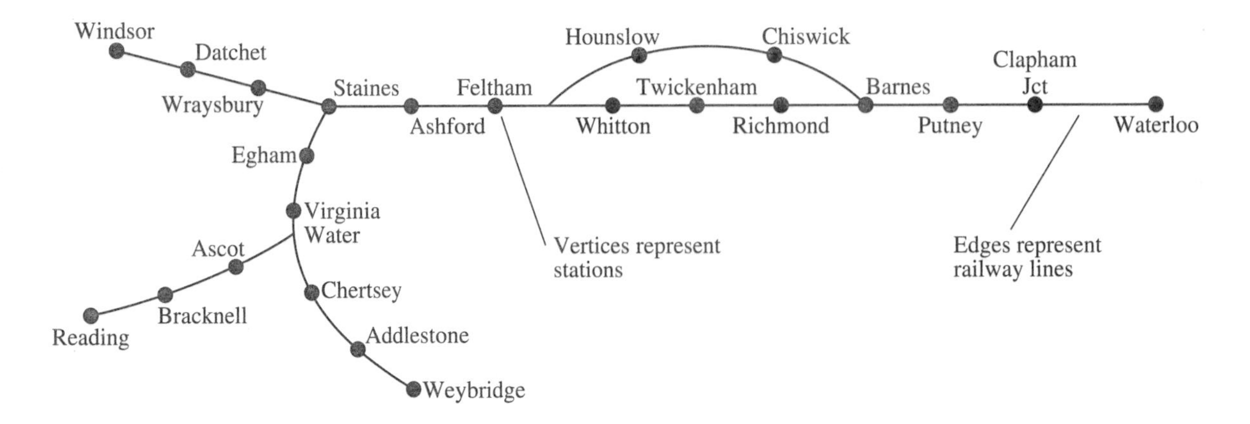

Example 2

In a college there are three faculties: Arts, Science and Management, each of which has a Dean. The heads of the departments in the faculties are responsible to the Dean of their faculty. The Deans are responsible to the Vice Principal (Academic). There is also a Vice Principal (Administration) who is assisted by the Accountant and the Registrar. The two Vice Principals are responsible to the Principal.

This administration can be modelled by the following graph:

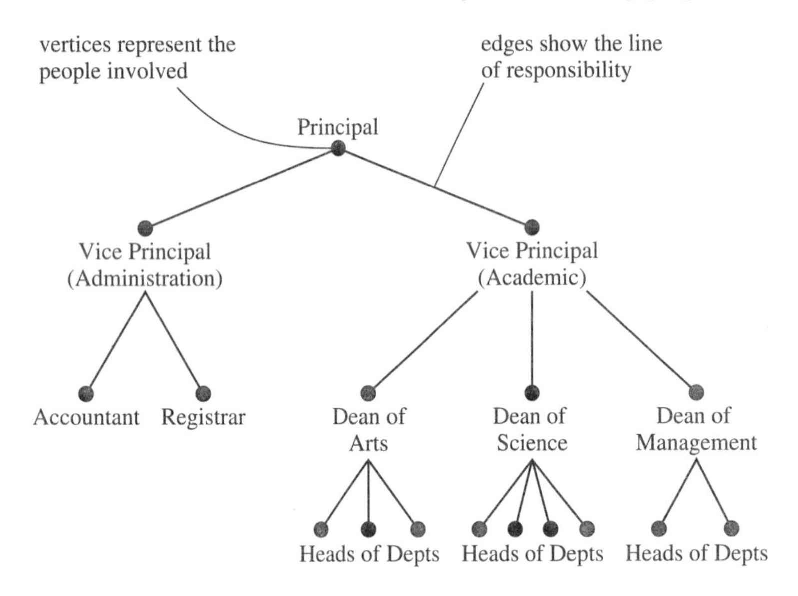

Example 3

A charity shop wishes to find one person per day to manage the shop from Monday to Friday. Five people, Mr Ahmed, Mr Brown, Ms Candy, Ms Davis and Mrs Evans, come forward. They each fill in a form and the following information is obtained:

Mr Ahmed is available Thursday and Friday
Mr Brown is available Tuesday and Wednesday
Ms Candy is available Tuesday and Thursday
Ms Davis is available Monday, Tuesday and Wednesday
Mrs Evans is available Tuesday and Friday

This information can be modelled by the following graph:

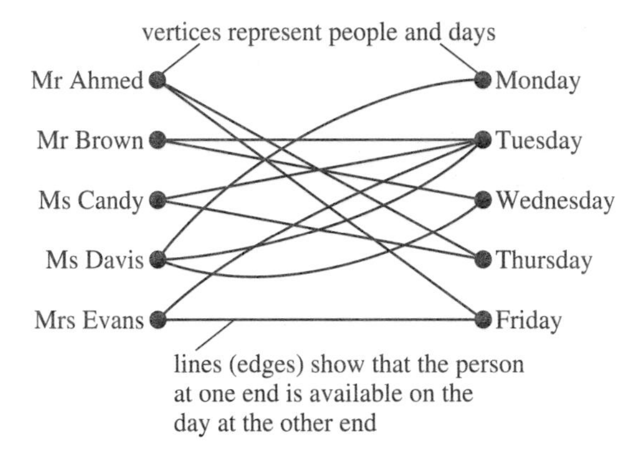

Example 4

A project consists of activities A, B, C...H which are not all independent. Activities A and B can start any time. Once A is complete C and D can start. When B is complete G can Start. E can start only when C is complete. F requires both D and E to be complete before it can start. H can start when F and G are complete. The project is complete when H is complete.

All this information can be modelled by the following graph:

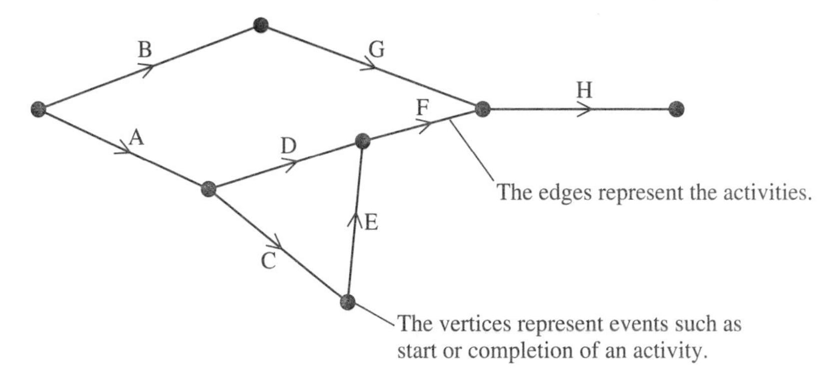

Example 5

A travelling showman (S) has with him a wolf (W), a goat (G) and a cabbage (C). He wishes to cross a river. He can use a small boat which can transport himself plus one only of his possessions at a time. Left unsupervised, the goat would eat the cabbage, and the wolf would eat the goat. How can he get all three safely across the river?

Leaving aside the desirability or otherwise of venturing into a small boat with a wolf (or a goat for that matter), we can represent the arrangements of the four participants on the banks of the river as vertices of a graph. For instance, let WC|SG represent the situation when the showman has transported the goat to the far bank, leaving the wolf and the cabbage safely behind.

Ignoring the safety aspects, list all of the possible vertices:

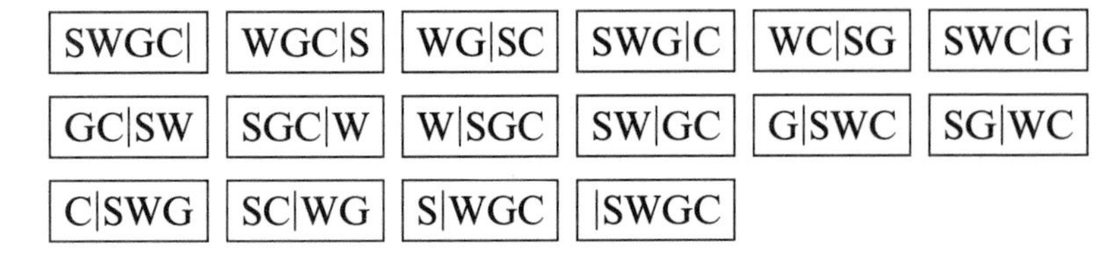

Delete from the list those vertices which represent situations in which eating would take place.

(For instance WG|SC would lead to W|SC !)

This leaves:

SWGC| SWG|C WC|SG SWC|G SGC|W W|SGC

G|SWC SG|WC C|SWG |SWGC

Now draw a graph in which pairs of vertices are connected by edges if one crossing of the river can transform one to the other. For instance SWC|G and W|SCG can be connected since the showman can transport the cabbage from the left bank of the river to the right bank, or vice versa, whilst the wolf is on the left bank and the goat is on the right bank.

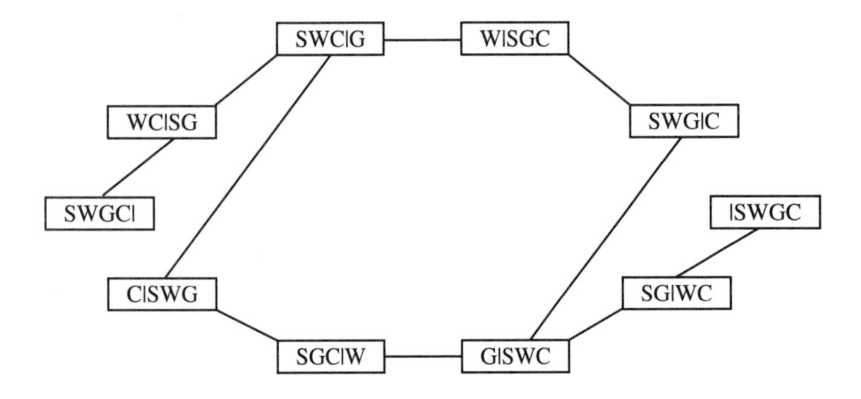

You can now produce a solution by travelling along arcs from the left of the graph to the right.

For example:

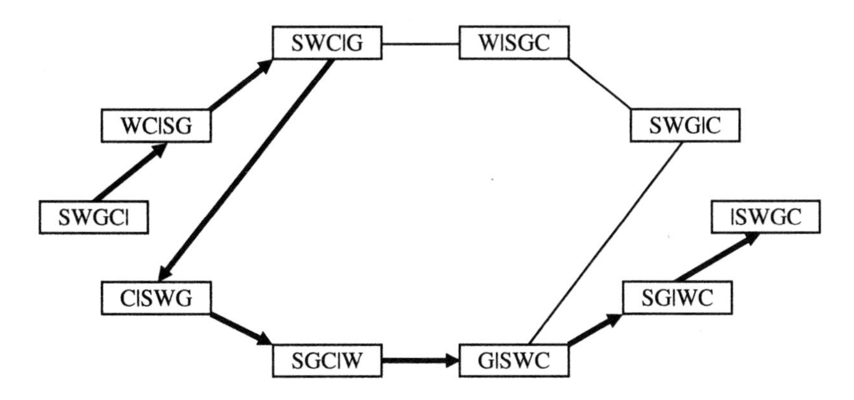

This represents the solution in which the showman takes the goat across and returns for the wolf. He then has to return the goat across the river (or the wolf would eat the goat) before transporting the cabbage and returning, once more, for the goat.

If you look at the graph you will see that there are two ways of working from left to right, so there are two solutions to the problem. And if you are allowed to 'cycle' round the path more than once there are infinitely many solutions.

2.2 Definitions of terms used in graph theory

- A **walk** is a finite sequence of edges such that the end vertex of one edge is the start vertex of the next.

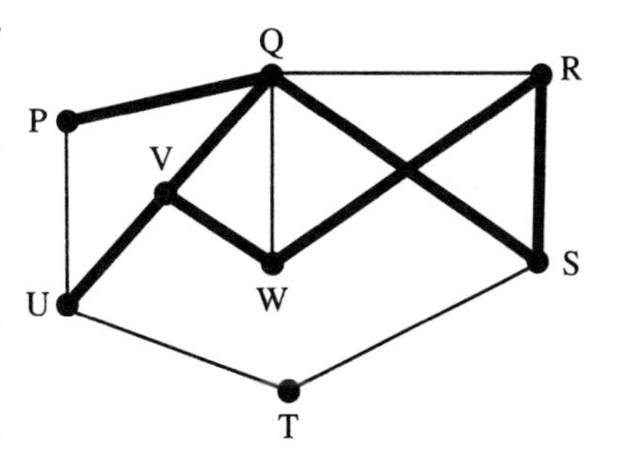

PQ QV VW WR RS SQ QV VU are the edges of a walk. The walk is described as PQVWRSQVU.

- A **trail** is a walk in which no edge appears more than once.

The walk PQVWRSQVU is not a trail because the edge QV appears twice.

- A **path** is a trail in which no vertex is visited more than once.

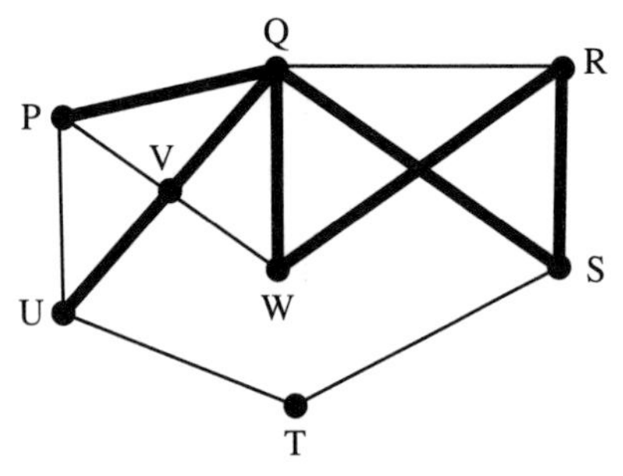

PQWRSQVU is a trail, but not a path, since Q is visited twice

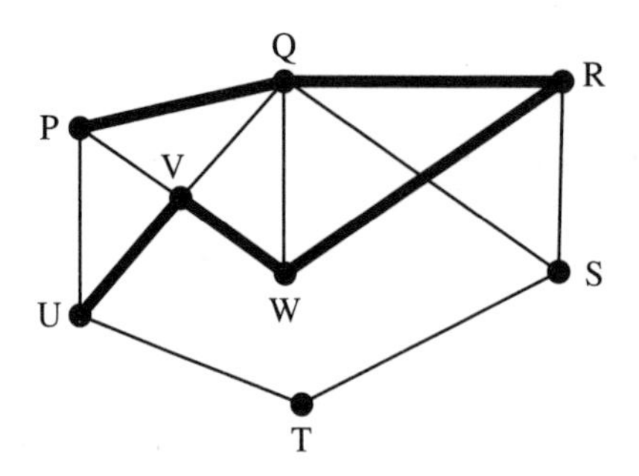

PQRWVU is a path

QS and RW are drawn as crossing, but there is no vertex there, so this is not significant in graph theoretic terms.
Thus there is no direct connection between S and W.

- A **cycle** (or circuit) is a closed path – the end vertex of the last edge is the start vertex of the first edge.

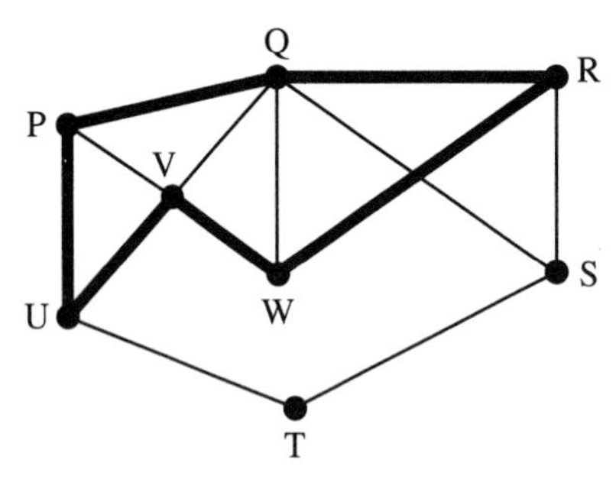

PQRWVUP is a cycle.

- An **edge set** is the set of all edges.
- A **vertex set** is the set of all vertices.
- A **subgraph** is a subset of the vertices together with a subset of the edges.
- Two vertices are **connected** if there is a path between them.
- A graph is **connected** if all of its vertices are connected.

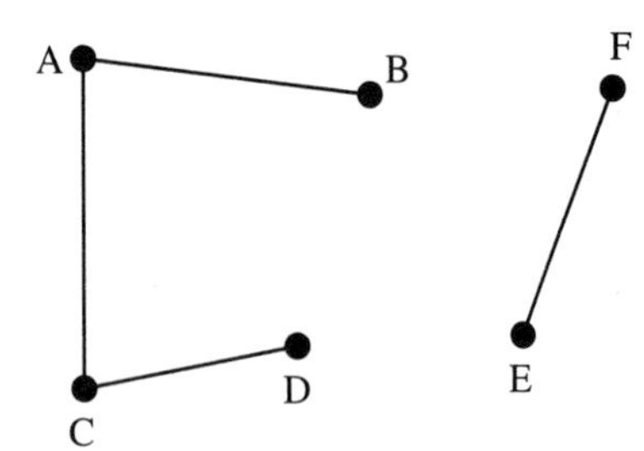

not connected

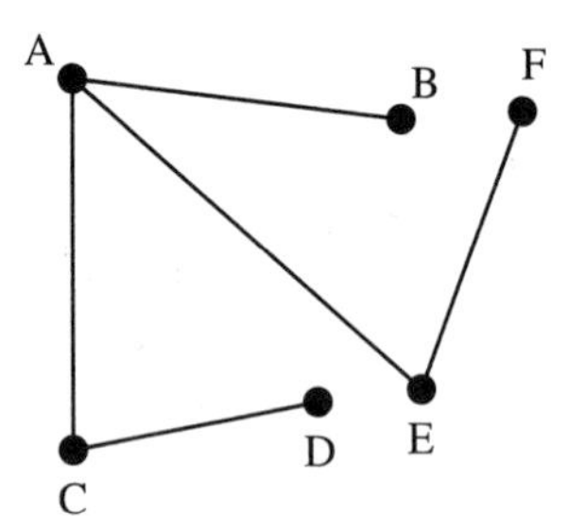

connected

- A **simple graph** is one in which there is no edge with the same vertex at each end, and no more than one edge connecting any pair of vertices.

An edge with the same vertex at each end is called a **loop.**

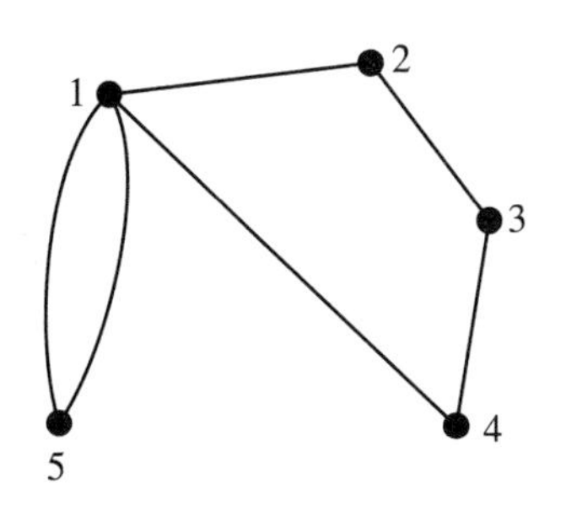

not simple
(two edges connecting 1 and 5)

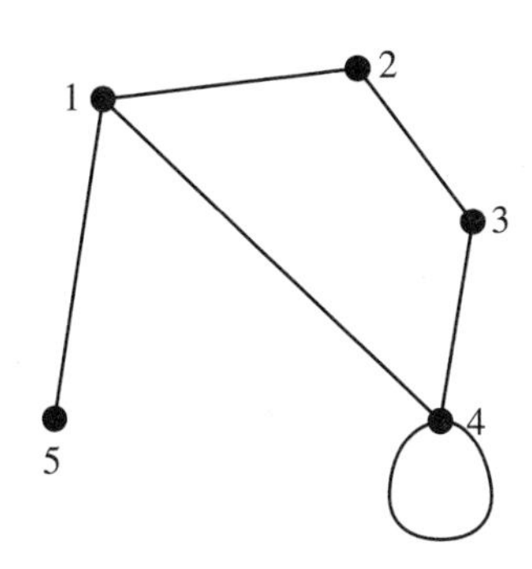

not simple
(a loop)

- The **degree** of a vertex is the number of edges connected to it.

The **degree** of a vertex is sometimes called its **order** or **valency**.

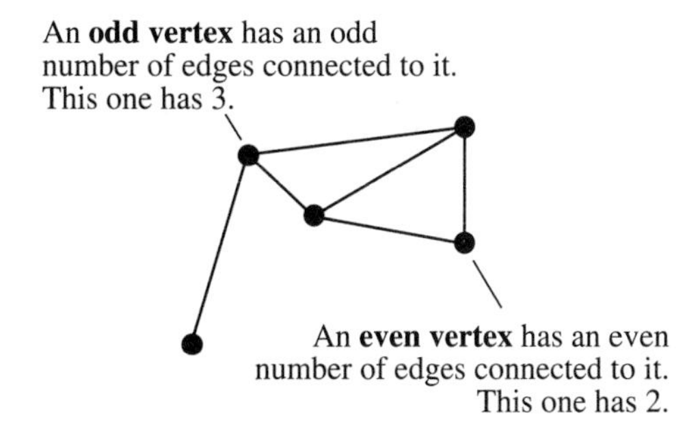

- If the edges of a graph have a direction associated with them they are known as **directed edges**, and the graph is known as a **digraph**.

The degree of a vertex is only defined for a graph, not for a digraph.

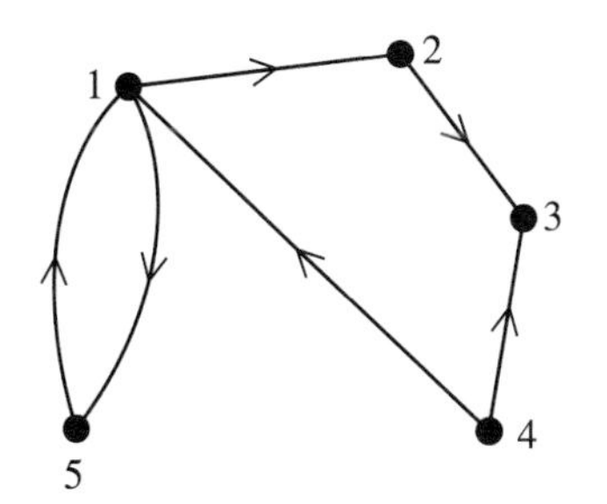

a digraph

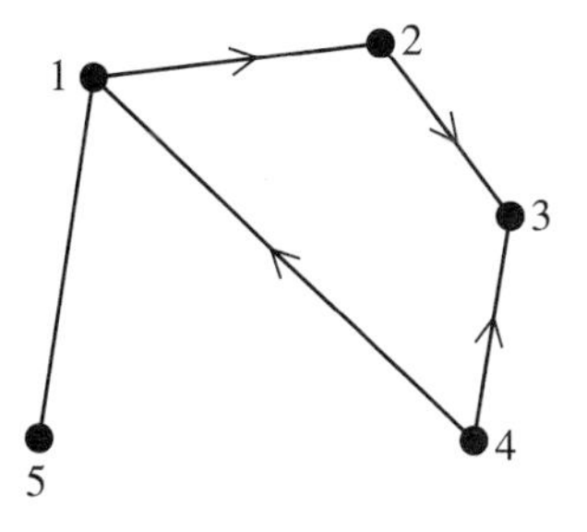

the same digraph shown differently

Here are some examples of graphs to illustrate the definitions:

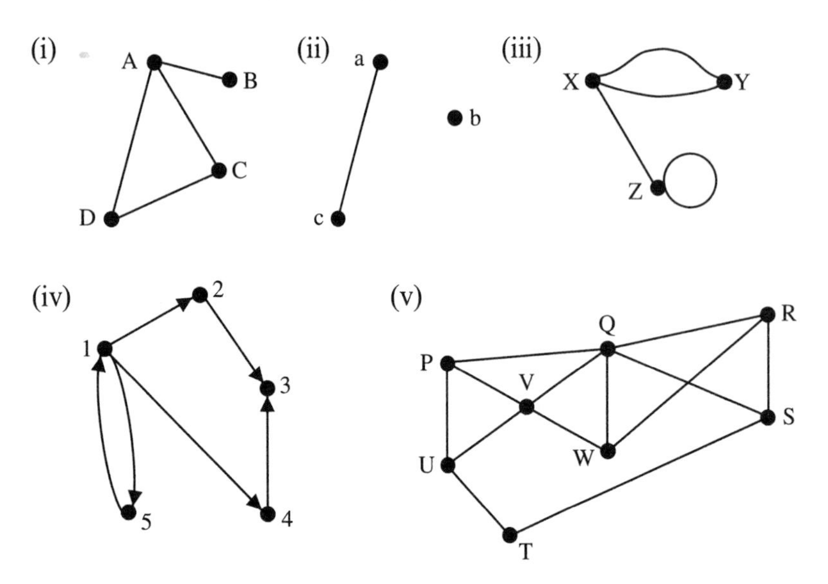

Graph (i) is simple and connected. Vertex A has order 3. Vertex B has order 1, and vertices C and D are both of order 2.

Graph (ii) is not connected. Notice that vertex b is of order 0.

Graph (iii) is not simple. Vertices X and Y are connected by two edges, and there is a loop on vertex Z.

Graph (iv) is a digraph.

In graph (v) PVWRS (for example) is a path, and PQWRSTUVP is a cycle. Notice that the edges QS and RW do not share a vertex.

Exercise 2A

1

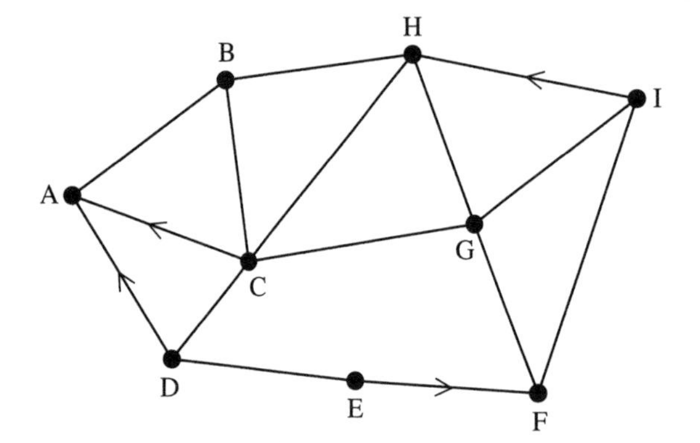

(a) Give the degrees of B and G.

(b) Give a path from A to I via H.

(c) Give a path from A to I via E.

(d) Is there a cycle which visits every vertex?

2 The plan of the ground floor of a house is shown below.

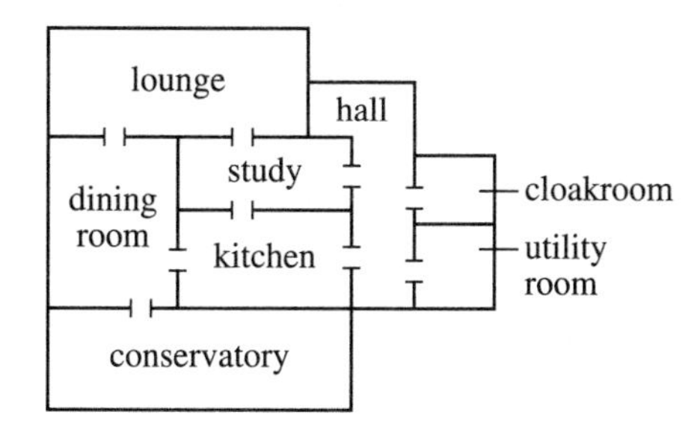

Draw a graph in which the vertices represent rooms, and in which two vertices are connected by an edge if there is a connecting door between the rooms. (This is called a *circulation graph*.)

3 The knight on a chessboard moves two squares in one direction and one square in a perpendicular direction like this:

The knight can move to any of the marked squares

(a) Given a 3×3 chessboard produce a graph in which the vertices represent squares of the chessboard, and in which two vertices are connected by an edge if a knight can move directly between them.

(b) Draw a graph representing a 4×4 chessboard, with the edges connecting vertices if a knight can move directly between them.

4 This graph has two vertices and one edge:

You can translate it in a direction perpendicular to the edge so that the vertices trace out new edges:

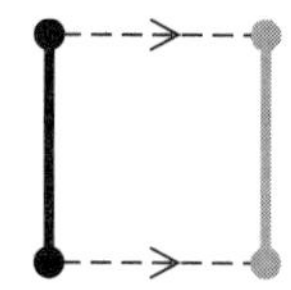

You cannot translate this graph in a direction perpendicular to *all* the edges in two dimensions, but it can be done in three dimensions and can be represented in two dimensions like this:

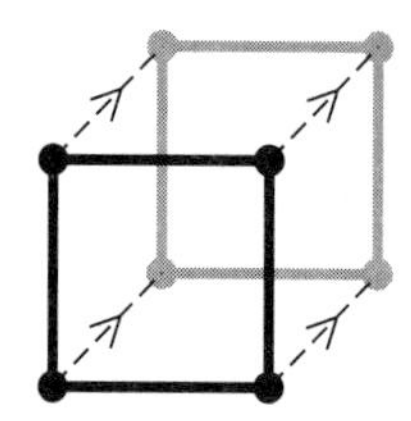

You cannot translate this graph in a direction perpendicular to all the edges in three dimensions, but you *can* represent the resulting graph on paper. Draw this graph.

5

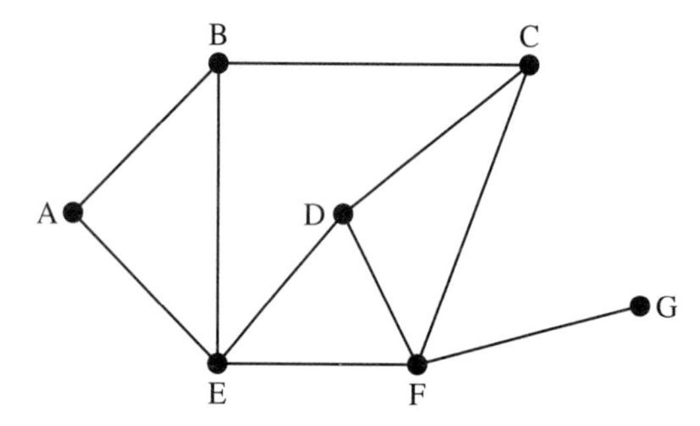

(a) Give a walk which includes every edge of this graph

(b) Find a trail containing the largest possible number of edges.

(c) Find a path containing the largest possible number of vertices.

6 The two graphs shown below are both subgraphs of the knight's move graph from question 3(b). Neither graph is connected. How many connected subgraphs are there in each graph?

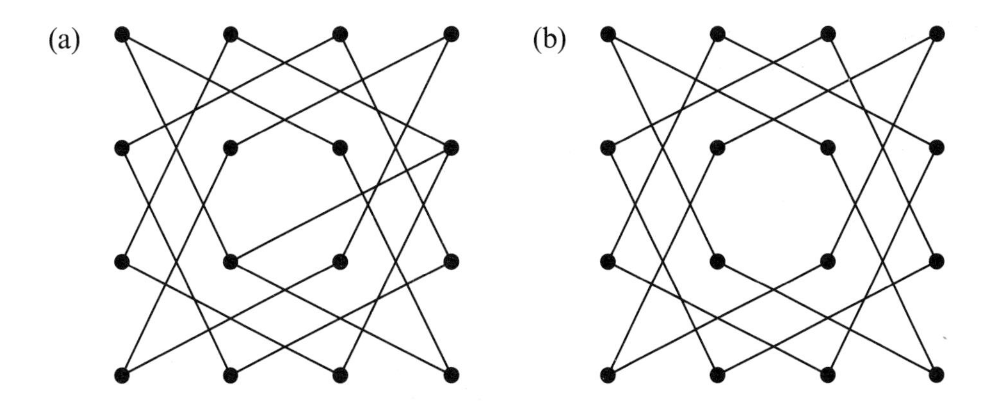

2.3 Other ways of representing graphs

Graphs are very useful for picturing relationships between objects. But graphs do not have to be represented by pictures. They can also be represented by lists or by matrices.

For example, this graph:

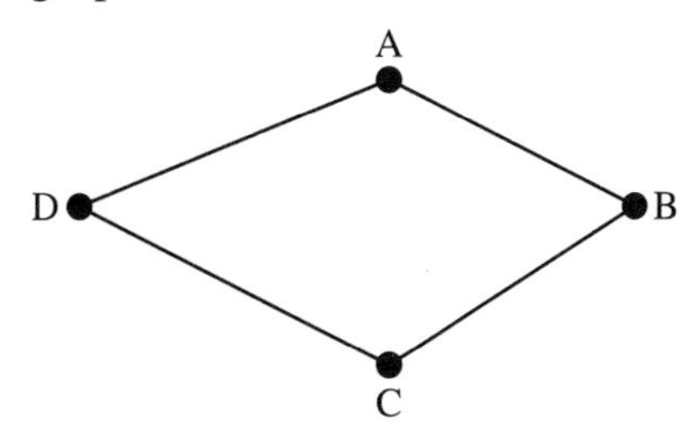

can be represented by this matrix:

		To			
		A	**B**	**C**	**D**
	A	0	1	0	1
From	**B**	1	0	1	0
	C	0	1	0	1
	D	1	0	1	0

The vertex set is {A,B,C,D,}
The edge set is {AB,BC,CD,DA}

Representing a graph by a matrix allows the information stored in the graph to be manipulated by a computer. Some graph problems can be solved much more quickly by running an algorithm on a computer.

The following are all equivalent representations of the same digraph (notice that the digraph is not connected):

(a)

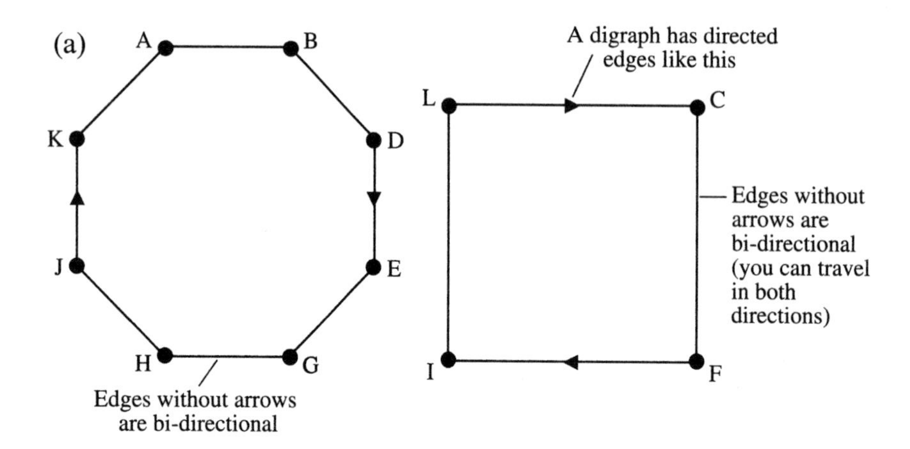

(b) Vertex set = {A, B, C, D, E, F, G, H, I, J, K, L}

$$\text{Edge set} = \left\{ \begin{array}{lllll} AB & BA & BD & DB & DE \\ EG & GE & GH & HG & HJ \\ JH & JK & KA & AK & LC \\ CF & FC & FI & IL & LI \end{array} \right\}$$

(c)

From \ To	A	B	C	D	E	F	G	H	I	J	K	L
A	0	1	0	0	0	0	0	0	0	0	1	0
B	1	0	0	1	0	0	0	0	0	0	0	0
C	0	0	0	0	0	1	0	0	0	0	0	0
D	0	1	0	0	1	0	0	0	0	0	0	0
E	0	0	0	0	0	0	1	0	0	0	0	0
F	0	0	1	0	0	0	0	0	1	0	0	0
G	0	0	0	0	1	0	0	1	0	0	0	0
H	0	0	0	0	0	0	1	0	0	1	0	0
I	0	0	0	0	0	0	0	0	0	0	0	①
J	0	0	0	0	0	0	0	1	0	0	1	0
K	1	0	0	0	0	0	0	0	0	0	0	0
L	0	0	①	0	0	0	0	0	①	0	0	0

This is called an **incidence matrix**.
A one indicates that there is a (directed) edge.
A zero indicates that there is no edge.

a directed edge from L to C

The vertices L and I are connected by a bidirectional edge from L to I and from I to L, so two 1s appear in the matrix

You need to make sure that your pictorial representations are unambiguous. For instance, a picture such as this:

would not be clear. We would not know whether the graph was:

	A	B	C	D
A	2	0	1	1
B	0	0	0	1
C	1	0	0	0
D	1	1	0	0

or

	A	B	C	D
A	2	1	0	1
B	1	0	0	0
C	0	0	0	1
D	1	0	1	0

Exercise 2B

1 Produce matrices to represent the following graphs and digraphs.

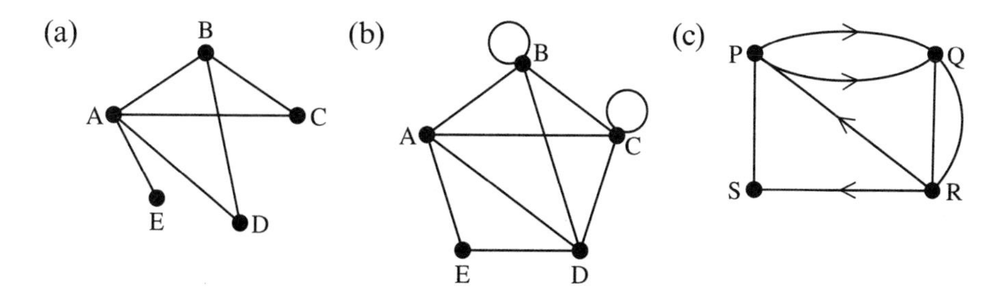

2 Draw a pictorial representation of the following graph.

	A	B	C	D	E
A	1	1	1	0	1
B	1	0	1	1	1
C	1	1	0	0	1
D	0	1	0	1	0
E	1	1	1	0	1

3 (a) Produce an incidence matrix for the circulation graph from question 2 of Exercise 2A

(b) Produce incidence matrices for the two graphs in question 6 of Exercise 2A.

4 Produce a matrix to represent the following digraph, and draw a picture of it:

Vertex set = {V, W, X, Y, Z}

$$\text{Edge set} = \left\{\begin{matrix} \text{VW} & \text{VX} & \text{VY} & \text{VZ} & \text{WV} \\ \text{WY} & \text{XV} & \text{XW} & \text{XX} & \text{XZ} \\ \text{YV} & \text{YW} & \text{YZ} & \text{ZW} & \text{ZY} \end{matrix}\right\}$$

5 (a) Why must the sum of the degrees of all the vertices in any graph always be even?

(b) Deduce a result concerning the number of odd vertices in a graph.

(c) Show that in a group of nine people it is not possible for each to be friends with exactly five others.

6 Draw all simple graphs with four vertices {A, B, C, D} and two edges, one of which is AB.

7 Indicate which of the following are possible, drawing any that are:

(a) a graph with four vertices of degrees 1, 1, 2 and 3

(b) a graph with four vertices of degrees 1, 1, 3 and 3

(c) a simple graph with four vertices of degrees 1, 1, 3 and 3.

8 Leonhard Euler (1707–1783) translated a simple practical problem into a graph theory problem. The problem was to find a route for a walk around Königsberg (now Kaliningrad) crossing each of its seven bridges once and only once.

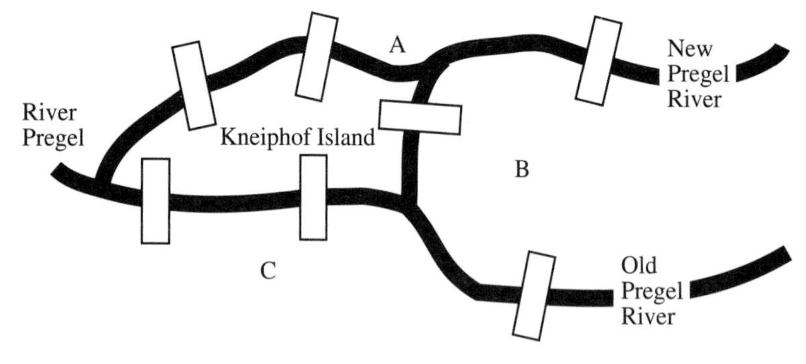

(a) Produce a graph in which the vertices represent areas A, B, C and Kneiphof Island, and the edges represent bridges.

(b) A graph in which it is possible to trace each edge once and only once without lifting the pencil from the paper, is called *traversable*, or *Eulerian*. A graph is Eulerian if it has no odd vertices.
A graph is called *semi-Eulerian* if it has just two odd vertices. Is a semi-Eulerian graph traversable?

(c) Use your graph from part (a) to show that there is no solution to the Königsberg problem.

9 A popular puzzle has four cubes with different colours on their faces. The aim of the puzzle is to stack the cubes into a $4 \times 1 \times 1$ cuboid so that there is no repetition of any colour on any of the long faces of the cuboid:

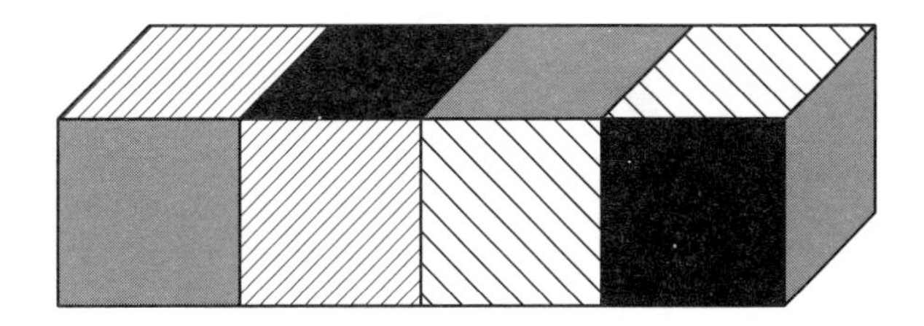

Suppose that the colours used are red, yellow, blue and green, and that the cubes are painted as follows:

Cube no 1	*Cube no 2*	*Cube no 3*	*Cube no 4*
r opposite y	b opposite y	r opposite y	b opposite g
g opposite y	b opposite g	r opposite b	r opposite r
b opposite g	g opposite r	g opposite y	y opposite b

Draw a graph with four vertices labelled r, y, b and g. The graph is to have twelve edges, each numbered 1, 2, 3 or 4. The vertices labelled r and y will be connected by two arcs numbered 1 and 3, because cubes 1 and 3 each have a pair of opposite faces coloured red and yellow.

Find how to split your graph into two separate subgraphs to solve the puzzle. (There are three possible solutions.)

2.4 Trees

- A **tree** is a connected graph with no cycles.

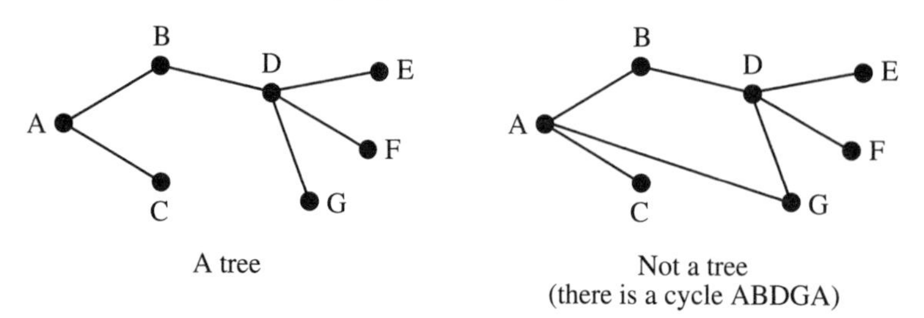

Trees are very useful graphs. You will probably have used computers in which the storage systems are organised in trees:

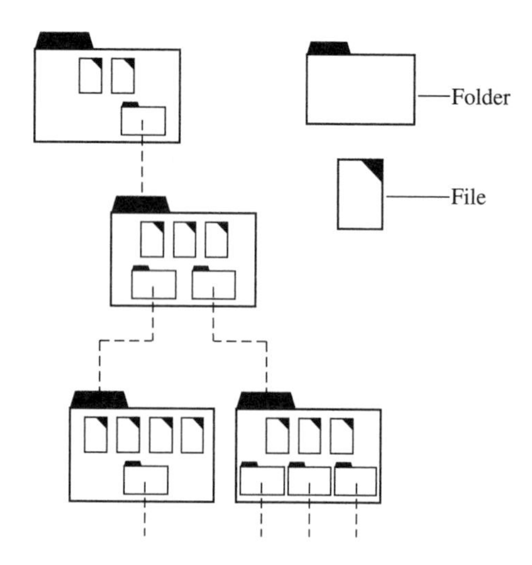

Here are some other examples of trees.

A family tree

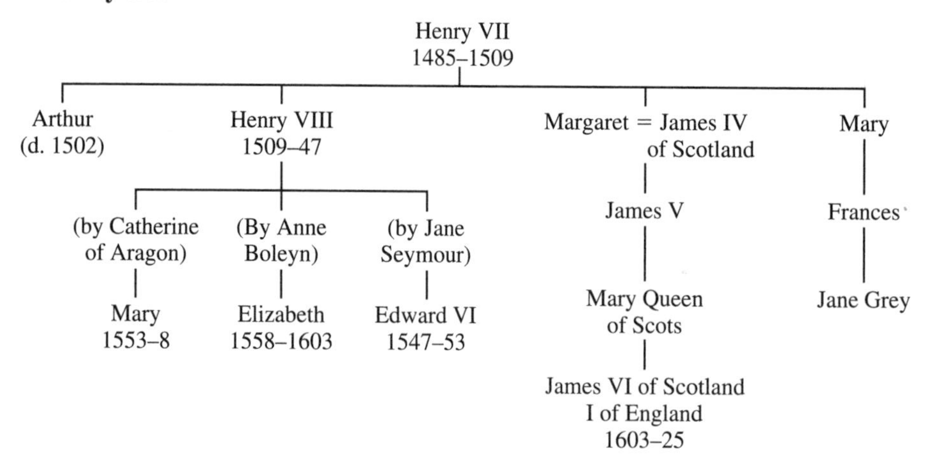

A probability tree.

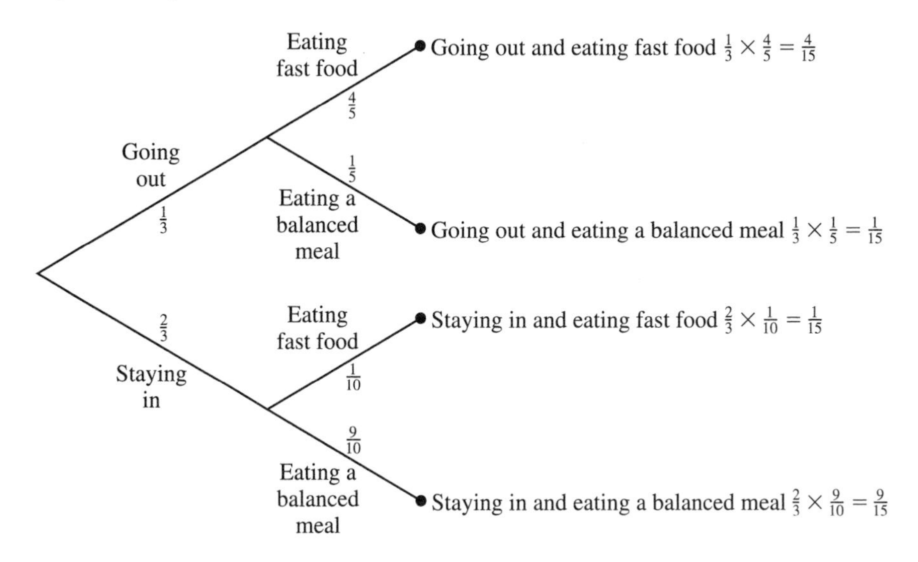

A river system

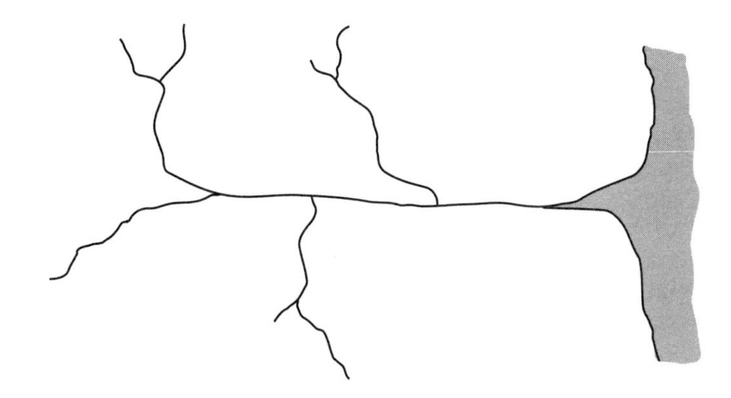

In some of the techniques which follow we shall be solving problems by finding subgraphs which are trees. However, for many problems we need to introduce the idea of a **weighted graph**.

2.5 Networks

To model many real-world problems we need to associate a number called a **weight** with each edge of a graph. To see the need for this, you only have to think about maps, in which the edges have distances associated with them. But the weights need not be distances. They could be times, or costs, or any other quantity that is associated with pairs of vertices.

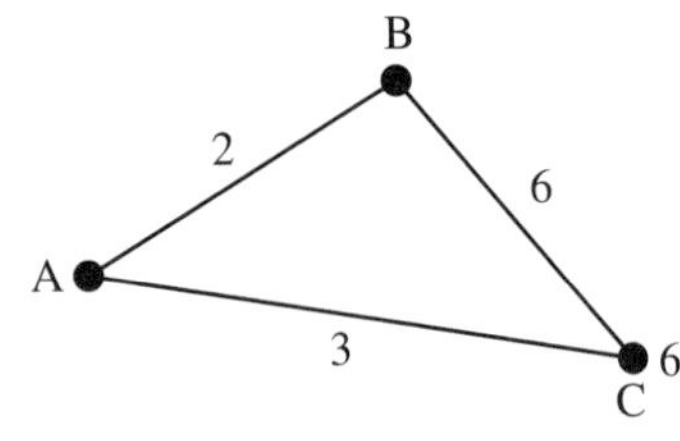

This example shows a weighting of 2 on the edge AB. It could represent the cost of travel between A and B, or the distance between A and B, or ...

Notice that to draw a triangle no one side can be longer than the sum of the lengths of the other two. This is called the triangle inequality. However, networks do not have to satisfy the triangle inequality. In our example:

$$\text{weight (BC)} > \text{weight (BA)} + \text{weight (AC)}$$

This can happen in many situations, even if the weights are lengths. For example, our network could be representing direct routes from B to A and from A to C, with BC representing a by-pass that misses out A but is actually longer than the route via A.

- A weighted graph is known as a **network**.

Here is another weighted graph in which the weights are distances (metres), but which includes one-way streets.

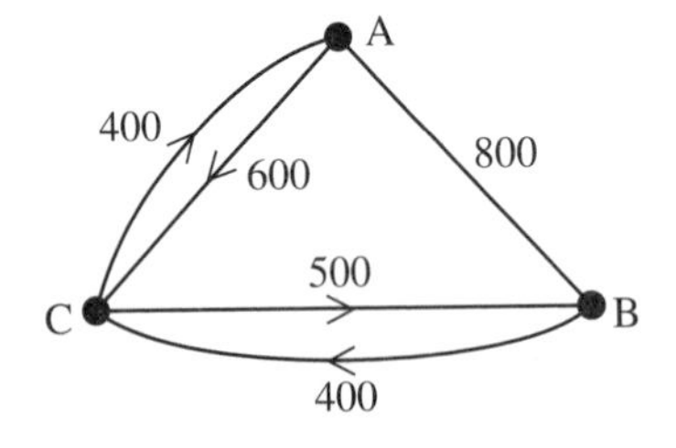

Exercise 2C

1 List all cycles, together with their total weights, in the following network:

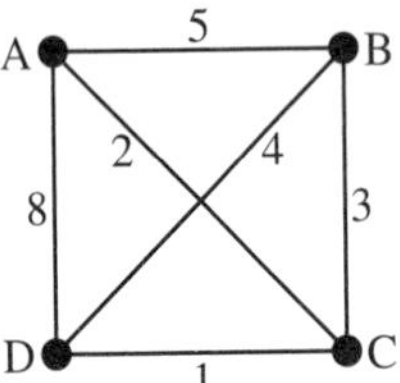

2 In the following network list all paths from A to B, together with their total weight:

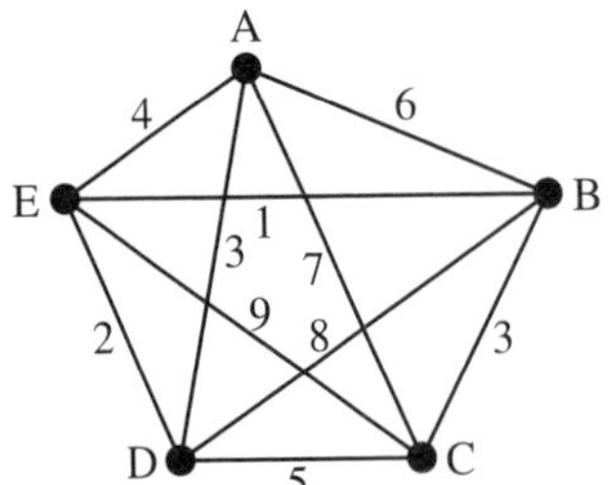

3 A network can be represented by a matrix in which the entries in the matrix represent the weights on the edges. Where there is no edge there is no entry in the corresponding cell of the matrix.

(a) Draw a pictorial representation of the following network:

	P	**Q**	**R**	**S**	**T**
P	12	23	13	—	14
Q	23	—	35	52	17
R	13	35	—	—	67
S	—	52	—	29	—
T	14	17	67	—	15

(b) Give the order of each vertex.

(c) Show that there is no cycle incorporating all the vertices of the graph.

(d) Find a walk of lowest total weight which visits every vertex and returns to the start vertex.

4 (a) Say which of the following graphs are trees.

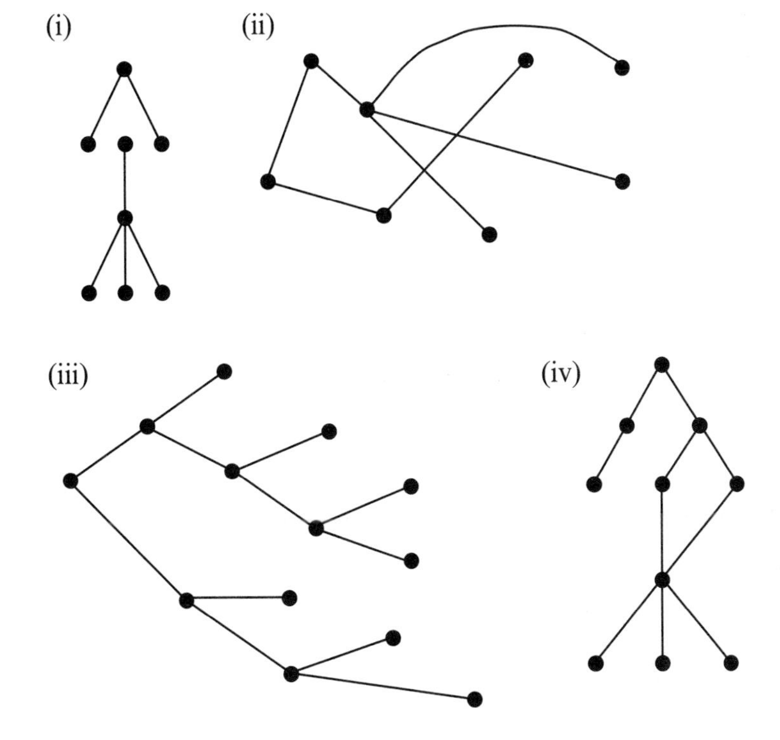

(b) What is the relationship between the number of vertices and the number of edges of a tree?

5 To transmit a message by digital means it first has to be encoded into a string of 0s and 1s. These are binary digits or 'bits'. A Huffman code is an efficient means of achieving this.

For example:

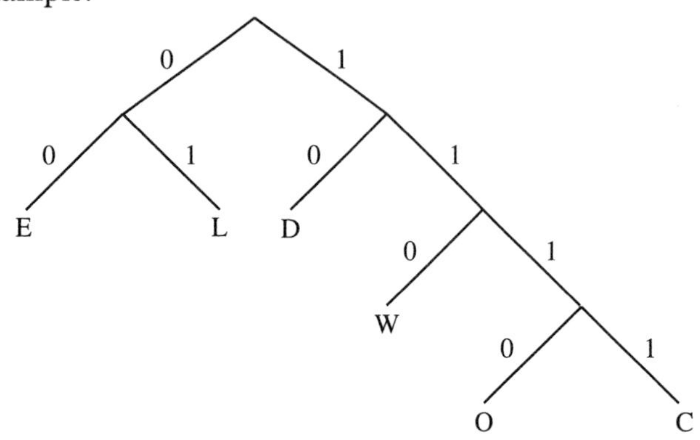

Using this particular Huffman code the letter 'C' would be coded as 1111. The word 'code' would be coded as 111111101000. Codes can be designed in which letters which are used frequently need only a few 'bits'.

(a) Decode the following message:
110000101111111110100010

(b) A language uses the following subset of letters with the given frequencies:

Letter	E	S	T	A	R	I	O	N	G	U
Relative frequency (%)	20	15	15	10	10	10	5	5	5	5

E should be near to the top vertex of the tree, since it occurs often. U can be further down the tree, since it occurs less often.

Design an efficient Huffman code to transmit messages in this language.

6 A walker can pack up to 8 kg in her knapsack. She can choose at most one of each item from the following list, and she wishes to carry items which give the maximum total value.

Item	A	B	C	D	E
Weight (kg)	2	7	2	3	6
Value	5	10	2	4	5

(a) Copy and complete the following tree diagram which represents the walker's problem. A branch of the tree is terminated when the associated weight exceeds 8 kg. This is known as a *bound* – hence the term '*branch and bound*' is used for this approach.
(In the tree diagram 'A' represents selecting A and '∼A' represents not selecting A.)

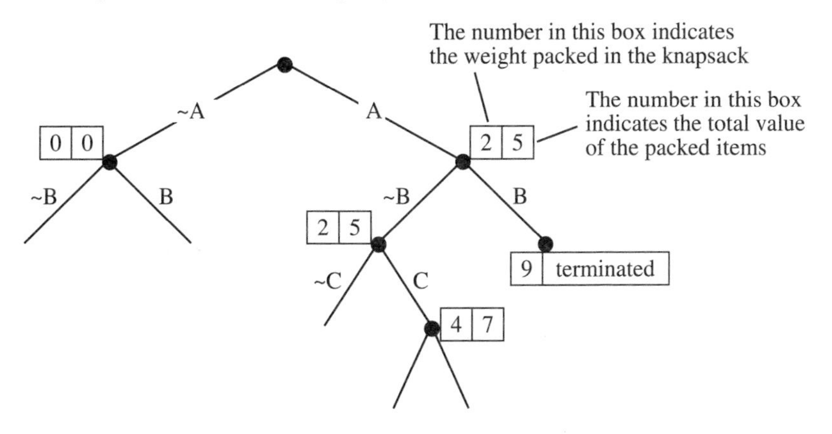

7 A charter aircraft can carry 1000 kg of cargo in addition to passengers. The following cargoes are awaiting shipment:

Cargo type	Units available	Weight per unit (kg)	Delivery fee (£)
A	2	400	440
B	2	250	240
C	4	150	140
D	3	100	60

(a) Find how many units of each type of cargo should be loaded so as to gain the largest total delivery fee.

(b) How should the loading plan be changed if an extra passenger is accepted, reducing the cargo carrying capacity to 900 kg?

2.6 Mathematical modelling

Mathematics is both a subject in its own right ('pure mathematics') and a very powerful problem-solving tool ('applied mathematics'). Graph theory mirrors these two aspects, and examples of both may be found in Exercises 2B and 2C.

The process of using the power of mathematics to solve real-world problems is known as **mathematical modelling**. Example 5 on page 23 (the showman, the wolf, the goat and the cabbage) provides a good example of this process in action. The process involves translating the essential elements of the real-world problem into a mathematical problem.

Real-world problem

In the case of example 5 the situation is modelled by a graph. The problem of transporting the party across the river becomes the problem of finding a path through the graph. The initial vertex of the path is the vertex representing the situation in which the entire party is on one bank of the river. The end vertex is the vertex representing the situation in which they are all on the other bank.

Mathematical problem

The next stage in the process is to solve the mathematical problem. In the case of the showman, the wolf, the goat and the cabbage this was easy – once written down, two paths could easily be seen. In most cases the pure mathematics is now put to use, providing techniques to solve the mathematical problem, and giving further insight into its nature and into any limitations of the solution.

Solution to the mathematical problem

The solution to the mathematical problem then has to be interpreted, in the hope that it provides a solution to the real-world problem. For our example that required the path through the graph to be interpreted as a sequence of river crossings for the showman.

Solution to the real-world problem

It is at this final stage that the adequacy of the modelling is revealed. The process of mathematical modelling involves essential simplification, and it may be that the first attempt fails adequately to capture the essence of the real-world problem. If so then the solution suggested by the solution to the mathematical problem will be inadequate, and it will be necessary to try again with a modified model.

The modelling process:

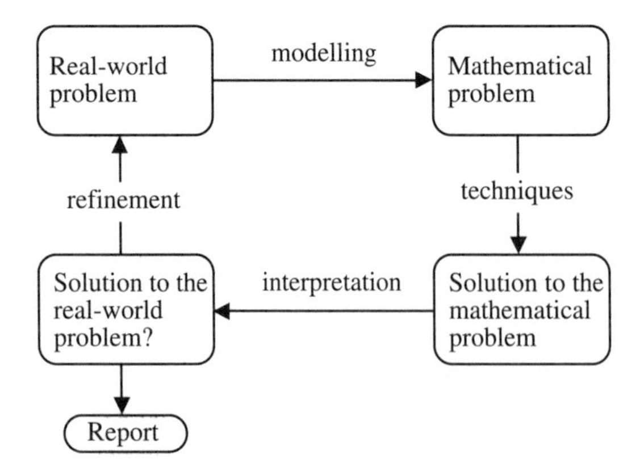

The problems tackled in sections 2.7 and 2.8 below relate to geographical networks, that is, networks in which the vertices represent locations, the edges represent routes connecting those places, and the weights represent the costs of using those routes (in terms of distance, time or expenditure). In these cases the 'modelling' part of the process is obvious and relatively easy. But the other parts of the process should not be forgotten. Nor should it be imagined that the modelling itself is always easy! (Would *you* have thought of modelling the showman/wolf/goat/cabbage problem as a graph?)

2.7 The minimum spanning tree (or minimum connector)

- **For a connected and undirected network, a minimum spanning tree (also known as a minimum connector) is a connected subgraph of minimum total weight incorporating every vertex of the network.**

That's a rather brief, and mathematical, definition of a mathematical problem. To help understand it, let's look at a few examples.

Consider the weighted graph (network):

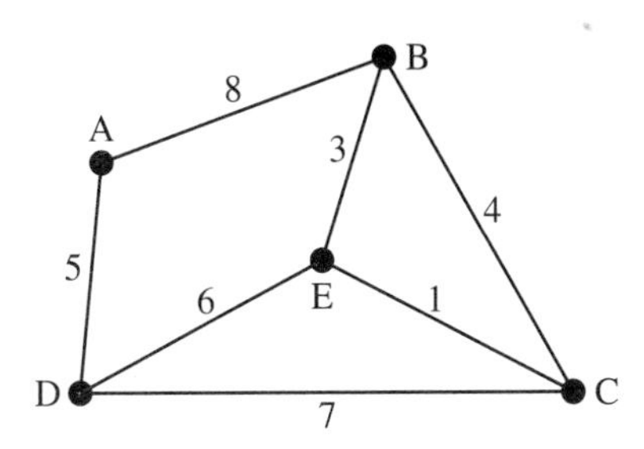

Some spanning trees and their weights are:

(i)

weight $(8 + 5 + 6 + 7) = 26$

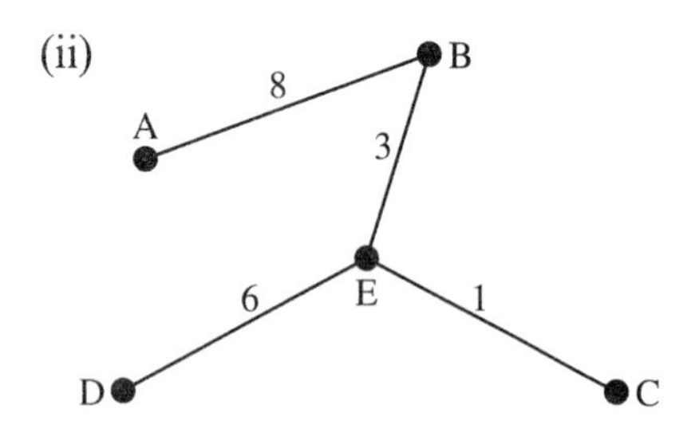

weight $(8 + 3 + 6 + 1) = 18$

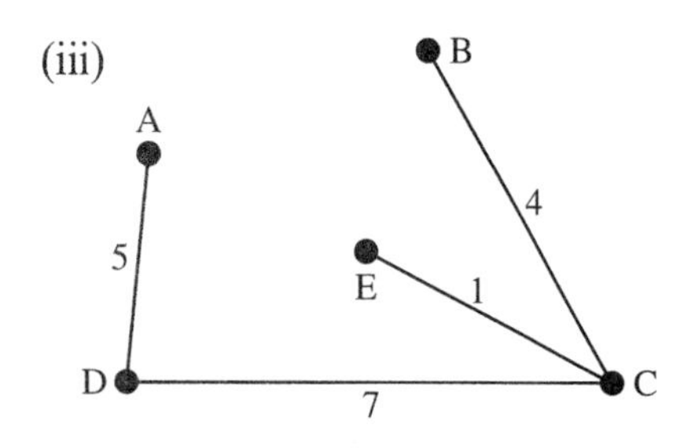

weight $(5 + 7 + 1 + 4) = 17$

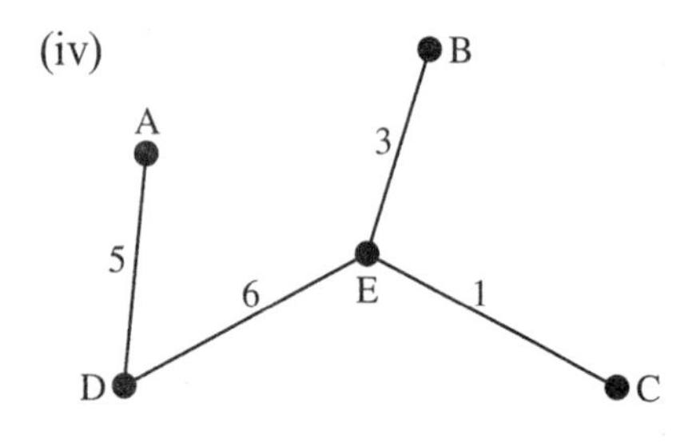

weight $(5 + 6 + 3 + 1) = 15$

Here is a real-world problem:

Imagine that you have to link together a number of computers using a network. The cost of making a connection between two machines is relatively low if they are in the same building, but more expensive the further apart they are. To complete the network every computer must be linked to every other computer, but not necessarily by a direct link – computers can communicate via other computers.

This type of problem can be solved using a mathematical network to represent the computer network. A mathematical network is a weighted graph. In this case the weights represent the costs of using different links.

To find the cheapest way of completing the network you need to find the **minimum spanning tree**, also known as a **minimum connector**. This is a subgraph (part of the whole graph) of minimum total weight.

Here are the costs of links for this computer network, in units of £100. The problem is to link the computers at minimum cost.

	A	B	C	D	E	F	G	H	I	J	K
A		5				5				3	
B	5				25	5				6.5	
C				2					1.5	27	17
D			2				20		2		
E		25					3	2			2
F	5	5						30		6	
G				20	3			4			4
H					2	30	4			31	3
I			1.5	2							20
J	3	6.5	27			6		31			
K			17		2		4	3	20		

First draw a pictorial representation of the network:

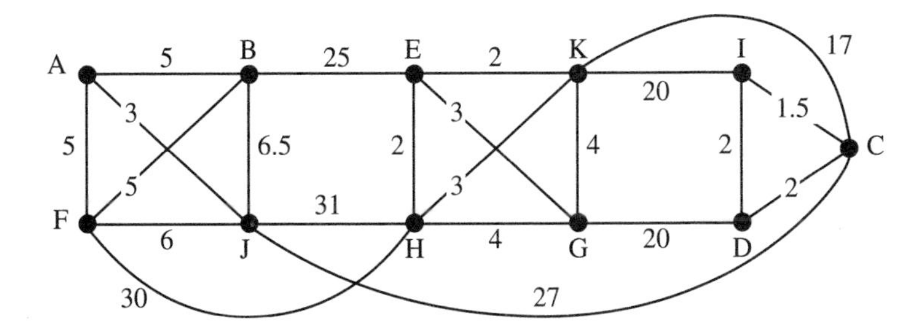

Now we have a picture of the network we need to find the minimum spanning tree. The next two sections show you two algorithms you can use to solve problems of this type.

The techniques for solving mathematical problems of this type consist of algorithms (see chapter 1). In practice the real-world problems are large, and the algorithms need to be efficient – even when we are implementing them on fast computers. Both of the algorithms which we give split into stages, and at each stage there is a choice to be made. Both end up with an optimal result. There may be more than one minimum spanning tree, so they may not produce the same tree, but they are certain to produce the same length.

2.8 Prim's algorithm: finding a minimum spanning tree from a graph

Step 1 Choose a starting vertex.

Step 2 Connect to it the 'nearest' vertex, that is, out of all the edges joined to the start vertex, choose (the) one with minimum weight. (Notice that there might be a choice here.)

Step 3 Connect to the tree of connected vertices that vertex which is 'nearest' to the connected set. (Again, there may be a choice.)

Step 4 Repeat step 3 until all vertices are connected.

> **Warning — a common error**
> Step 3 is quite explicit, but is often misread/misunderstood to read:
> **Step 3** Connect to the tree of connected vertices that vertex which is 'nearest' to the last vertex connected.

- **Prim's algorithm builds a minimum spanning tree by adding one vertex at a time (and an associated edge) to a connected subgraph. It has two forms, graphical and matrix.**

Example 6

Use Prim's algorithm to find a minimum spanning tree for this network

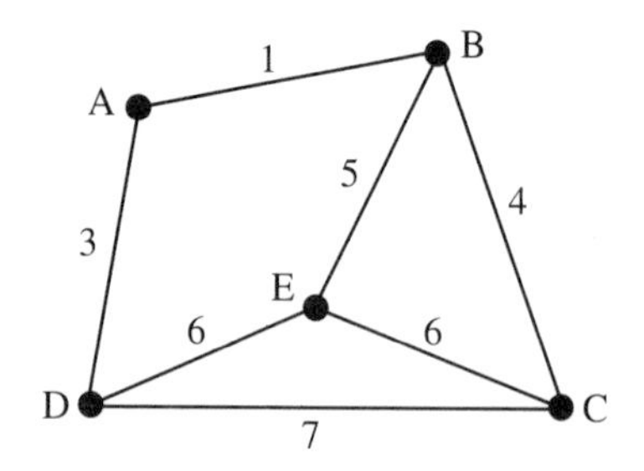

Step 1 Choose vertex A.

Step 2 Nearest vertex to A is vertex B

Step 3 (i) Vertex nearest to either A or B is vertex D.

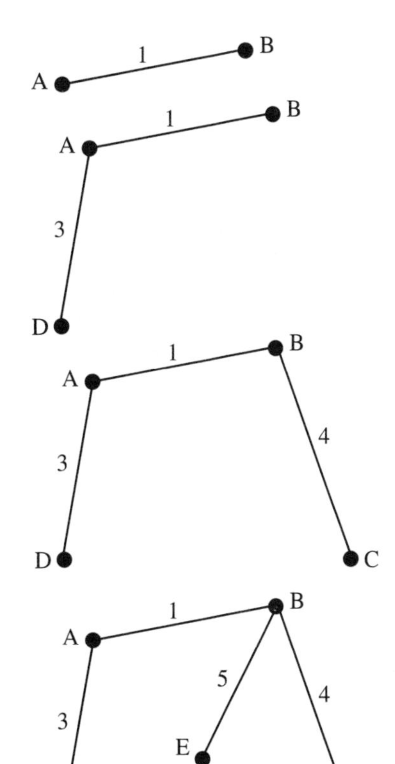

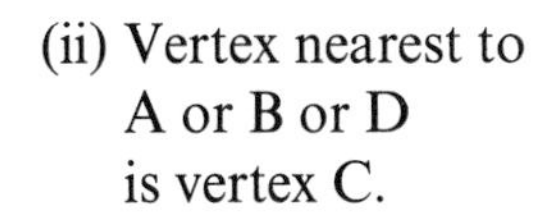
(ii) Vertex nearest to A or B or D is vertex C.

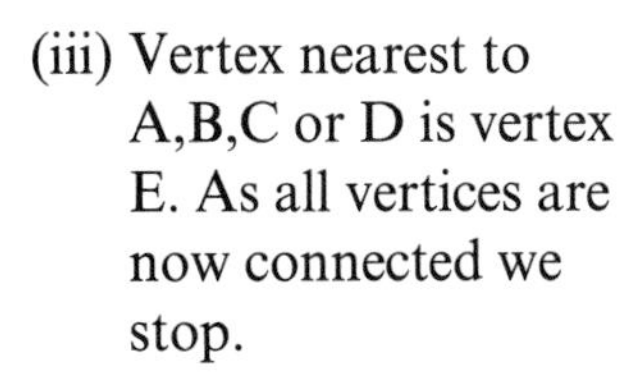
(iii) Vertex nearest to A,B,C or D is vertex E. As all vertices are now connected we stop.

All vertices are now connected and we have a minimum spanning tree (MST) for the given network. It is of weight 13. In this case it is unique since at no stage was there a choice of which vertex to include.

Prim's algorithm applied to the computer network

Here is the result of applying the algorithm to the problem on page 45:

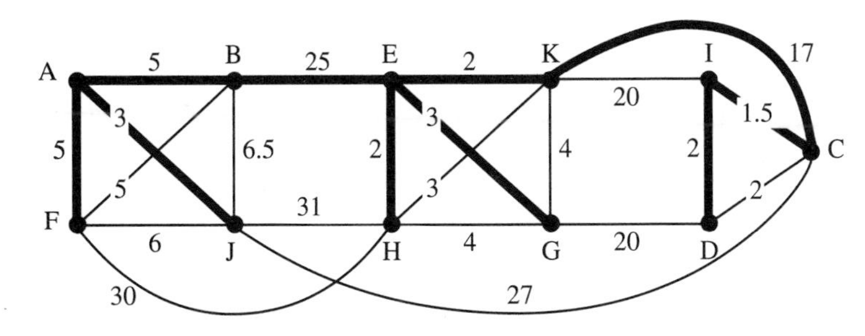

Length of minimum connector = 65.5 (representing £6550).

The minimum connector highlighted above was started from vertex A. The order in which edges were included was as follows:

AJ, AB, AF, BE, EK, EH, EG, KC, CI, ID.

Applying Prim's algorithm to a matrix

Here is how to use Prim's algorithm for a matrix representation of a graph.

Step 1 Start with the matrix representing the network and choose a starting vertex. Delete the *row* corresponding to that vertex.

Step 2 Label with a '1' the *column* corresponding to the start vertex, and ring the smallest undeleted entry in that *column*. (There may be a choice of entry to ring.)

Step 3 Delete the *row* corresponding to the entry that you have just ringed.

Step 4 Label (with the next label number) the *column* corresponding to the vertex which you have just ringed.

Step 5 Ring the lowest undeleted entry in *all* labelled columns. (There may be a choice here.)

Step 6 Repeat instructions 3, 4 and 5 until all rows are deleted. The ringed entries represent the edges in the minimum connector.

Example 7

Use Prim's algorithm to find a minimum spanning tree for the network represented by this matrix.

	A	B	C	D	E
A	—	8	—	5	—
B	8	—	4	—	3
C	—	4	—	7	1
D	5	—	7	—	6
E	—	3	1	6	—

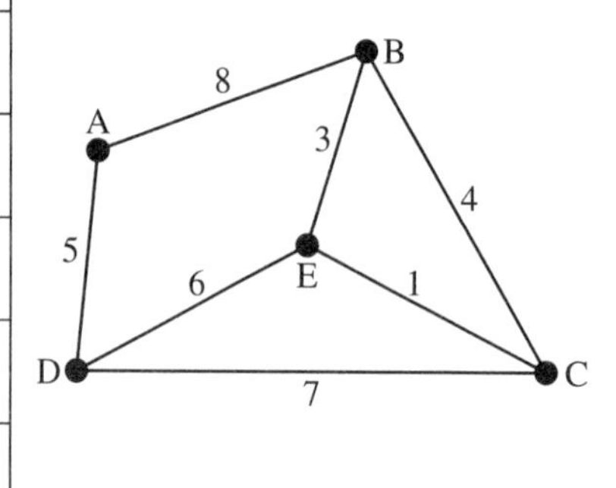

Step 1 Choose vertex A and delete the row corresponding to A.

Step 2 Label the A column with 1. Ring the smallest entry in column A. This is the 5 in row D.

Step 3 Delete row D.

	1				
	A	B	C	D	E
A	~~—~~	~~8~~	~~—~~	~~5~~	~~—~~
B	8	—	4	—	3
C	—	4	—	7	1
D	(5)	~~—~~	~~7~~	~~—~~	~~6~~
E	—	3	1	6	—

Step 4 Label D column with 2.

Step 5 Ring smallest entry in either A column or D column.
This is the 6 in row E.

	1			2	
	A	**B**	**C**	**D**	**E**
A	~~—~~	~~8~~	~~—~~	~~5~~	~~—~~
B	8	—	4	—	3
C	—	4	—	7	1
D	(5)	~~—~~	~~7~~	~~—~~	~~6~~
E	—	3	1	(6)	—

Step 3 Delete the row corresponding to E.

Step 4 Label the E column as 3.

Step 5 Ring smallest entry in columns A, D or E.
This is the 1 in row C.

	1			2	3
	A	**B**	**C**	**D**	**E**
A	~~—~~	~~8~~	~~—~~	~~5~~	~~—~~
B	8	—	4	—	3
C	—	4	—	7	(1)
D	(5)	~~—~~	~~7~~	~~—~~	~~6~~
E	~~—~~	~~3~~	~~1~~	(6)	~~—~~

Step 3 Delete the row corresponding to C.

Step 4 Label the C column as 4.

Step 5 Ring smallest entry in columns C, D or E.
This is the 3 in row B.

Step 3 Delete the row corresponding to B.

	1		4	2	3
	A	**B**	**C**	**D**	**E**
A	~~—~~	~~8~~	~~—~~	~~5~~	~~—~~
B	~~8~~	~~—~~	~~4~~	~~—~~	(3)
C	~~—~~	~~4~~	~~—~~	~~7~~	(1)
D	(5)	~~—~~	~~7~~	~~—~~	~~6~~
E	~~—~~	~~3~~	~~1~~	(6)	~~—~~

All rows are now deleted. The edges of the minimum connector or minimum spanning tree are AD, DE, EC, and EB, with total weight of $5 + 6 + 1 + 3 = 15$.

Applying the matrix form of Prim's algorithm to the computer network

Here is the result of applying the matrix form of the algorithm to the computer networking problem on page 45. To help you to follow it the rows have been labelled with the order in which they were deleted:

		1	3	9		5	4	8	7	10	2	6
		A	**B**	**C**	**D**	**E**	**F**	**G**	**H**	**I**	**J**	**K**
1	**A**		5				5				3	
3	**B**	(5)				25	5				6.5	
9	**C**				2					1.5	27	(17)
11	**D**			2				20		(2)		
5	**E**		(25)					3	2			2
4	**F**	(5)	5						30		6	
8	**G**				20	(3)			4			4
7	**H**					(2)	30	4			31	3
10	**I**			(1.5)	2							20
2	**J**	(3)	6.5	27			6		31			
6	**K**			17		(2)		4	3	20		

The ringed entries in the table show the edges that have been included in the minimum spanning tree. The ⑤ in the second row and first column shows that the edge BA is included. Thus the minimum spanning tree produced by this method is:

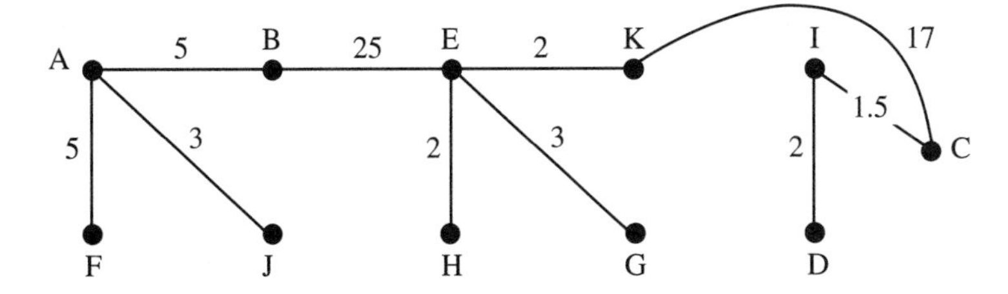

How this method works

The method works by building up a connected set of vertices, recognised by the labelled columns. At each stage the unconnected vertex nearest to the connected set is chosen at step 5. Deleting the row corresponding to that vertex stops us from trying to include it again.

Exercise 2D

1 Use the matrix form of Prim's algorithm to find a minimum spanning tree for the network:

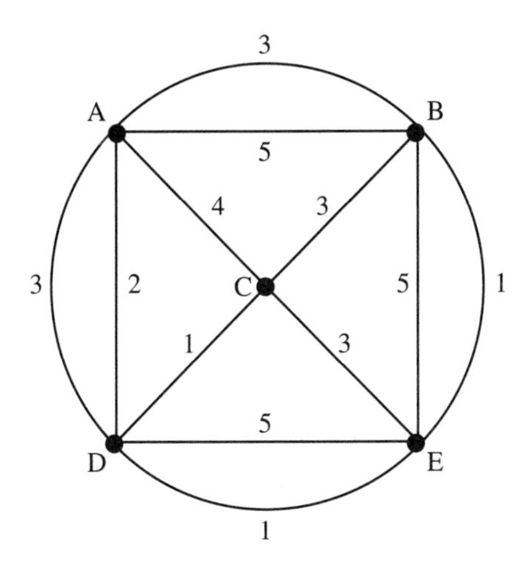

2 Use both forms of Prim's algorithm to find minimum spanning trees for the network shown below.

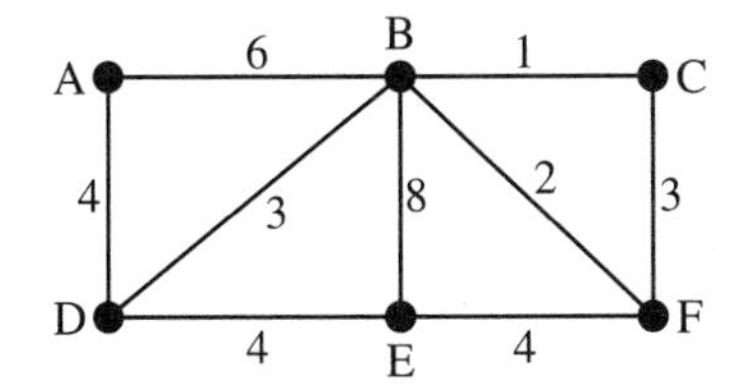

3 The table gives the distance (in miles) between six places in Ireland. Use the matrix form of Prim's algorithm to find a minimum spanning tree connecting these places.

	Athlone	Dublin	Galway	Limerick	Sligo	Wexford
Athlone	—	78	56	73	71	114
Dublin	78	—	132	121	135	96
Galway	56	132	—	64	80	154
Limerick	73	121	64	—	144	116
Sligo	71	135	80	144	—	185
Wexford	114	96	154	116	185	—

4 The table gives the distances (in miles) between six places in Scotland. Draw the network and use the graphical form of Prim's algorithm to find a minimum spanning tree.

	Aberdeen	Edinburgh	Fort William	Glasgow	Inverness	Perth
Aberdeen	—	125	147	142	104	81
Edinburgh	125	—	132	42	157	45
Fort William	147	132	—	102	66	105
Glasgow	142	42	102	—	168	61
Inverness	104	157	66	168	—	112
Perth	81	45	105	61	112	—

2.9 Kruskal's algorithm for finding a minimum spanning tree

At each stage Prim's algorithm adds an edge to the connected set by finding the vertex which is 'nearest' to the set. Kruskal's algorithm focuses entirely on the edges – and is happy to work with a disconnected set.

To follow Kruskal's algorithm you work through the graph gradually building up a tree by trying out the edges that have the smallest available weight. If the next available edge would create a cycle you don't add it to the tree.

At the start all the edges are assumed to be available for inclusion in the tree. As the tree is gradually built up some edges will be included in it and others will be excluded.

- **Kruskal's algorithm builds a minimum spanning tree by adding an edge at a time (and associated vertices) to a subgraph. The subgraph need not be connected at intermediate stages, though the final subgraph is connected.**

Step 1 Choose an available edge with least weight. (Clearly there may be choices available.) If, when taken together with some of the edges labelled 'included', it forms part of a circuit, then label it 'excluded'. Otherwise label it 'included'.

Step 2 Repeat step 1 until no edges are labelled 'available'.

Example 8

Apply Kruskal's algorithm to this network:

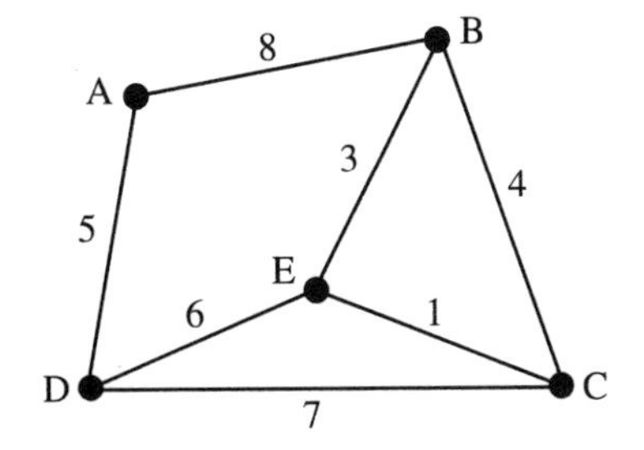

First choice — Edge of smallest weight is edge EC of weight 1.
EC is now 'included'.

Second choice — Edge of smallest weight 'available' is edge EB of weight 3.
EB is now 'included'.

Third choice — Edge of smallest weight 'available' is edge BC of weight 4.
However, this would form a cycle with 'included' edges so is 'excluded'.

Edge of smallest weight now 'available' is edge AD of weight 5.
This does not form a cycle with 'included' edges and so is included.
Notice that because the parts of the graph added to the tree so far are not connected we now have a disconnected graph – this does not matter.

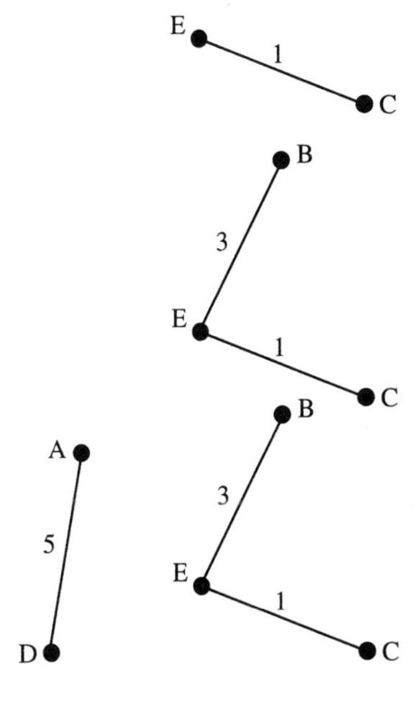

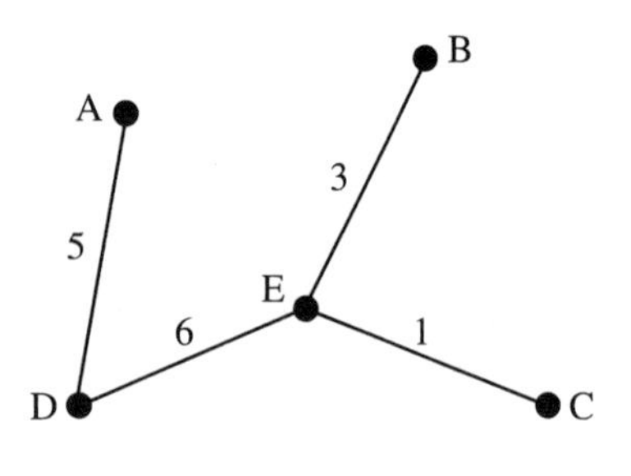

Fourth choice Edge of smallest weight 'available' is edge DE of weight 6. This does not form a cycle with 'included' edges and so is included.

Fifth choice Edge of smallest weight 'available' is edge DC of weight 7. However, this would form a cycle with 'included' edges so is 'excluded'.

Edge of smallest weight 'available' is edge AB of weight 8. However, this would form a cycle with 'included' edges so is 'excluded'.

There are no further edges available, and so we have the minimum spanning tree.

You might have noticed in the previous worked example that the minimum spanning tree was completed after the fourth choice had been made, and that subsequent work was not needed. In fact you can use a more efficient stopping criterion at step 2.

A better stopping condition ...

Step 2 Repeat step 1 until $n - 1$ edges are 'included', where n is the number of vertices.

Kruskal's algorithm *appears* to be (even) easier than Prim's. However, step 2 is actually quite complex. It is quite difficult in a large and complex network to check for circuits. It is necessary to construct an algorithm (within the algorithm!) so that the check can be incorporated in a computer program.

Example 9

Using Kruskal's algorithm find as many minimum spanning trees as possible for the following network:

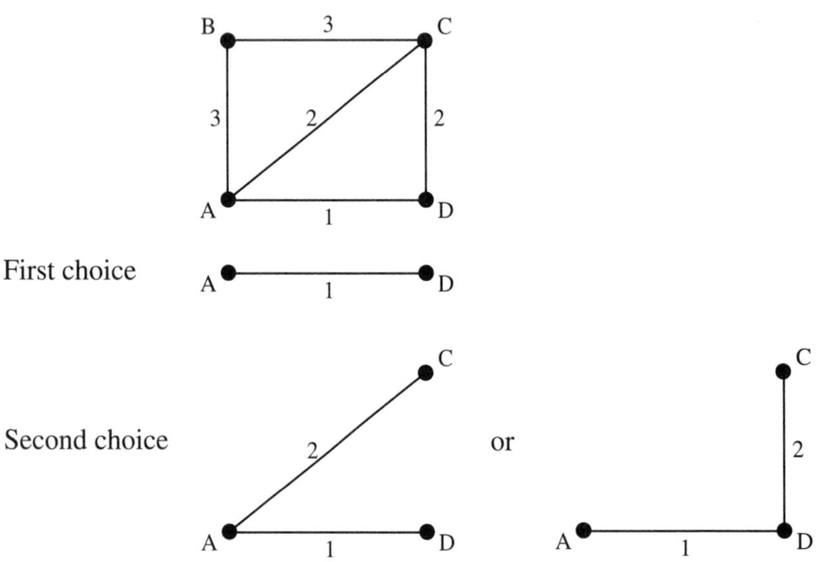

Third choice

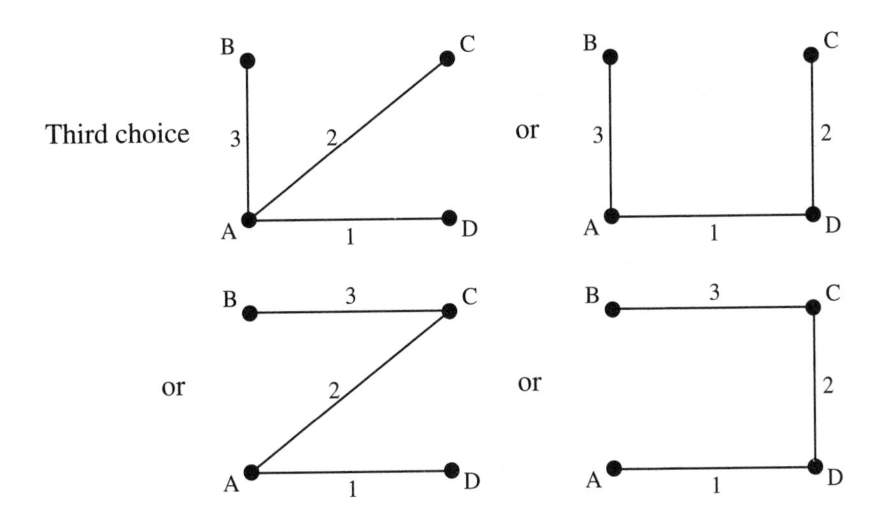

Notice that all four minimum spanning trees have weight 6.

Why is Kruskal's algorithm greedy?

Kruskal's algorithm is often called a 'greedy algorithm'; in fact both Prim and Kruskal are greedy. They each split into stages, and at each stage there is a choice to be made. Greedy algorithms take the best available at each stage, and by doing so end up with the optimal results. In chapter 3 we shall meet a problem for which greediness is not the best policy.

Exercise 2E

1 Draw a pictorial representation of the network defined by the following matrix:

$$\begin{pmatrix} 0 & 12 & 23 & 15 & 18 & 10 \\ 12 & 0 & 20 & 21 & 24 & 13 \\ 23 & 20 & 0 & 29 & 31 & 14 \\ 15 & 21 & 29 & 0 & 9 & 25 \\ 18 & 24 & 31 & 9 & 0 & 17 \\ 10 & 13 & 14 & 25 & 17 & 0 \end{pmatrix}$$

(a) Use Kruskal's algorithm to find the minimum connector, and give its total weight.

(b) Use the matrix form of Prim's algorithm to find the minimum connector.

2 The map shows six cities, A, B, C, D, E and F, together with the costs involved (£ million) in building pipelines to connect them.

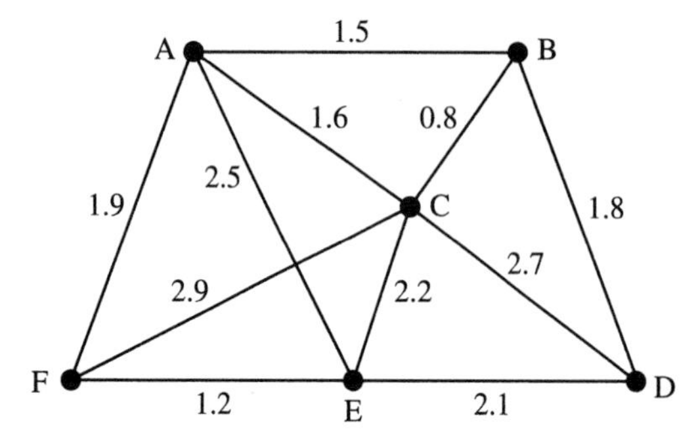

Use Prim's algorithm (graphical form) to find the least cost pipe network which connects all six cities together.

3 (a) Produce the matrix which represents the network in question 2.
(b) Use the matrix form of Prim's algorithm to find a minimum spanning tree.

4 Five new houses, A, B, C, D and E, are to be connected to a drainage system. Each is to be connected to the sewer at the point S on the diagram, either directly or via another house. Alternatively, houses may be connected to an intermediate manhole at M, directly or via another house. This manhole must in turn be connected to S, either directly or via another house. All connecting pipes must be such that water can drain downhill, and all connections are to be made in a straight line. The direction of drainage is shown on the diagram.

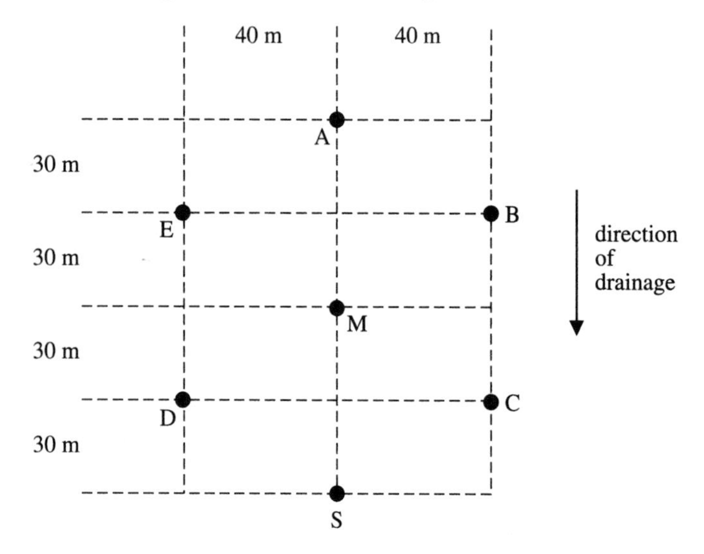

(a) Starting from S, use Prim's algorithm to find a minimum connector for A, B, C, D, E, M and S.
Show which pipes are in your connector, indicating the order in which they were included, and give their total length.
Does your connector represent a system which drains correctly?

(b) Investigate whether or not the provision of the intermediate manhole at M is worthwhile. Justify your conclusions.

5 The vertices of the following graph are at the vertices of a regular hexagon with sides of length 2. The hexagon has two sides missing and has four interior edges. The weights on the edges are equal to their lengths as drawn.

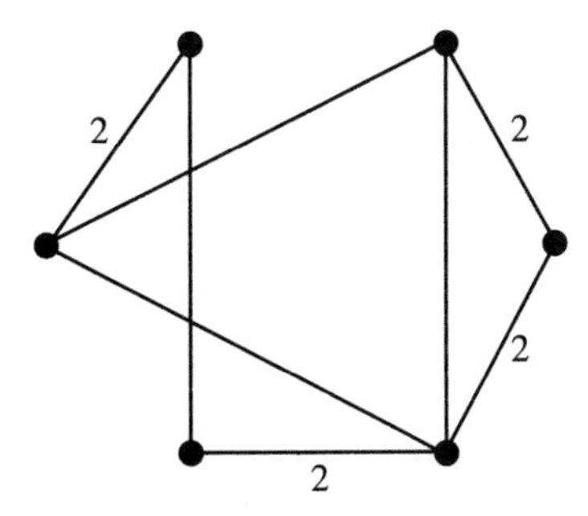

(a) Find all the minimum spanning trees.

(b) Give their length.

6 (a) The points on the graph below may be connected in pairs by straight lines.

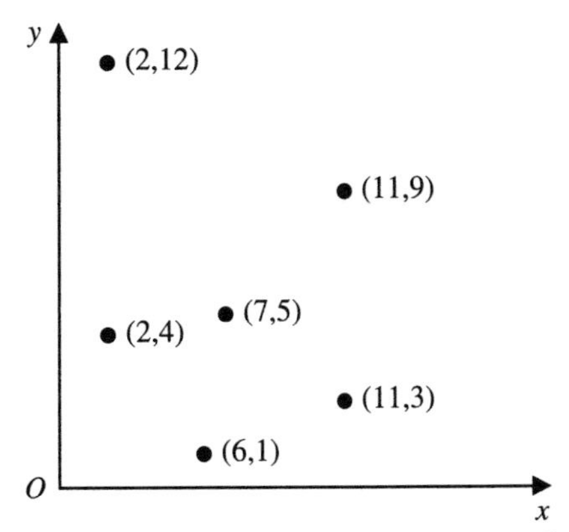

Use Prim's algorithm to find a connector of minimum length for the six points.

(b) The points on the graph below may also be connected in pairs by straight lines.

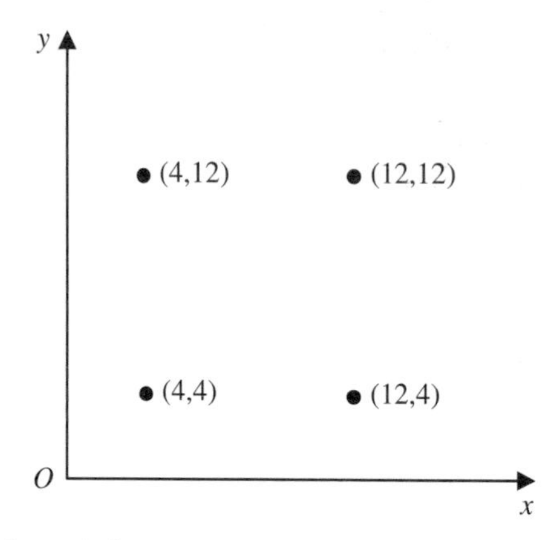

(i) Find a minimum connector for the four points.

(ii) Locate a fifth point so that the minimum connector for the five points is shorter than the minimum connector for the original four points. Show your minimum connector and give its length.

7 The *complete* graph on n vertices is the simple graph in which every vertex is connected to every other vertex.
The complete graph on five vertices is:

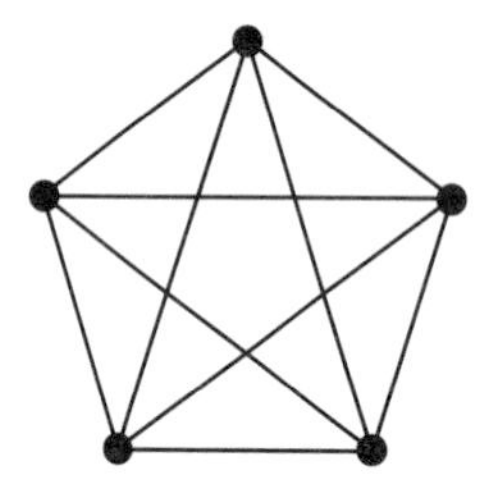

Compare the time that you would expect it to take to execute Prim's algorithm on a complete network on six vertices, compared with that on a complete network on five vertices.
Repeat the above for a complete network on 50 vertices.

8 An international publisher employs authors to produce books in their native languages. Translations are then made so that the books can be sold in other countries, although not all books are suitable for all countries. The costs (£00) of making translations between 10 principal languages, where translators are available, are as follows:

From language

To language	A	B	C	D	E	F	G	H	I	J
A	—	3	3	10	3	4	—	2	1	2
B	3	—	2	4	—	1	1	4	2	1
C	2	1	—	2	—	4	1	1	—	3
D	10	4	3	—	5	—	3	6	2	1
E	3	—	—	5	—	6	4	5	3	5
F	4	1	6	—	6	—	2	3	1	1
G	—	1	2	3	4	4	—	2	4	5
H	2	4	2	6	5	5	2	—	3	1
I	1	2	—	2	3	2	4	3	—	2
J	2	1	4	1	5	1	5	1	2	—

(a) A book has been written in language F, and the publisher wishes to have it translated into *all* other languages (a worldwide translation). Find the minimum cost method of achieving this.

(b) The cost matrix is not symmetric. What effect does this have? Would the cost of making a worldwide translation be the same for a book originally written in language C?

(c) Why would translating from language D to language A never be used in making a worldwide translation? Is it worth retaining the services of the A↔D translator?

2.10 Finding the shortest path through a network

Why is the walk of least weight a path?

How can you find the shortest path (or the path of least total weight) between any two vertices connected by a network?

There are many real-world problems for which this mathematical problem is a model. The most obvious are those which involve finding the shortest distance/time/cost route between two points in a geographical network.

In this directed network edge weights represent distances. The problem is to find the shortest route from A to G.

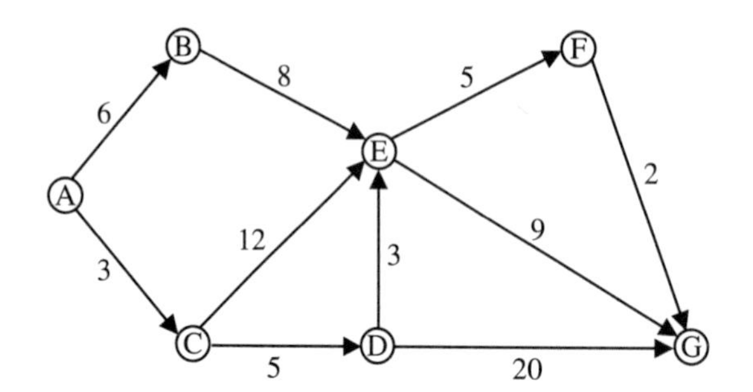

Here are some paths from A to G with their lengths:

ABEFG	$(6+8+5+2)=21$
ACDG	$(3+5+20)=28$
ACEG	$(3+12+9)=24$
ACDEFG	$(3+5+3+5+2)=18$

This is a fairly simple network so you could find the shortest path fairly easily by trying out different routes. Sometimes, though, networks can be very complex. Here is an example to make the point. *You are not expected to be able to solve this problem.*

A complex network problem

A production plan is required for a factory for the next six months. The factory can make four different products, but can make only one product in a given month.

The factory is currently set up to make product A, and it must be returned to that state at the end of the six month period. The costs of switching production from one product to another are (in £000):

Cost of switching production

		To			
		A	**B**	**C**	**D**
From	**A**	0	10	12	6
	B	8	0	8	4
	C	10	8	0	10
	D	5	6	9	0

The profits to be made from each product vary from month to month, partly due to fluctuations in the prices of raw materials. Those profits are (£000):

	Month					
Product	**1**	**2**	**3**	**4**	**5**	**6**
A	30	35	30	40	35	45
B	20	30	40	50	60	50
C	10	20	30	40	50	50
D	40	30	20	60	30	20

We can model this problem as a shortest path problem through a directed network. The network is:

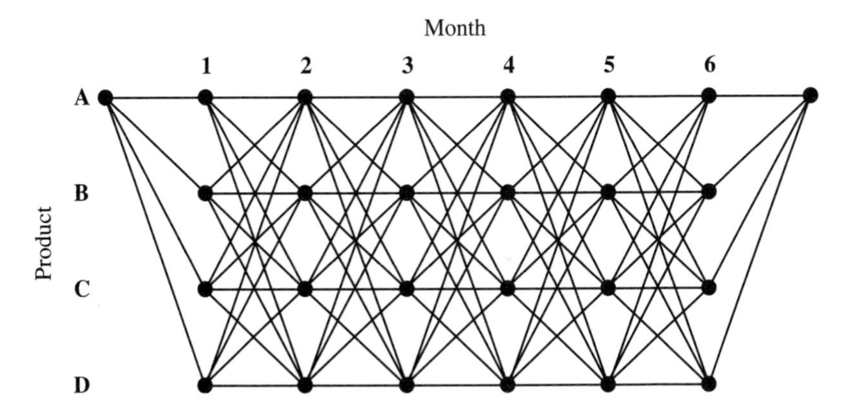

Each vertex represents a product being made in a particular month, except that the left-hand vertex represents the fact that we start with the factory set up to produce A, and the right-hand vertex shows that it must be returned to that state at the end of the six month production run.

The edges are directed from left to right, but there are so many of them that there is no room for the arrow heads! Each edge has a weight, but again, there are too many to put all of them on the diagram. Some of the weights are shown below to show how they are computed:

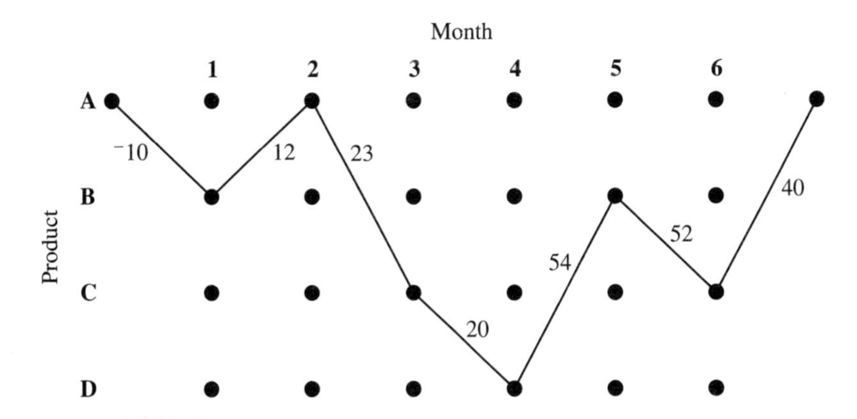

The -10 represents the cost of switching from product A to product B for the first month's production.

The 12 represents the profit of 20 from B in month 1, less the cost of 8 for switching from B to A for month 2.

The 23 represents the profit of 35 from A in month 2, less the cost of 12 for switching from A to C for month 3.

The 20 represents the profit of 30 from C in month 3, less the cost of 10 for switching from C to D for month 4.

The 54 represents the profit of 60 from D in month 4, less the cost of 6 for switching from D to B for month 5.

The 52 represents the profit of 60 from B in month 5, less the cost of 8 for switching from B to C for month 6.

The 40 represents the profit of 50 from B in month 6, less the cost of 10 for switching from C back to A at the end of the production run.

Thus the diagram shows one out of the many possible paths through the network, and represents one out of the many possible production plans. The plan is to produce B in month 1, then A in month 2, then C, then D, then B and finally C. The total weight of the path represents the profit from the production plan, and in this case is $^{-}10 + 12 + 23 + 20 + 54 + 52 + 40 = 191$ (£000s).

Algorithmic complexity

Here is an algorithm for finding the minimum weight path between two vertices in a network.

1. List all the paths between the two vertices.
2. Find the total weight of each path.
3. Choose the path of least total weight.
   (There may be more than one.)

This is known as **complete enumeration**. However, for the factory problem there are 4096 paths to list, corresponding to 4096 possible production plans! In real-life problems the number of possibilities can easily mushroom to an extent that can defeat even the capabilities of fast computers. The extent to which the number of possibilities that need to be examined increases as the size of problem increases is known as the **complexity** of the algorithm. So the complexity of an algorithm gives an indication of the time it will take to solve a problem using it. (There is more about algorithmic complexity on page 233.)

Can you see that there are $4^6 = 4096$ possible production plans?

To solve problems like this you need an algorithm which is more efficient than complete enumeration. Dijkstra's algorithm is one such algorithm.

2.11 Dijkstra's algorithm

Dijkstra's algorithm is an example of a **labelling algorithm**. It finds the shortest route from an initial vertex to any other vertex in a network. At each iteration a fresh vertex is assigned a label. This label gives the shortest 'distance' from the start or source vertex. Since problems are often geographical, 'distance' is used to mean the weight of a path.

The algorithm works gradually through the network, labelling one vertex at each stage – when the shortest distance is known for certain – and improving working values as it goes. These working values give the shortest distances found so far to unlabelled vertices.

Dijkstra is pronounced 'Dike stra'

It is convenient, when applying the algorithm by hand, to record working values and labels in boxes (one for each vertex) thus:

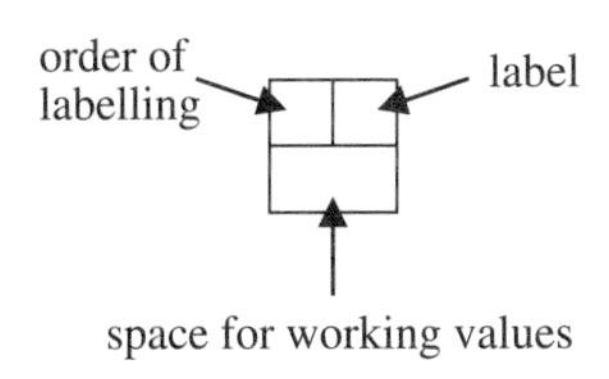

But remember that it is designed, as are all these algorithms, for computer use on large networks.

Step 1 Label the start vertex '0' (in the top right-hand box) and mark it as the first vertex labelled (in the top left-hand box).

Step 2 Update working values for all vertices Y which can be reached directly from the vertex X which has just been labelled. The rule for updating the working value of a vertex Y is:

Take the weight of the arc connecting the vertex Y to the vertex X that has just been labelled and add it to that label. If the vertex Y has no working value then the result becomes the working value. If the vertex Y already has a working value then the result replaces it if it is lower.

It seems odd on the first pass to refer to the vertex just labelled, but the reason for doing so will become clear later.

Step 3 Out of all unlabelled vertices with working values, choose that with the lowest working value. (There may be a choice here, in which case it is a free choice.) Label it with that working value and record the order in which it has been labelled (second, third, fourth, etc.).

Step 4 Repeat steps 2 and 3 until the destination vertex is labelled.

Step 5 The label on the destination vertex is the shortest distance to that vertex.
The shortest route is found by 'tracing-back' as follows:
If vertex N lies on the route, then vertex M is the previous vertex *if*:

label at N − label at M = weight of arc MN

Repeat until the route is traced back to the initial vertex.

It is tempting to try to avoid tracing-back by keeping a record of the route whilst labelling, but this is much less efficient.

Example 11

Use Dijkstra's algorithm to find the shortest route from A to G in this network.

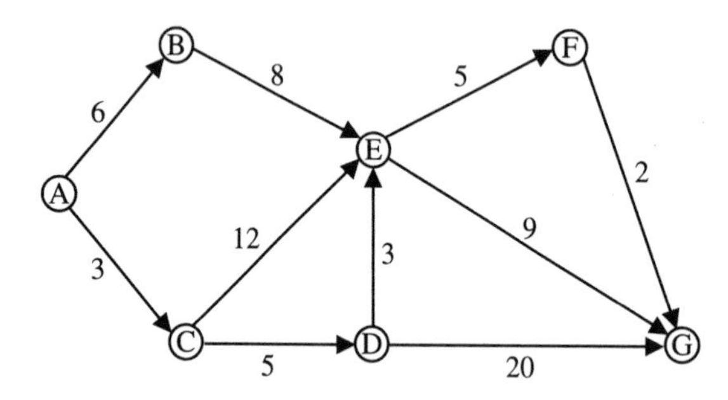

Step 1 Label start vertex A with 0 and number it 1 as the first vertex labelled.

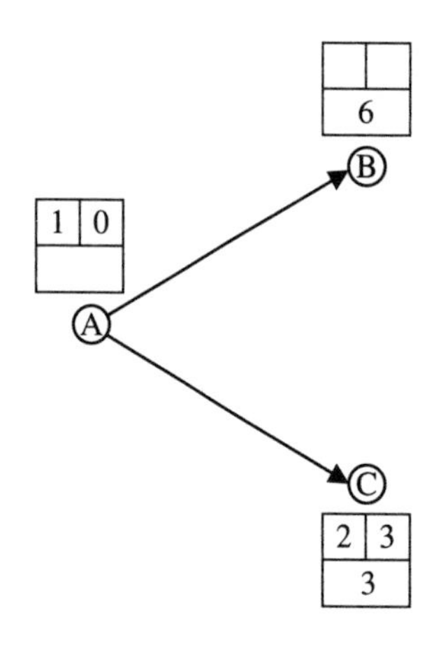

Step 2 Vertices B and C can be reached directly from A.
Working value for the distance along the path to B is $0 + 6 = 6$.
Working value for the path to C is $0 + 3 = 3$.

Step 3 Smallest working value is 3 at C, so label C with 3 and number it as 2, the second vertex labelled.

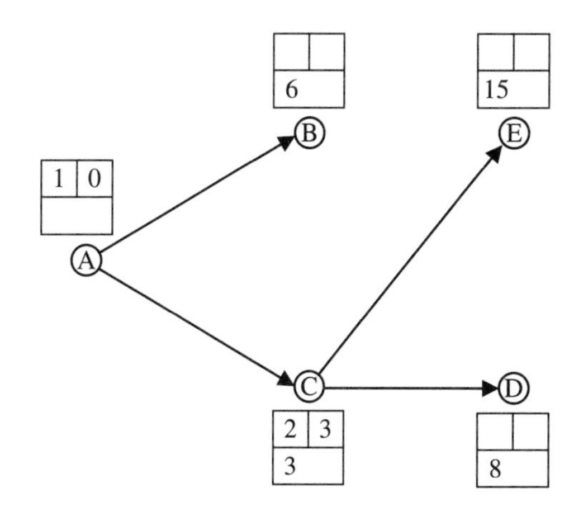

Step 2 Vertices D and E can be reached directly from C, just labelled. Working value for D is $3 + 5 = 8$ (label of C + weight of CD). Working value of E is $3 + 12 = 15$.

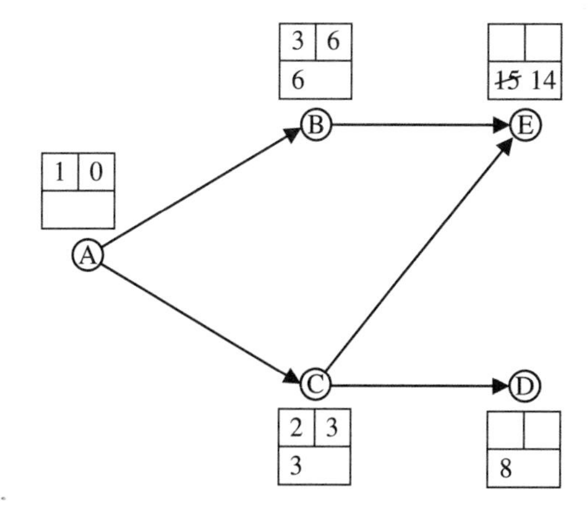

Step 3 The working values are now B (6), D (8) and E (15). The smallest is 6 at B, so label B with 6 and number it as 3.

Step 2 Vertex E can be reached directly from B, just labelled. Now (label B) + (weight BE) $= 6 + 8 = 14$. As this is smaller than 15, E's existing working value, we replace the 15 by 14.

E: ~~15~~ 14

Step 3 The working values are now D (8) and E (14). The smallest is 8 at D, so label D with 8 and number it as 4.

D: 4 | 8; 8

Step 2 Vertices E and G can be reached directly from D, just labelled. E already has a working value of 14, but (label D) + (weight DE) $= 8 + 3 = 11$, and so 14 is replaced by 11. G has no working value, and so we give it one calculated from (label D) + (weight DG) $= 8 + 20 = 28$.

E: ~~15~~ ~~14~~ 11

G: 28

Step 3 The working values are now E (11) and G (28). The smallest is 11 at E, so label E with 11 and number it as 5.

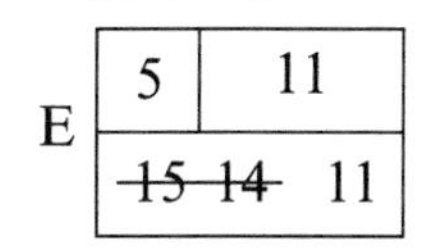

Step 2 Vertices F and G can be reached directly from E, just labelled. G already has a working value of 28, but (label E) + (weight EG) = 11 + 9 = 20, and so 28 is replaced by 20. F has no working value, and so we give it one calculated from (label E) + (weight EF) = 11 + 5 = 16.

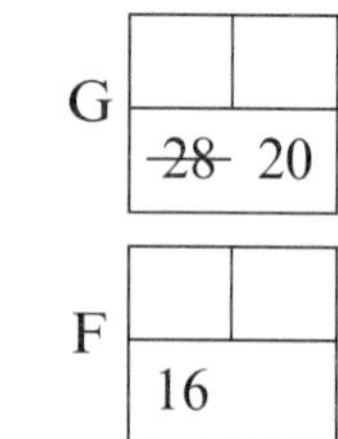

Step 3 The working values are now F (16) and G (20). The smallest is 16 at F, so label F with 16 and number it as 6.

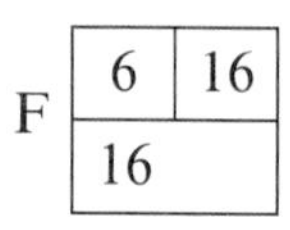

Step 2 Only G remains to be reached. G already has a working value of 20, but (label F) + (weight FG) = 16 + 2 = 18, and so 20 is replaced by 18.

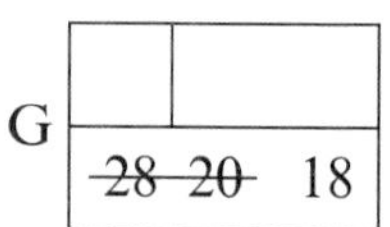

Step 3 The only (and therefore the smallest) unlabelled vertex is G, and this is now labelled and numbered 7. Since we have labelled the destination vertex we now stop repeating steps 2 and 3, and go to step 5.

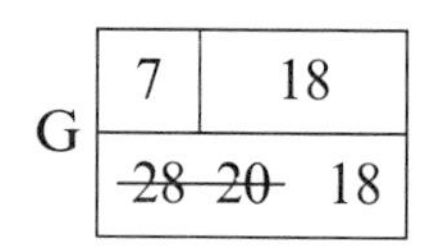

Step 5 The length of the shortest route from A to G is 18, which is the label of G.

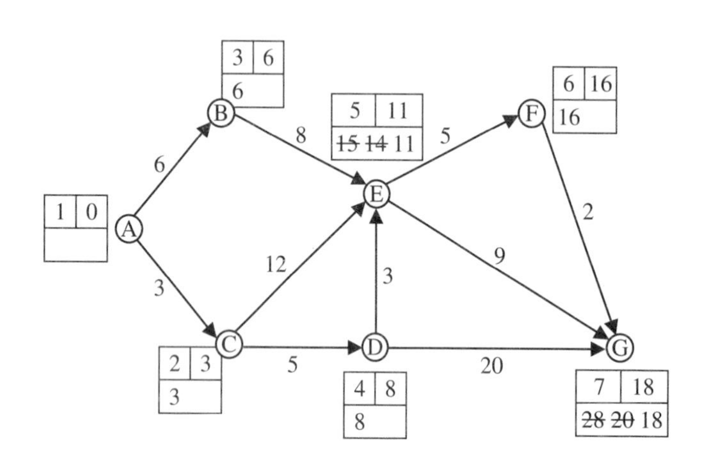

To find the shortest route work backwards from G:
(label G) – (label F) = 18 – 16 = 2 = (weight FG)
(label F) – (label E) = 16 – 11 = 5 = (weight EF)
(label E) – (label D) = 11 – 8 = 3 = (weight DE)
(label D) – (label C) = 8 – 3 = 5 = (weight CD)
(label C) – (label A) = 3 – 0 = 3 = (weight AC)

So FG, EF, DE, CD and AC are on the shortest route, which is therefore A → C → D → E → F → G.

For all other edges: (label Y) – (label X) ≠ (weight XY).

For example, take DG: (label G) – (label D) = 18 – 8 = 10, but the weight of DG is 20.

Exercise 2F

1 In the following network the edge weights represent distances:

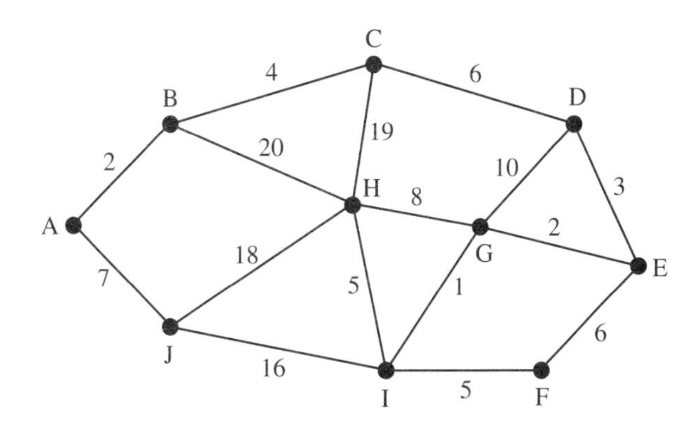

(a) Use Dijkstra's algorithm to find the shortest route from A to E.

(b) Give the shortest route from E to H.

(c) Use Dijkstra's algorithm to find the shortest route from A to H.

2 The matrix below gives fares for direct bus journeys between towns, P, Q, R, S, T, U and V, in a tourist area. Blanks indicate no direct service.

Fares in pence

	P	Q	R	S	T	U	V
P		67		45			80
Q	67		63	71		170	
R		63			59		
S	45	71			40		54
T			59	40		58	62
U		170			58		50
V	80			54	62	50	

(a) Draw a network to represent this information.

(b) Using Dijkstra's method, find the minimum cost route from P to U.

3 The diagram represents the roads joining 10 villages, labelled A to J. The numbers give distances in km.

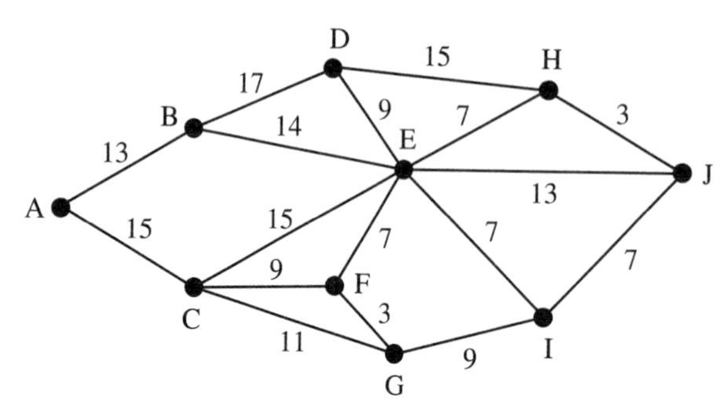

(a) Use Dijkstra's algorithm to find a shortest route from A to J. Explain the method carefully, and show all your working. Give a shortest route and its length.

A driver usually completes this journey driving at an average speed of 60 km/h. The local radio reports a serious accident at village E, and warns drivers of a delay of 10 minutes.

(b) Describe how to modify your approach to (a) to find the quickest route, explaining how to take account of this information. What is the quickest route, and how long will it take?

4 The diagram shows road connections and distances (in km) between some major archaeological sites in Crete.

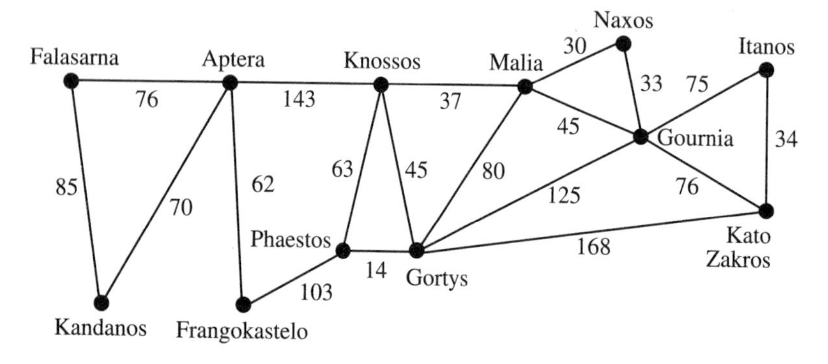

(a) Use Dijkstra's algorithm to find the shortest distance and route from Frangokastelo to Itanos.

(b) Crete is very mountainous and most of the marked connections involve mountain roads on which an average speed of 50 km/h is all that can be achieved. Faster roads,

including the State Highway, run along the north coast, connecting the sites of Falasarna, Aptera, Knossos, Malia and Gournia. Average speeds of 100 km/h are attainable on these roads.
Use Dijkstra's algorithm to find the fastest route from Frangokastelo to Itanos.

(c) There is no direct connection between Kandanos and Frangokastelo, but there is a small car ferry operating along the south coast. This allows a direct journey to be made from Kandanos to Frangokastelo, at extra cost in terms of time and money. What distance on the journey from Kandanos to Itanos would be saved by making such a connection?

5 The diagram shows a directed network in which the weights on the arcs represent costs for journeys in the directions indicated.

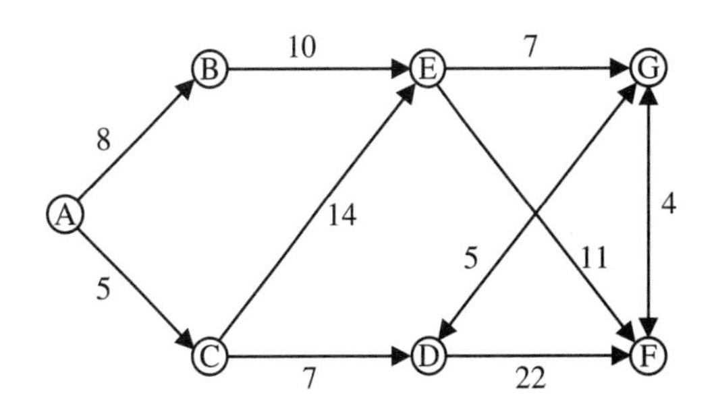

(a) Apply Dijkstra's algorithm to find the smallest cost, and the associated route, to get from A to F.

(b) Suppose now that a new route is introduced from B to C, and that a profit of 6 may be made by making a delivery from B to C when using that route. This is shown as a negative cost in the network.

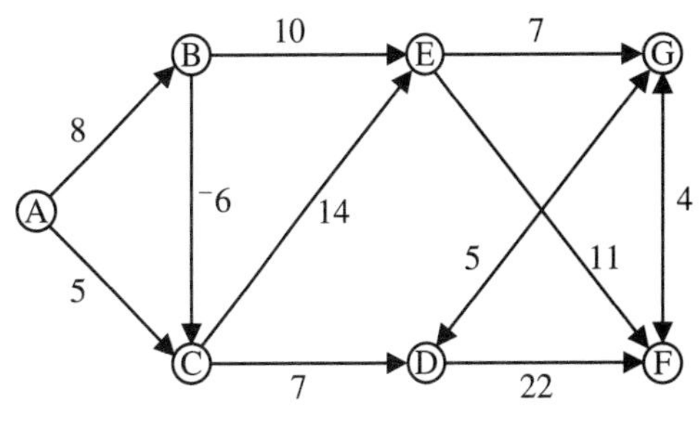

Work through Dijkstra's algorithm to find the best route from A to C in this modified network.
Say what is the best route from A to C, and explain why Dijkstra's algorithm fails to find it.
What extra problem would be faced if there were also a route from C to B at a cost of less than 6?

6 The following network represents an electronic information network. The vertices represent computer installations. The arcs represent direct links between installations. The weights on the arcs are the rates at which the links can transfer data (in units of 100 000 bits per second).

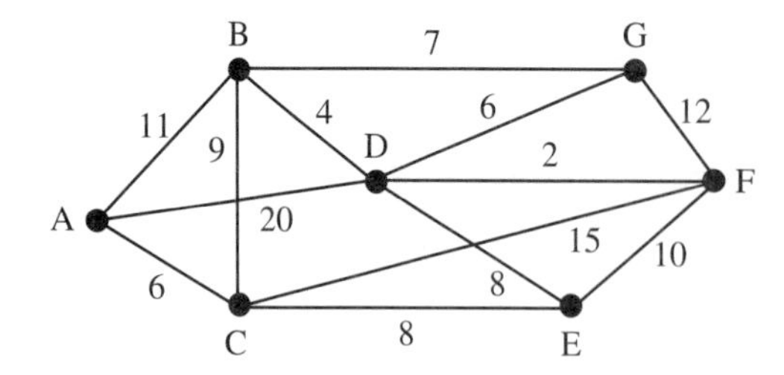

A chain of links may be established between any two installations via any other installations. For such chains the rate at which the data can be transferred is given by the speed of the slowest link in the chain.

(a) The fastest chain linking A to G is needed. Devise a way of adapting Dijkstra's algorithm to produce an algorithm to achieve this.
Apply your algorithm to the network. State the fastest chain and its speed of transferring data.

(b) How could you find a slowest chain linking A to G? Give such a chain.

7 Colin uses Dijkstra's algorithm to find a shortest path in a complete network on five vertices:
It turns out that he has to label all the other vertices before he can label his destination vertex. (This is an example of a worst-case scenario.) It takes Colin 10 minutes to do this.

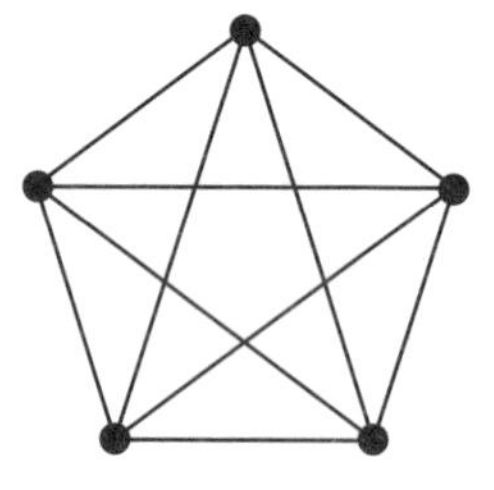

(a) Approximately how long would it take Colin to complete a similar worst-case task in a complete network on ten vertices? Roughly how long for 100 vertices?

(b) Joan can do the five vertex problem in 5 minutes. How long would it take her to do the ten vertex problem?

(c) Using Dijkstra's algorithm, roughly how long would it take each of Colin and Joan to find the shortest routes between each pair of vertices in a complete network on five vertices? How long for a complete network on ten vertices? How long for 100 vertices?

8 Use Dijkstra's algorithm to solve the production planning problem on page 60.

SUMMARY OF KEY POINTS

1. A graph consists of a finite number of vertices connected by a set of edges.
2. A walk is a finite sequence of edges such that the end vertex of one edge is the start vertex of the next.
3. A trail is a walk in which no edge appears more than once.
4. A path is a trail in which no vertex is visited more than once.
5. A cycle (or circuit) is a closed path – the end vertex of the last edge is the start vertex of the first edge.
6. Two vertices are connected if there is a path between them.
7. A graph is connected if all its vertices are connected.
8. A simple graph is one in which there is no edge with the same vertex at each end and no more than one edge connecting any pair of vertices.
9. The degree of a vertex is the number of edges connected to it.
10. If the edges of a graph have a direction associated with them they are known as directed edges, and the graph is called a digraph.
11. A tree is a connected graph with no cycles.
12. A weighted graph is known as a network.
13. For a connected and undirected network, a minimum spanning tree (also known as a minimum connector) is a connected subgraph of minimum total weight incorporating every vertex of the network.
14. Prim's algorithm builds a minimum spanning tree by adding a vertex at a time (and an associated edge) to a connected subgraph. It has two forms, graphical and matrix.
15. Kruskal's algorithm builds a minimum spanning tree by adding an edge at a time (and associated vertices) to a subgraph. The subgraph need not be connected at intermediate stages, though the final subgraph is connected.
16. Dijkstra's algorithm works through a network, labelling one vertex at each stage and improving working values as it goes. The working values give the shortest distances found so far to unlabelled vertices.

Decision making in graphs

3

In chapter 2 we introduced some network problems for which there are **viable** algorithmic solutions. By 'viable' we mean that the algorithms can be implemented on reasonably large networks in acceptable times on fast computers. We will make this **definition of viability** a little more precise in this chapter and will examine two well known network problems for which no such viable algorithms have been developed. It is widely suspected, but not yet proved, that such algorithms cannot exist.

Nevertheless, the problems are real enough and good methods of solution are needed, even if viable algorithms leading to optimal solutions are not available. We will define and develop **heuristic** algorithms for solving these problems. Heuristic means that the algorithms work in an intuitively obvious or 'common sense' way, without guaranteeing optimality. A heuristic algorithm usually produces good results, but isn't guaranteed to find the best result.

The two problems that we will be examining are the Travelling Salesman Problem (TSP) and the Route Inspection Problem (also known as the Chinese Postman Problem).

The salesman has to visit every **vertex** of a network. The postman has to travel along every **edge**.

3.1 The travelling salesman problem (TSP)

The problem is easy to state and understand:

Find a route of minimum weight which visits every vertex in an undirected network.

The vertices usually represent cities. As in chapter 2 we will use 'distance' or 'length' instead of weight, whilst remembering that there will be applications which are concerned with costs, time, or with other edge weightings.

There are several ways of visiting every vertex of a network like this one.

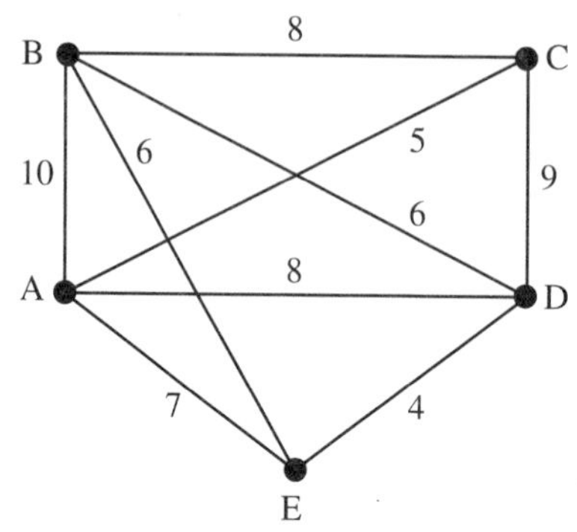

- **A cycle which visits every vertex, returning to the starting vertex, is called a tour.**

Here are some possible tours for the network:

(i) ABCDEA of length 38.
(Notice that this is essentially the same as AEDCBA, or BCDEAB, or BAEDCB, etc. Depending on the choice of starting vertex, and the direction taken, there are 10 ways of writing it down.)

(ii) ACBDEA of length 30.

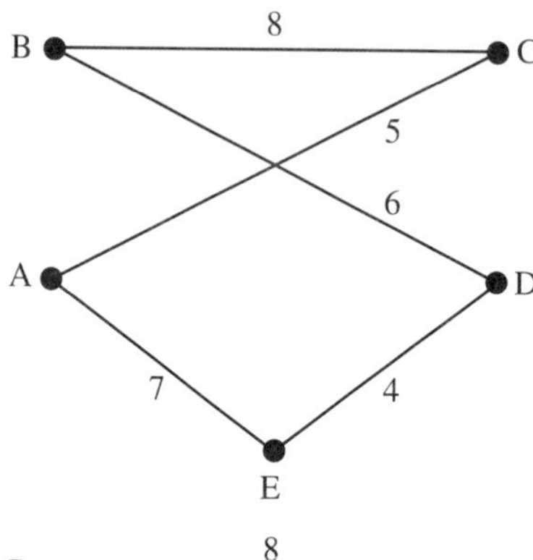

(iii) ACBEDA of length 31.

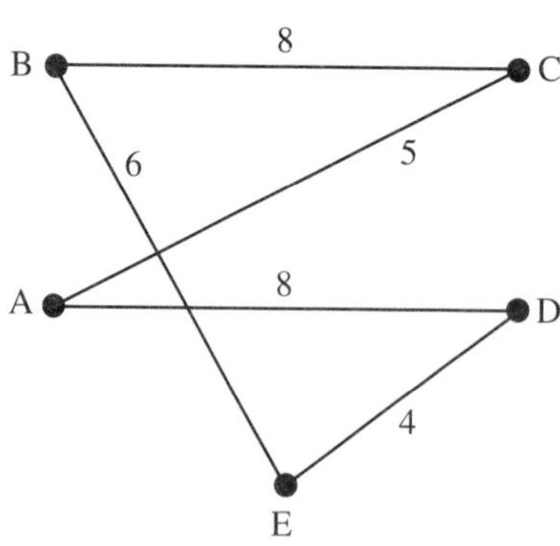

(iv) ACDBEA of length 33.

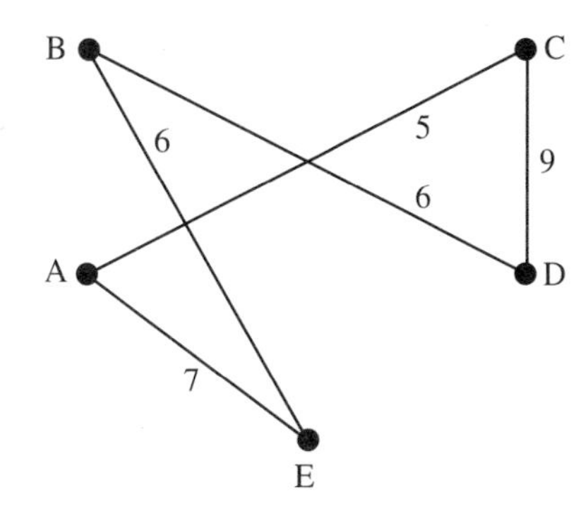

In a real-life travelling salesman problem the salesman would not mind backtracking through a previously visited city. However, in mathematics you will sometimes be asked to solve this problem:

Find a tour of minimum weight which visits each vertex once only.

This is usually described as the 'classical' travelling salesman problem to distinguish it from real-life or 'practical' problems in which backtracking is allowed. The classical problem was first studied by Sir William Hamilton in the 19th century, so a cycle which includes all the vertices of a network is known as a Hamilton cycle. The travelling salesman problem is to find the Hamilton cycle of least weight.

- **A tour which visits each vertex once only is called a Hamilton cycle.**

You can transform a practical problem into the corresponding classical problem by creating a complete network in which the edge weights are the shortest distances in the original network. If you are interested in this look at Appendix 1. Your syllabus (London 1997 and 1998) will only consider travelling salesman problems in networks which are complete, and in which the triangle inequality holds (see chapter 2, page 38). In these networks the classical and practical problems are the same, so you will not *need* to study the material in Appendix 1.

> A complete network is one in which every vertex is connected to every other vertex.

3.2 The difficulty with the TSP

A complete enumeration algorithm is one which systematically tries every possibility. We can consider such an algorithm to solve the classical travelling salesman problem:

1. List all Hamilton cycles.
2. Find the total length (weight) of each cycle.
3. Choose the cycle of least total length. (There may be more than one such cycle.)

However, there is a problem with this algorithm. For the TSP on a complete network on 5 vertices there are $4! = 4 \times 3 \times 2 \times 1 = 24$ cycles to consider (discounting any differences due to choosing different starting vertices). They are:

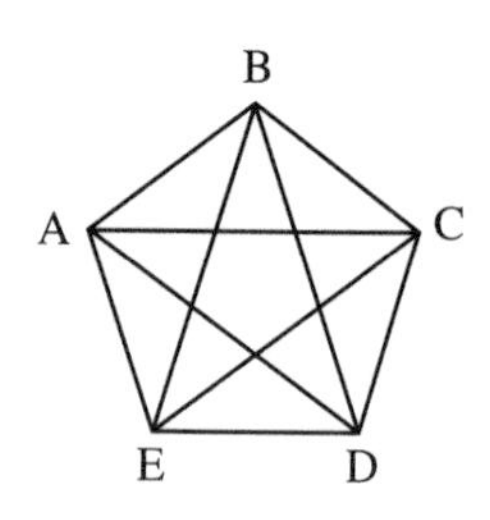

4! is 4 factorial $= 4 \times 3 \times 2 \times 1$

ABCDEA; ABCEDA; ABDCEA; ABDECA; ABECDA;
ABEDCA; ACBDEA; ACBEDA; ACDBEA; ACDEBA;
ACEBDA; ACEDBA; ADBCEA; ADBECA; ADCBEA;
ADCEBA; ADEBCA; ADECBA; AEBCDA; AEBDCA;
AECBDA; AECDBA; AEDBCA; AEDCBA.

(Notice that there are effectively only 12 different cycles since the 24 come in pairs, such as ABCDEA / AEDCBA, but it is easier to list them all if we think of them as being different.)

Imagine that we have access to a machine which can list and find the lengths of 10 million Hamilton cycles per second, and let us ignore the time that it would take to find the shortest one once the list is complete. A travelling salesman problem could well involve a network of 50 vertices (or towns). But there are 49! Hamilton cycles in such a network (counting 'clockwise' cycles as different from 'anticlockwise'). So to complete the task our fast computer would need:

$$\frac{49!}{10\,000\,000} \text{ seconds}$$

That's more than $1.9 \times 10^{48} =$

1 900 000 000 000 000 000 000 000 000 000 000 000 000 000 000 000 years!

You can read more about this in Appendix 1 (Complexity of the TSP, page 233). Whilst there will be no examinable material on this topic, it is useful to appreciate that we need to be able to put bounds on the total length of the solution to a TSP. We might not be sure of having the best solution, but we might be able to say that the best solution must be of length at least x, or of length at most y. If we can do this, then x is said to be a **lower bound**, and y is an **upper bound**. The larger we can make our lower bound the more useful it will be. The smaller we can make our upper bound the more useful it will be. If x and y are close together, and if we can find a tour with total length between x and y, then we would probably be satisfied with that tour, even though we might not be sure that it was the *best* tour.

Of course, the total length of *any* tour is an upper bound for the total length of the solution to a TSP, since the solution is a tour of lowest total length.

3.3 An upper bound to the practical problem

Given a connected network, a tour can be constructed from a minimum spanning tree (MST). This involves travelling twice along each edge of the minimum spanning tree. A tour constructed in this way will visit some vertices more than once, so it will be appropriate for the practical problem, not the classical. So:

- **Twice the length of the minimum spanning tree is an upper bound for the total length of the solution to the practical travelling salesman problem.**

In your examination, networks in travelling salesman problems will be constructed so that the practical and classical problems are identical and this technique *can* be used.

Example 1

Let's consider this network again:

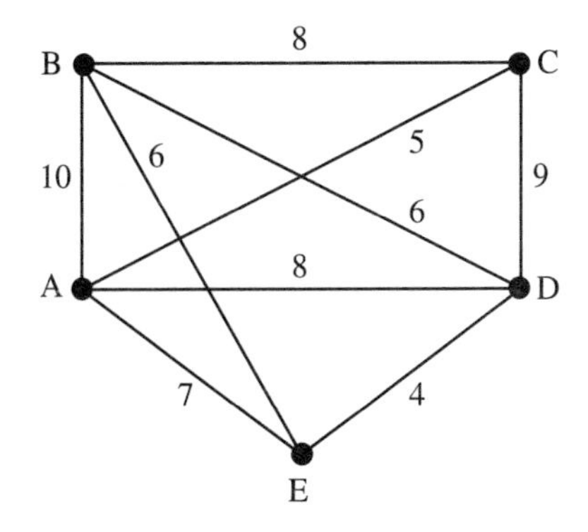

The first step is to use Prim's algorithm or Kruskal's algorithm to produce a minimum spanning tree. In this example there are two possible solutions:

(i) Total length = 22 units

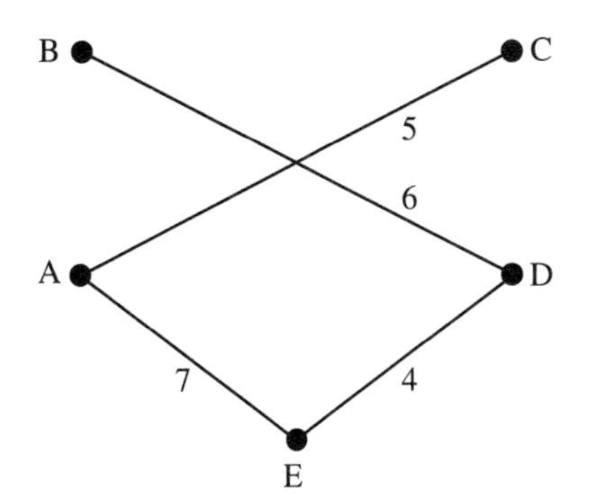

(ii) Total length = 22 units

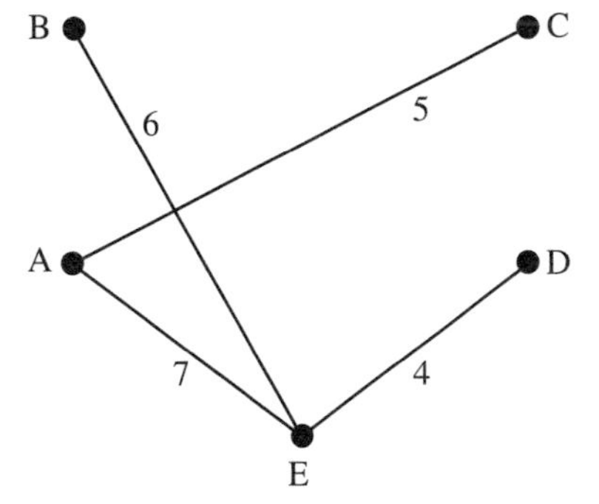

Starting at A, we can now construct a tour by travelling to the end of each branch of the minimum spanning tree, and back as far as is necessary.

So from (i) we can produce AEDBDEACA or ACAEDBDEA.
From (ii) we can produce ACAEDEBEA or ACAEBEDEA or AEDEBEACA or AEBEDEACA.

All these tours are of total length $2 \times 22 = 44$ units.

However, we can do better than this by using **shortcuts**. Using (i) and travelling outwards from A we come to the end of a branch at B. Instead of retracing our steps back along the tree we can look back to the original network to see if we can move across to the end of a different branch

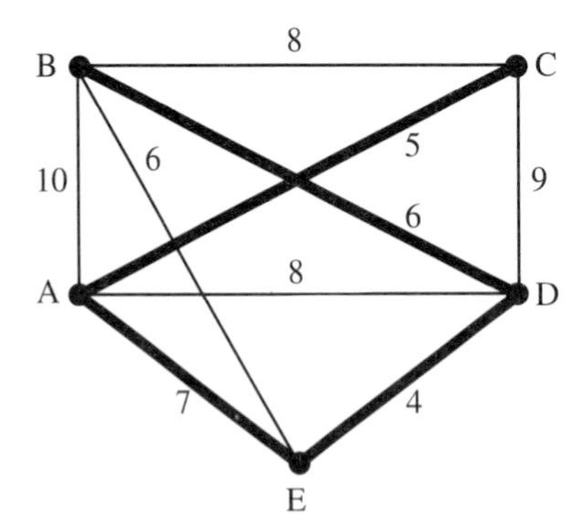

We can move from B direct to C without using edges on the minimum spanning tree. This represents a shortcut, since we can now return to A along the minimum spanning tree, giving a shorter tour.
The tour is AEDBCA. Its total length is 30 units.

Shortcutting in (ii):

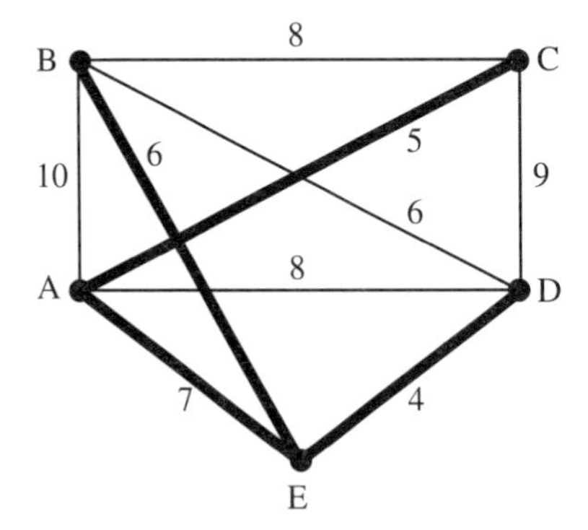

On arriving at B we can go directly to D without back-tracking to E. We can then go directly to C, to return along the minimum spanning tree to A.
The tour is AEBDCA, and its total length is 33 units.

We prefer the lowest of the upper bounds that we have constructed, since that will be closer to the length of the best cycle. This is AEDBCA, with a total length of 30 units. So we can say that the best solution to the practical travelling salesman problem in this network has total length less than or equal to 30 units.

The ‘twice the minimum connector plus shortcuts’ approach can produce good (i.e. low) upper bounds, but it is difficult to make it algorithmic – in a large network it would be difficult to find a good set of shortcuts.

Exercise 3A

1

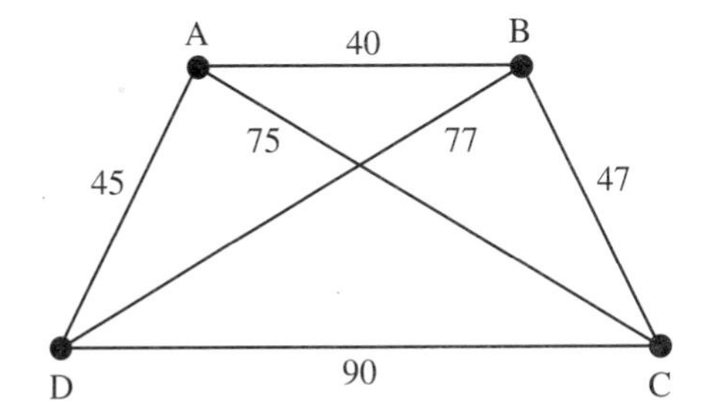

(a) Find a minimum spanning tree for the above network and hence find an upper bound for the total length of the solution of the travelling salesman problem.

(b) Obtain a better upper bound by using a shortcut.

2

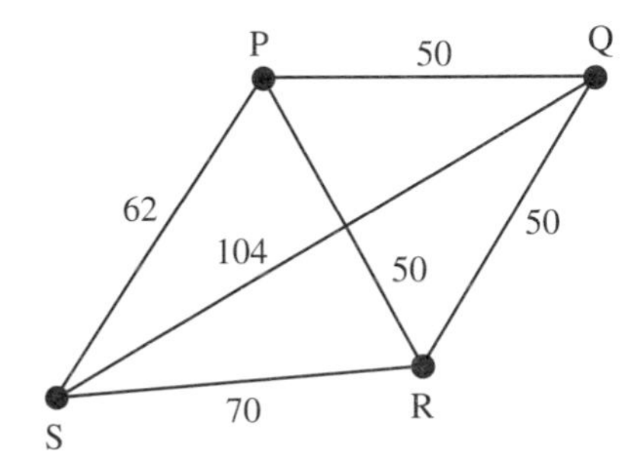

For the above network obtain three minimum spanning trees. Hence obtain an upper bound for the total length of the travelling salesman problem. By using shortcuts obtain an upper bound which is less than 240.

3

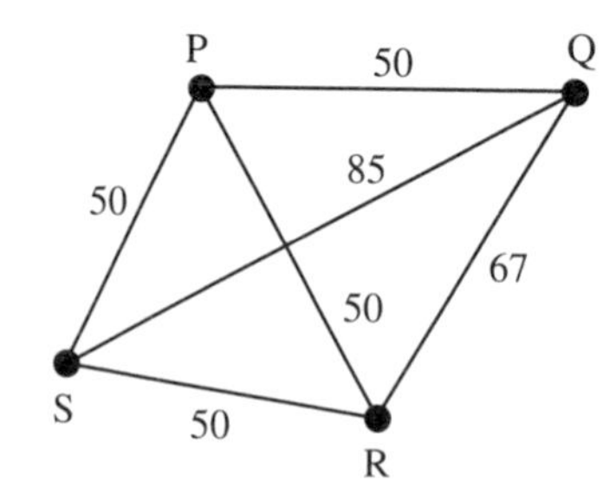

(a) Obtain three minimum spanning trees for the above network and hence obtain an upper bound for the total length of the solution of the travelling salesman problem.

(b) Consider each of these trees in turn and obtain in each case a better bound by using shortcuts.

4

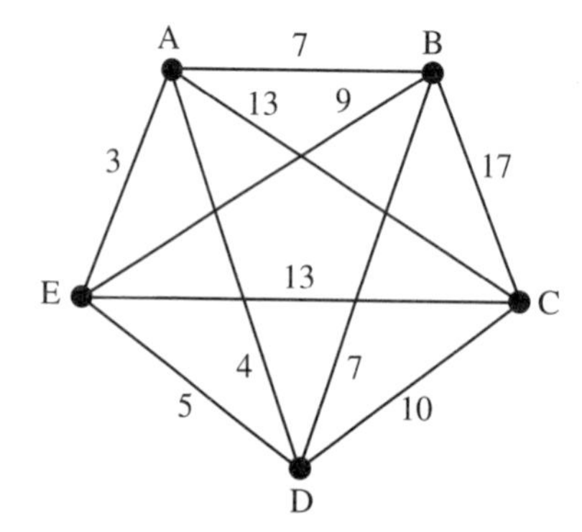

A delivery van based at A is required to deliver goods in towns B,C,D,E shown in the diagram. The numbers on the arcs are distances in miles.

(a) Find two minimum spanning trees for the above network and hence find an upper bound for the distance travelled by the van.

(b) Improve this upper bound by using shortcuts.

5 A computer engineer lives in town A and needs to visit each of the towns B,C,D and E to service various installations. He must return to town A and visit each town once. The distances in miles between the towns are shown in the table.

	A	B	C	D	E
A	—	17	10	9	12
B	17	—	8	14	5
C	10	8	—	7	11
D	9	14	7	—	11
E	12	5	11	11	—

(a) Use Prim's algorithm to find the length of a minimum spanning tree that connects the five towns.
(b) Hence find an upper bound for the total distance that must be travelled by the engineer.
(c) Improve the upper bound by using shortcuts.

6 The table shows the distances, in miles, between some cities. A politician has to visit each city once, starting and finishing at A. She wishes to minimise her total travelling distance.

	A	B	C	D	E	F	G	H
A	—	47	84	382	120	172	299	144
B	47	—	121	402	155	193	319	165
C	84	121	—	456	200	246	373	218
D	382	402	456	—	413	220	155	289
E	120	155	200	413	—	204	286	131
F	172	193	246	220	204	—	144	70
G	299	319	373	155	286	144	—	160
H	144	165	218	289	131	70	160	—

(a) Find a minimum spanning tree for the network.
(b) Hence find an upper bound for the politician's problem.
(c) Reduce this upper bound to a value below 1400 miles by using shortcuts.

3.4 A lower bound for the classical problem

There is a useful method of finding a lower bound for the classical problem. This method can be used for the practical problem when the practical and classical problems are identical.

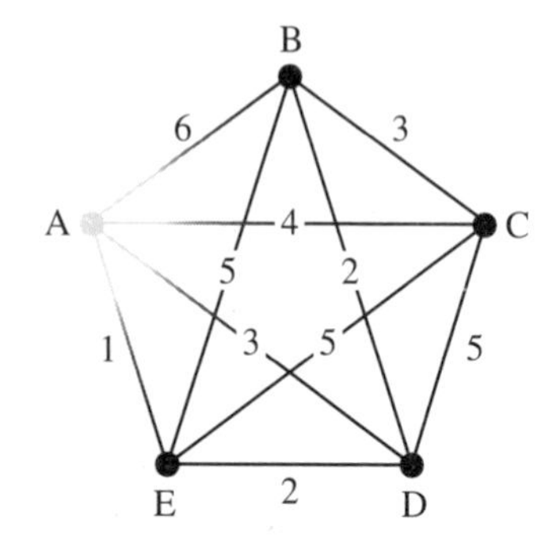

The method is based on deleting a vertex *from the original network*, together with all the edges that are connected to it. When you do this, the minimum Hamilton cycle will also lose a vertex and the two edges connected to it. The remaining vertices of the Hamilton cycle will still be connected.

You will not know which of the edges that you have deleted *from the original network* are the two edges of the *minimum* Hamilton cycle that pass through the vertex you have deleted. You only know that two of the deleted edges will be part of the minimum Hamilton cycle. There are only two since you can't pass through a vertex more than once in a Hamilton cycle.

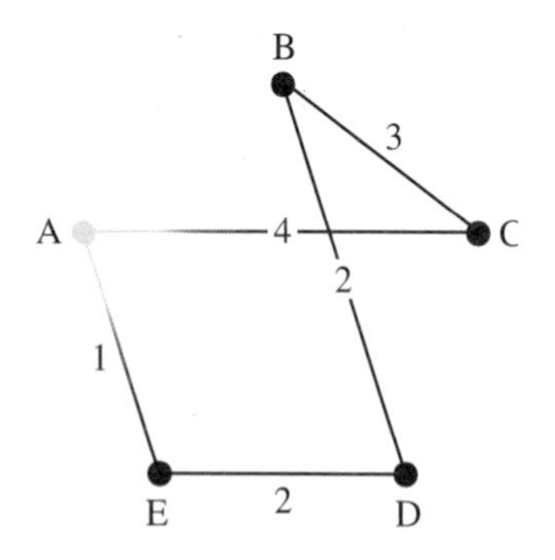

Now construct the minimum connector for the remaining vertices. If you then add back the deleted vertex in the most economical way, you can be sure that the length of the connector will not be greater than that of the minimum Hamilton cycle.

Here is the algorithm:

Step 1 Choose any vertex and delete it and all edges that are connected to it.

Step 2 Use a minimum connector algorithm to find the length of a minimum connector for the remaining vertices.

Step 3 Add to that length the sum of the lengths of the two shortest deleted edges.

A set of lower bounds can be constructed by deleting each vertex in turn. The largest of the lower bounds is then chosen.

Step 4 Repeat, choosing each vertex in turn at step 1. Choose the largest result.

Notice that this method of producing a lower bound often does not produce a tour.

Applying the algorithm to the problem in example 1 produces the following results.

Here is the original network: Deleting A and edges connected to A:

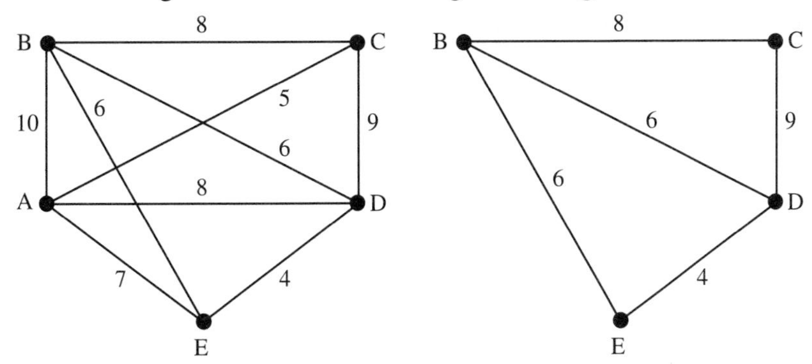

There are two possible minimum spanning trees for the remaining vertices, both of length 18:

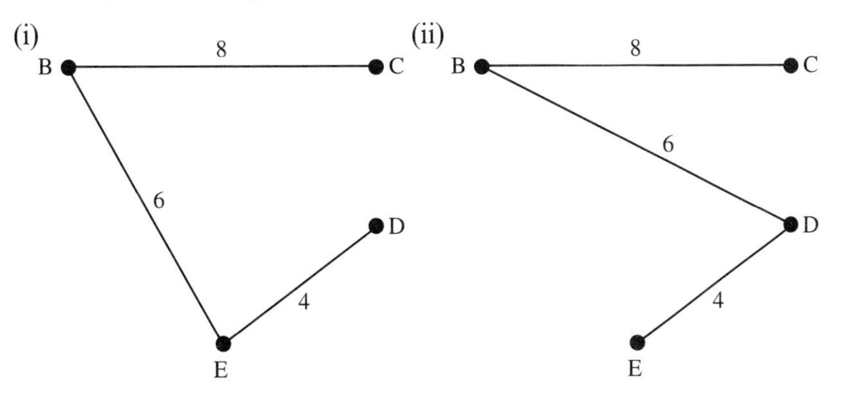

We now know that a minimum spanning tree of B, C, D and E has length 18. But the undeleted part of the minimum Hamilton cycle connects vertices B, C, D and E. So the total length of that undeleted portion cannot be less than 18.

We now add back in vertex A together with two edges. The edges which we deleted were AB (length 10), AC (length 5), AD (8) and AE (7). We do not know which are members of the minimum Hamilton cycle, but they cannot be of shorter total length than the sum of the lengths of AC and AE, the two shortest. Adding in those gives:

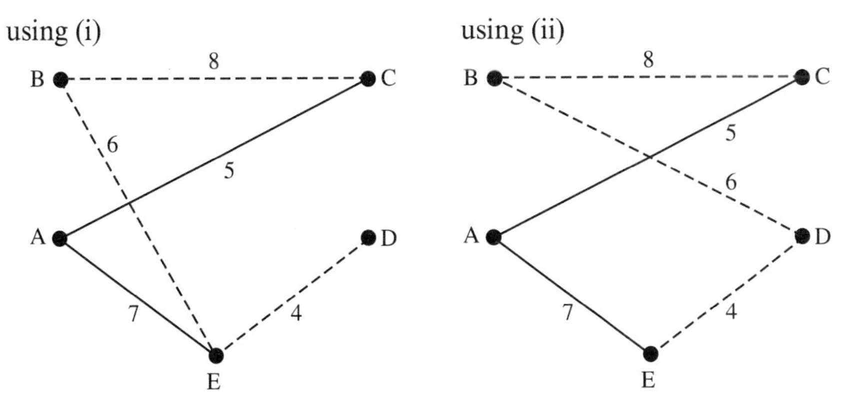

Only one of these sets of edges forms a tour, but in both cases the sum of their lengths gives a lower bound. In this case that lower bound is 30 units.

Applying step 4, the following results are produced:

Deleted vertex	A	B	C	D	E
Lower bound	$18 + (5 + 7) = 30$	$16 + (6 + 6) = 28$	$17 + (5 + 8) = 30$	$18 + (4 + 6) = 28$	$19 + (4 + 6) = 29$

Choosing the largest and using the upper bound constructed earlier, we can conclude that for the classical problem:

(lower bound) $30 \leqslant$ length of best tour $\leqslant 30$ (upper bound).

So in this small example we have lower bound = upper bound. We therefore have a solution. This will not happen in large scale problems.

Example 2

Find lower bounds for the classical travelling salesman problems in the following complete network:

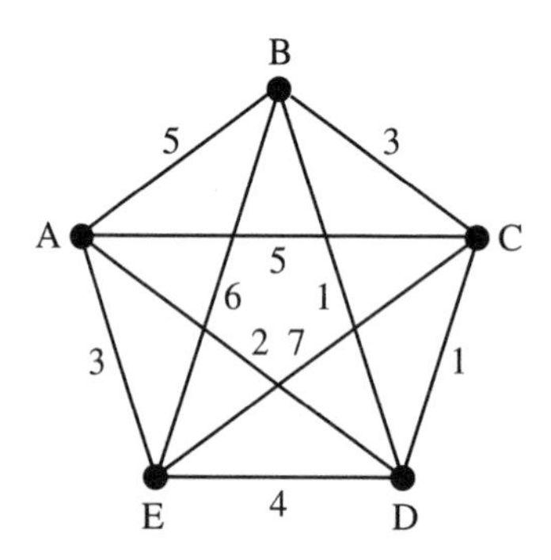

Removing A:

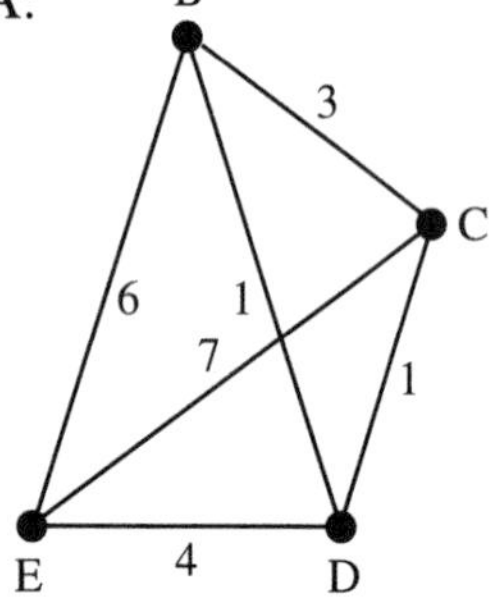

Minimum connector:

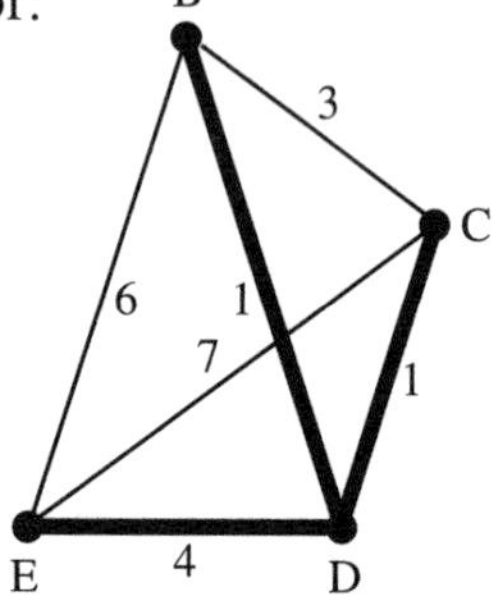

Connecting A back in:

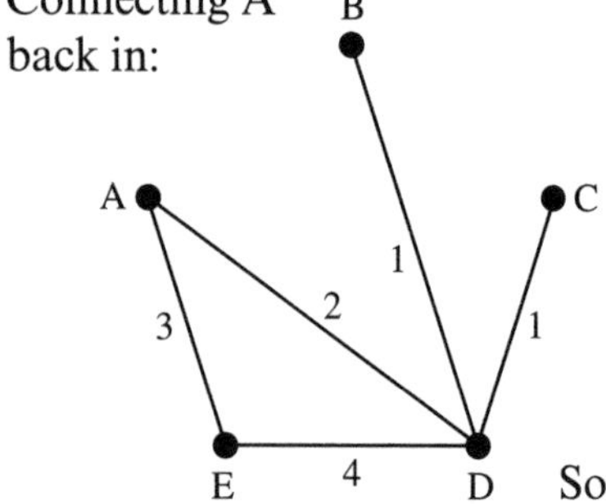

Applying step 4:

vertex deleted	*lower bound*
A	11
B	10
C	10
D	13
E	11

So we end up with a lower bound of 13.

Using the matrix

You can also calculate lower bounds using the matrix that represents the network. This is often easier when the number of vertices is not small.

The network in example 2 is represented by the matrix:

	A	B	C	D	E
A	—	5	5	2	3
B	5	—	3	1	6
C	5	3	—	1	7
D	2	1	1	—	4
E	3	6	7	4	—

Deleting A means you delete the row labelled by A and the column labelled by A.
So you obtain:

	B	C	D	E
B	—	3	1	6
C	3	—	①	7
D	①	1	—	4
E	6	7	④	—

Now you can use Prim's algorithm to obtain a minimum connector for this network. The ringed entries are the result, and give the minimum connector as BD, CD and DE of weight $1 + 1 + 4 = 6$.

To find the two edges needed to reconnect A to the network, look for the two smallest entries in the A row of the original matrix. The two edges to be used are therefore AD(2) and AE(3).

So by deleting the vertex A you obtain a lower bound of:

$$\text{MST}(6) + 2 + 3 = 11$$

Repeat the process for the vertex D. The matrix when D is removed is:

	A	B	C	E
A	—	5	5	3
B	⑤	—	3	6
C	5	③	—	7
E	③	6	7	—

Use of Prim's algorithm identifies the ringed entries as forming a minimum connector. The weight of this minimum connector is:

weight of AE(3) + weight of AB(5) + weight of BC(3) = 11

The two edges to be used to reconnect D are DB(1) and DC(1). So by deleting D you obtain a lower bound of 13.

The other lower bounds given above may be obtained by deleting B,C and E in turn.

For this small problem we can find the shortest tour by complete enumeration (working out all the possible tours to see which is shortest). In this case it is ADCBEA, of length 15.

Cycle	Length	Cycle	Length
ABCDEA	16	ACBDEA	16
ABCEDA	21	ACBEDA	20
ABDCEA	17	ACDBEA	16
ABDECA	22	ACDEBA	20
ABECDA	21	ACEBDA	21
ABEDCA	21	ACEDBA	22
ADBCEA	16	AEBCDA	15
ADBECA	21	AEBDCA	16
ADCBEA	15	AECBDA	16
ADCEBA	21	AECDBA	17
ADEBCA	19	AEDBCA	16
ADECBA	21	AEDCBA	16

Exercise 3B

1 In question 1 of Exercise 3A the following network was considered:

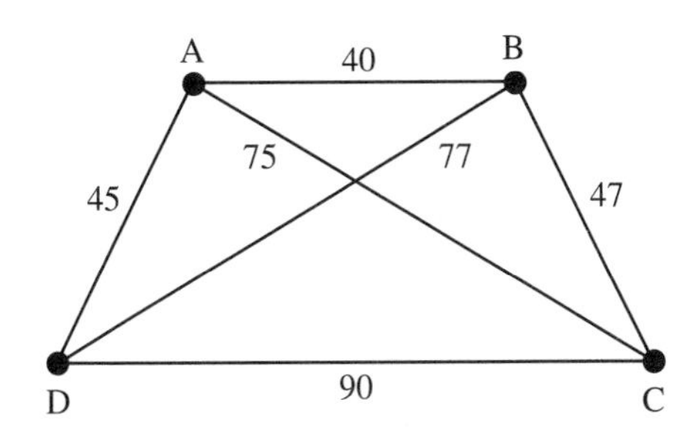

Obtain lower bounds for the length of the travelling salesman problem

(a) by deleting A (b) by deleting B (c) by deleting C
(d) by deleting D.

2 In question 2 of Exercise 3A the following was considered:

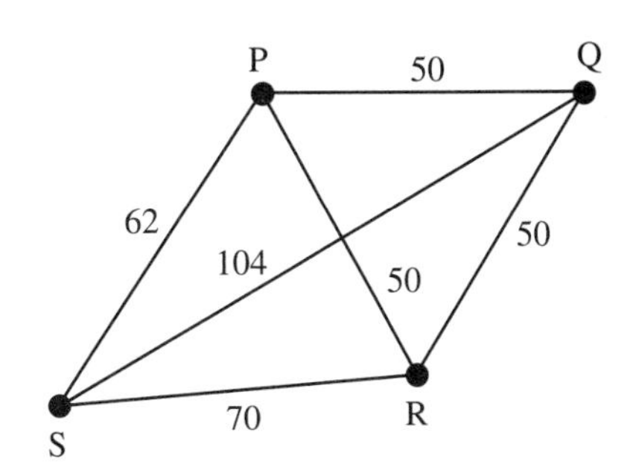

Obtain lower bounds for the length of the travelling salesman problem

(a) by deleting Q (b) by deleting S.
(c) Use your results for (a) and (b), together with your answers to question 2 of Exercise 3A, to obtain the tour in this network, of minimum weight.

3 In question 4 of Exercise 3A you had the following network.

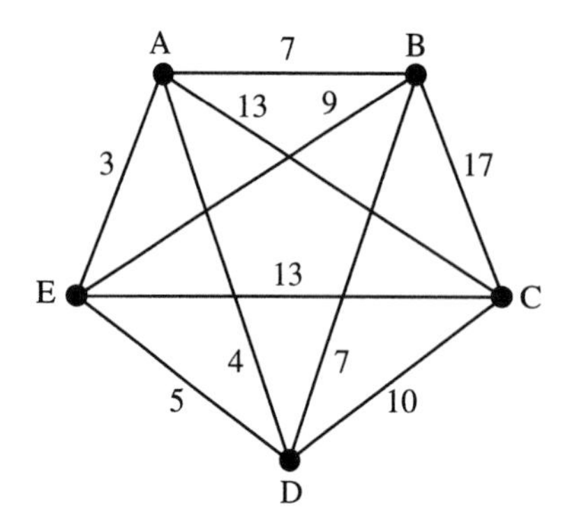

Obtain lower bounds for the length of the travelling salesman problem

(a) by deleting A (b) by deleting B (c) by deleting C

(d) by deleting D (e) by deleting E.

(f) Using your results for (a)–(e) and your answer to question 4 of Exercise 3A, write down an inequality for L, the minimum length of a tour.

(g) Write down a tour that satisfies this inequality.

4 In question 5 of Exercise 3A a matrix giving the distances between towns A, B, C, D and E was considered:

	A	**B**	**C**	**D**	**E**
A	—	17	10	9	12
B	17	—	8	14	5
C	10	8	—	7	11
D	9	14	7	—	11
E	12	5	11	11	—

Obtain lower bounds to the distance travelled by the computer engineer by

(a) deleting A (b) deleting B (c) deleting C

(d) deleting D (e) deleting E.

(f) From your answer to (e) and your answer to question 5 of Exercise 3A, obtain the minimum tour of length 41.

5 Below is shown the distance, in miles, between some cities as used in question 6 of Exercise 3A.

	A	B	C	D	E	F	G	H
A	—	47	84	382	120	172	299	144
B	47	—	121	402	155	193	319	165
C	84	121	—	456	200	246	373	218
D	382	402	456	—	413	220	155	289
E	120	155	200	413	—	204	286	131
F	172	193	246	220	204	—	144	70
G	299	319	373	155	286	144	—	160
H	144	165	218	289	131	70	160	—

Use the minimum spanning tree that you found for it to obtain lower bounds to the distance to be travelled by the politician

(a) by deleting B (b) by deleting C (c) by deleting D.

(d) Write down an inequality satisfied by L, the minimum length of the tour.

6 A lorry driver has to deliver milk every day to village shops. He starts and finishes at A and must visit villages B,C,D,E and F. The distances between A, B, ... F are shown in the table:

	A	B	C	D	E	F
A	—	11	13	8	15	13
B	11	—	10	16	5	6
C	13	10	—	17	8	8
D	8	16	17	—	17	16
E	15	5	8	17	—	4
F	13	6	8	16	4	—

(a) Obtain an upper bound for the distance the lorry must travel.

(b) Obtain a lower bound for the distance travelled.

(c) Using the best upper bound and lower bound you can obtain, write down an inequality for L, the minimum distance travelled.

(d) Write down the shortest possible route.

3.5 The route inspection problem

The following problem is often called the Chinese Postman Problem. The 'Chinese' refers to the nationality of the mathematician who posed the problem rather than to that of the postman. The problem can be stated as:

In a given undirected network a route of minimum weight is to be found which traverses every edge at least once, returning to the start vertex.

Again, many applications are geographical so the weights on the edges often represent real lengths.

A graph is **traversable** (or traceable) if it is possible to use a pencil to draw over each edge of the graph once and only once, without lifting the pencil from the paper (see Exercise 2B, question 8). There are two circumstances in which a graph may be traced – when there are no odd vertices, and when there are exactly two odd vertices. If there are no odd vertices then the tracing can start anywhere, but must end at the same point. If there are two odd vertices then it must start at one and end at the other.

So unless a network has no odd vertices it will be necessary to traverse some edges more than once in order to traverse all edges and return to the start. We can imagine that by repeating them we are effectively adding extra arcs to our network, and that by doing so we end up with all vertices of even order.

Why must the number of odd vertices in a graph always be even?

Example 3

Find a minimum weight route starting and finishing at A and traversing every edge of the network at least once.

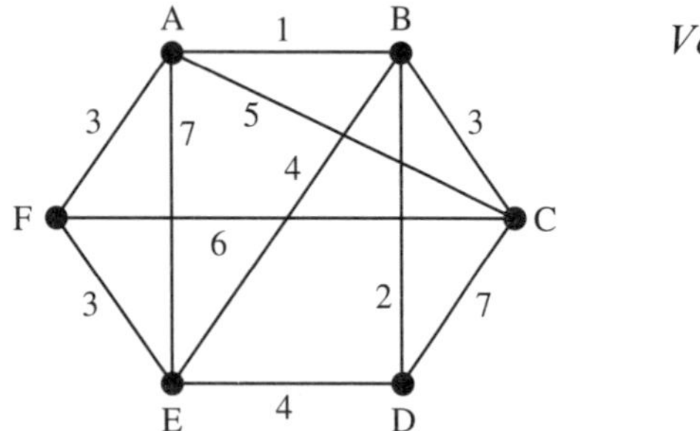

Vertex	*Order*
A	4
B	4
C	4
D	3
E	4
F	3

Vertices D and F are odd. Edges joining D and F will therefore have to be repeated. If necessary we can use Dijkstra's algorithm to find the shortest route between D and F, but in this simple case we can see that it is DBAF.

So edges DB, BA and AF will have to be repeated, and we can show this on the network:

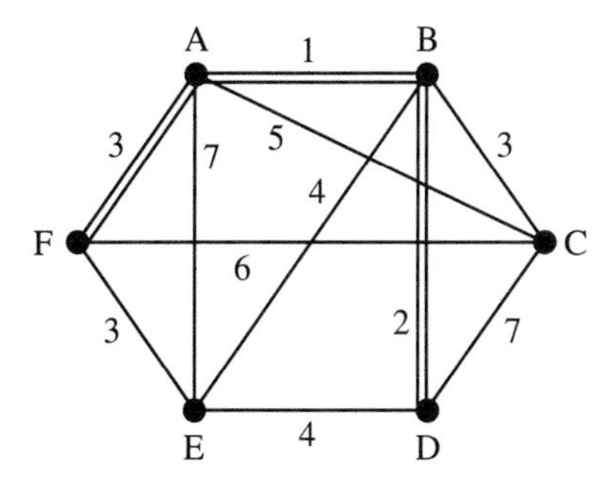

A route is:
ABCAFEAFCDBEDBA, with
total length $= 45 + 6 = 51$.

Alternative ways of pairing vertices

When there are more than two odd vertices then there will be alternative ways of pairing them up to form a **pairing**. For instance, suppose that there are four odd vertices, W, X, Y and Z. Then there are three possible pairings as the following example shows.

Example 4

Find a minimum weight route starting and finishing at U and traversing every edge of the network at least once.

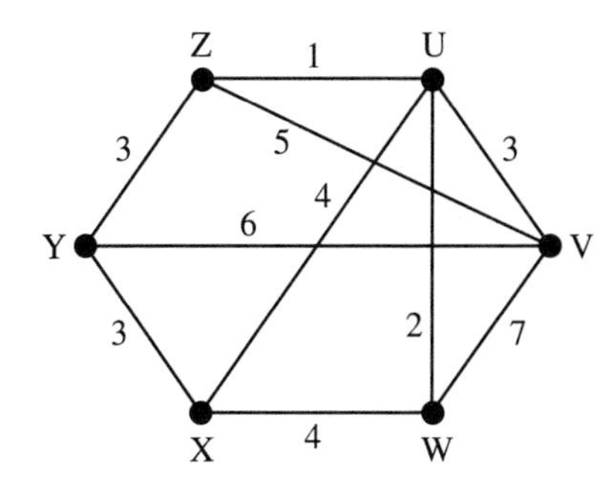

Vertex	*Order*
U	4
V	4
W	3
X	3
Y	3
Z	3

The three pairings are:

(i) W with X and Y with Z
(ii) W with Y and X with Z
(iii) W with Z and X with Y

For each pair in a pairing the shortest route between the two vertices has to be found. The edges on those shortest routes then have to be repeated:

Pairing	Shortest routes	Lengths of repeated edges
WX / YZ	WX and YZ	$4 + 3 = 7$
WY / XZ	WUZY and XUZ	$2 + 1 + 3 + 4 + 1 = 11$
WZ / XY	WUZ and XY	$2 + 1 + 3 = 6$

So the best pairing is the WZ / XY pairing:

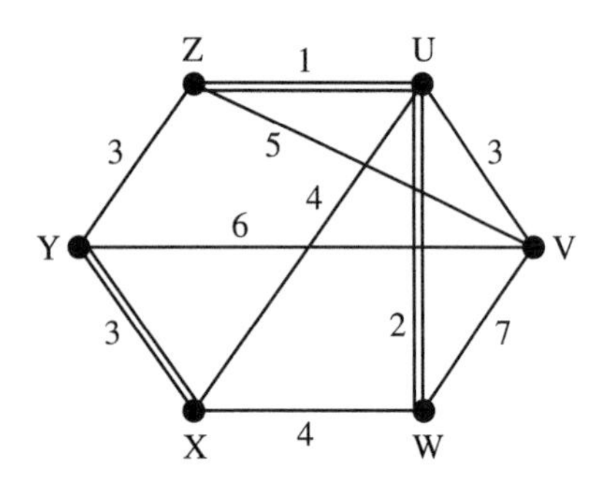

There are many possible routes, such as UVWUZVYXWUXYZU.

They all have length $38 + (2 + 1 + 3) = 44$.

The route inspection algorithm

To find a minimum weight route which traverses every edge of a given connected network at least once the algorithm is:

Step 1 List all odd vertices.

Step 2 Form all possible pairings.

Step 3 For each pairing find the edges that are best to repeat, and find the sum of the lengths of those edges.

Step 4 Choose the pairing with minimum sum and construct a route which repeats those edges. (This last step is easy since the graph with repeated edges will be traversable.)

Example 5

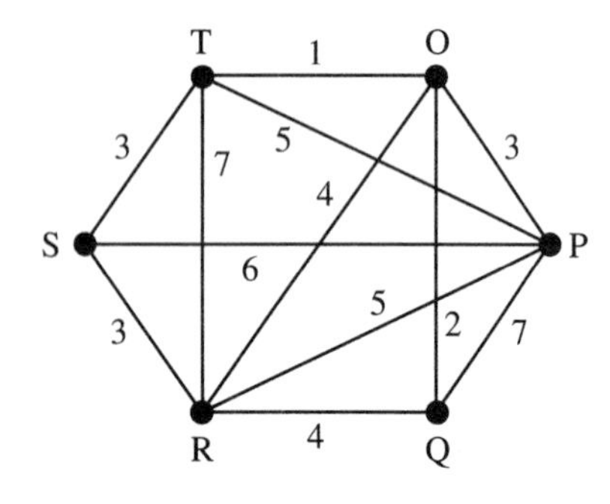

Vertex	*Order*
O	4
P	5
Q	3
R	5
S	3
T	4

Vertices P, Q, R and S are odd.

There are three possible pairings to consider, PQ / RS, PR /QS and PS /QR.

Shortest routes (by inspection in this simple example, but you would need to use Dijkstra's algorithm in practice):

from P to Q is POQ with length 5
from R to S is RS with length 3

So the PQ / RS pairing involves repeating edges of total length 8.

from P to R is PR with length 5
from Q to S is QOTS with length 6

For PR / QS the length is 11.

from P to S is PS with length 6 *For PS / QR the length is 10.*
from Q to R is QR with length 4

So the best choice is to repeat edges PO, OQ and RS, and we can show this on the diagram:

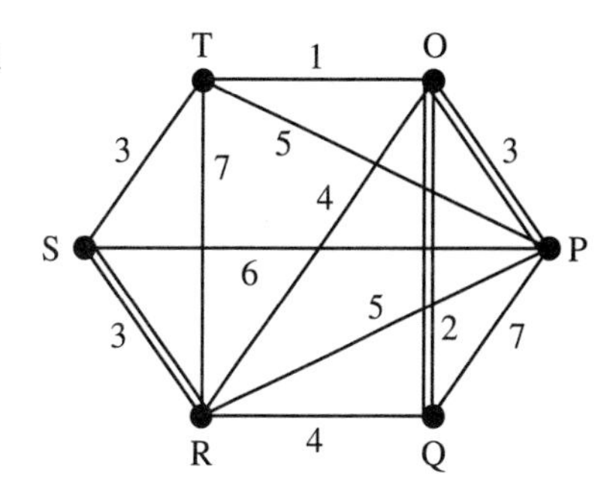

A route is: POQPOQROTPSTRSRP,
with total length $= 50 + 8 = 58$.

Exercise 3C

1

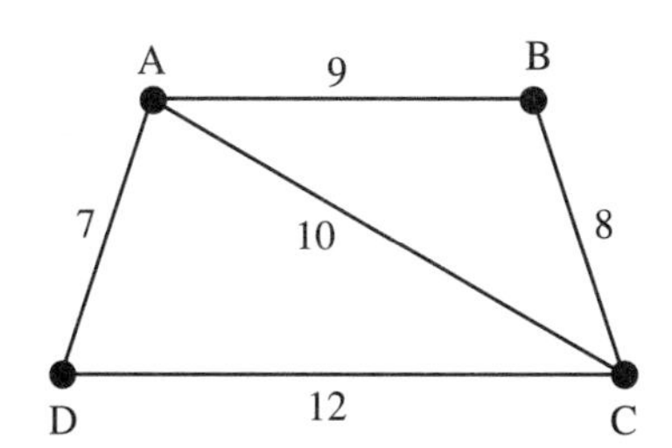

For the network shown above find a minimum weight route starting and finishing at A and traversing every edge of the network at least once. Give the weight of this route.

2 The table shows the distances in kilometres between the five locations A, B, C, D and E.

	A	B	C	D	E
A	—	27	40	30	20
B	27	—	28	45	—
C	40	28	—	—	—
D	30	45	—	—	22
E	20	—	—	22	—

(a) Draw a network representing this information.
(b) Find a minimum distance route starting and finishing at A and traversing every road at least once.

3

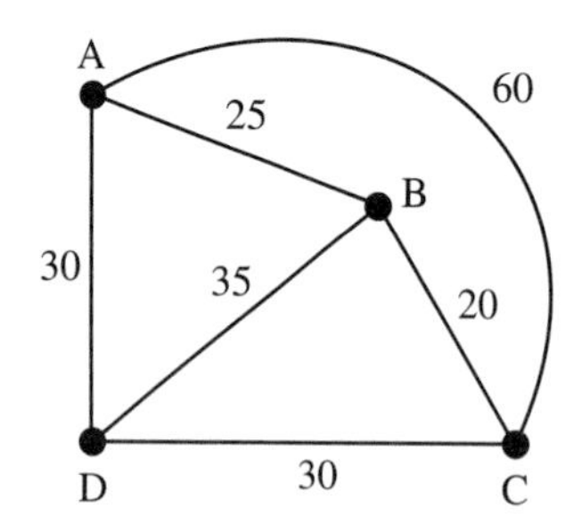

The diagram shows the layout of footpaths in a garden. The lengths of the paths are given in metres. The paths are all to be swept. Starting at A and finishing at A find a route of minimum length. Give the length of this route.

4 Following a storm, all the roads in a village have to be checked for fallen trees. The streets link five locations, A, B, C, D and E. The lengths of the streets, in metres, are given in the table.

	A	B	C	D	E
A	—	20	60	—	45
B	20	—	50	70	—
C	60	50	—	50	—
D	—	70	50	—	40
E	45	—	—	40	—

Find the minimum length route starting and finishing at A. Give the length of this route.

5 The matrix opposite gives fares for direct bus journeys between towns, P, Q, R, S, T, U and V, in a tourist area. Blanks indicate no direct service.

Fares in pence

	P	Q	R	S	T	U	V
P		67		45			80
Q	67		63	71		170	
R		63			59		
S	45	71			40		54
T			59	40		58	62
U		170			58		70
V	80			54	62	70	

Find one of the cheapest ways for a tourist to use every route.

6 A clothing company is to produce a logo incorporating the letters NX. A diagram of the logo is shown below. Vertices have been labelled ABCDEFG for convenience. The numbers indicate distances in cm between vertices.

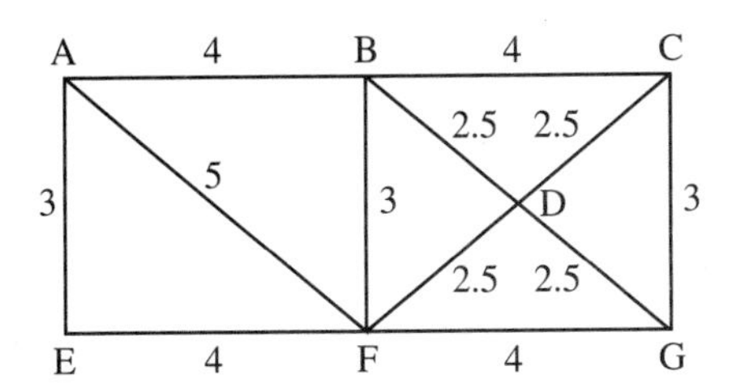

Plans involve the production of large numbers of garments, so it is important to produce the logo with the minimum of stitching. Machinists will start and end stitching at A, with no breaks.

(a) Explain why it will be necessary to sew some edges more than once.

(b) Use an appropriate algorithm to determine the order of stitching which minimises the length to be sewn. Give your minimum length, and a corresponding order in which vertices should be stitched.

7 The map shows a number of roads in a housing estate. Road intersections are labelled with capital letters and the distances in metres between intersections are shown. The total length of all the roads in the estate is 2300 m.

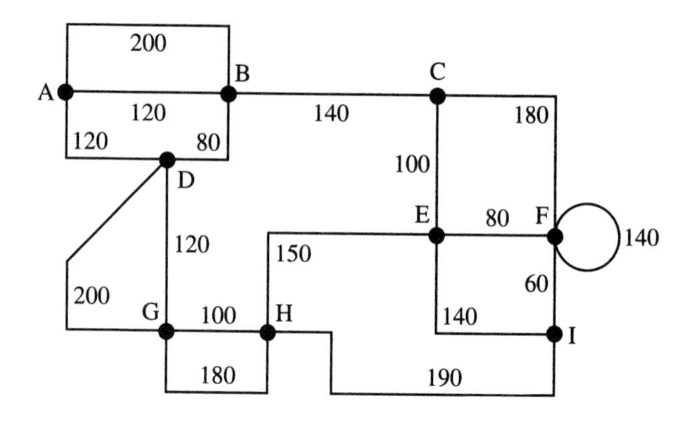

(a) A newspaper deliverer has to walk along each road at least once, starting and ending at A. By investigating all possible pairings of odd intersections, find the minimum distance which the newspaper deliverer has to walk.
(You need not apply a shortest distance algorithm to solve this. You should show the distance computations which lead to you choosing a particular pairing of odd intersections.)
For each intersection other than A, give the number of times that the newspaper deliverer must pass through that intersection whilst following the shortest route.

(b) The newspaper deliverer only calls at a proportion of houses. The postwoman has to call at most houses, and since the roads are too wide to cross continually back and forth, she finds it necessary to walk along each road twice, once along each side. She requires a route to achieve this in the minimum distance.
Describe how to produce a network to model this problem. Without drawing such a network, say why it will be traversable and calculate the minimum length of road along which the postwoman will have to walk.

(c) The streetcleaner needs to drive his vehicle along both sides of each road. He has to drive on the correct side of the road at all times. He too would like a shortest route.
Explain how the streetcleaner's problem differs from the postwoman's, and say how the network would have to be modified to model this.

8 A skier can travel across a section of mountain country by using a system of drag lifts. She has to queue for each lift in order that she can be towed up a hillside. She can then ski to the bottom of another lift, or, in some cases, back to the bottom of the same lift.

The bottoms of the lifts are labelled A to H, and the average queue plus tow time for each lift is listed in the table below.

Lift	A	B	C	D	E	F	G	H
Time for queue plus tow (min)	5	10	7	25	14	15	7	10

The time in minutes to ski from the top of a lift to the bottom of a lift, where a route exists, is given in the following table:

To \ From	A	B	C	D	E	F	G	H
A	6	8		10				
B	10			5				
C	8							7
D	11	5		5		15		
E		12			4			
F				17	5		4	
G						6		7
H			7					

(a) Explain why the total time from the bottom of B to the bottom of A is 18 minutes. Draw a network incorporating total times, its vertices representing the bottoms of lifts.

(b) Use Dijkstra's algorithm to find the shortest time for the skier to get from the bottom of lift A to the bottom of lift F.

(c) A skier wishes to ski all of the runs, starting and ending at the bottom of lift A.

State those runs that must be repeated, giving your reason.

Find a suitable route and the time it would take.

SUMMARY OF KEY POINTS

1 A cycle which visits every vertex, returning to the starting vertex is called a tour.

2 A tour which visits each vertex only once is called a Hamilton cycle.

3 The total length of any tour is an upper bound for the total length of the solution to a travelling salesman problem.

4 Twice the length of the minimum spanning tree is an upper bound for the total length of the solution to the practical travelling salesman problem.

Critical path analysis 4

In this chapter we consider the modelling of complex projects by a network and the analysis of the resulting model. This is sometimes called network analysis or network planning but more usually is now referred to as the **critical path method** (CPM) or **critical path analysis** (CPA).

Serious application of the method was first made in the mid-1950s. In Britain a team working for the CEGB (Central Electricity Generating Board) developed a method for scheduling the work of overhauling a generating plant. By 1957 they had devised a technique for identifying 'the longest irreducible sequence of events'. Using this technique, in 1958, they carried out an overhaul of a power station which reduced the overall time to 42% of the previous average time for the same work.

At about the same time a group of researchers working in the US for the E.I. du Pont Chemical Company used the critical path method to schedule the construction of a $10 million chemical plant. It was credited with saving the company $1 million.

Since that time CPM or CPA has been used in a large number of areas, including: overhaul, construction, civil engineering, town planning, marketing, ship building and design.

4.1 Precedence tables

The first step in scheduling a complex project is to break it down into a set of sub-projects known as **activities**. For example, if the project were building an extension to a house then this could be broken down into the following activities:

A Prepare the foundations
B Have foundations passed by inspector
C Obtain bricks
D Erect walls
E Construct roof
F Install plumbing
G Install wiring
H Plaster walls
I Decorate
J Landscape garden

It is clear that not all these activities are independent. Some of them are related in the sense that certain activities cannot start until others have been completed. The next step in the scheduling process is to identify which activities depend on which others being completed first. Some activities must precede (come before) others. The way in which the activities depend on each other can be summarised in a **precedence table** like this:

Activity	Depends on
A	—
B	A
C	—
D	B, C
E	D
F	D
G	E
H	F, G
I	H
J	E

A precedence table is sometimes called a **dependence table.**

The heading 'depends on' is a shorthand for 'the activities which must be completed before it can be started'. A dash — indicates that there are no such activities. Notice that the 'depends on' column only shows activities *immediately* preceding each entry in the activity column. For example, we have written E depends only on D. The fact that E depends on B, C is therefore implied.

The relationships between the activities may not be given in such a direct form. When this is the case it is suggested that you first produce a dependence table from the given information.

Example 1

A project has been broken down into the activities A, B, C, D, E, F and G. After a committee meeting the following information was produced. Draw up a precedence table which summarises this information.

Activities A, C and D do not depend on the completion of any other activity.

Activity A must be completed before activity B can start.

Both activities B and C must be completed before Activity E can begin.

Activity F can only start when A, B, C, D and E are completed.

The project is completed when G is finished. G requires all other activities to be completed before it can start.

The table below is the precedence table. The reasons for the entries are given immediately following the table.

Activity	Depends on
A	—
B	A
C	—
D	—
E	B, C
F	D, E
G	F

The entries for A, C and D follow from their independence of other activities.

The entry for B follows from the given statement directly.

The entry for E follows as *both* B and C must be complete before it can start.

The entry for F requires some explanation. Since E depends on B and C, and also B depends on A we need only include E here together with activity D.

As F depends on D and E, and E depends on B and C, and B depends on A the given statement implies G depends on all the others as required.

Exercise 4A

1 The project 'write and post a letter' may be broken down into the following activities.

A Purchase a pad of paper.
B Purchase a packet of envelopes.
C Purchase a stamp.
D Write the letter.
E Address the envelope.
F Stick the stamp on the envelope.
G Place the letter in the envelope.
H Seal the envelope.
I Post the letter.

Draw up a precedence table for this project.

2 Mrs Brown decides that the lounge needs a total change of paint, wallpaper and curtains. She identifies the following activities.

A Buy new curtains.
B Buy the paint.
C Buy the wallpaper.
D Take up the carpet.
E Remove the curtains.
F Paint the woodwork.
G Hang the wallpaper.
H Hang the new curtains.
I Replace the carpet.

Draw up a precedence table for this project.

3 A project consists of activities A, B, C, D and E. These activities must satisfy the following conditions:

Activities A and B are independent of the others.
Only when activity A is completed can activity C start.
Activity D requires activity B to be completed before it can start.
Activity E can only start when all other activities are finished.

Draw up a precedence table for this project.

4 A project consists of activities A, B, C, ..., J. The following information has been obtained about these activities.
Activities A, B and C do not depend on the completion of any other activity.
Activities A and B must be completed before activity D can begin.
Before activities E and F can start, activity D must be completed.
Activity H can only start when A, D and F are completed, whilst activity G can only start when activity E has also been completed.
One must wait until activities C, D, E, F and G are complete before activity I can begin.
The final activity J requires the completion of all the others.
Draw up a precedence table for this project.

4.2 Activity networks

Having produced a precedence table the next step in the critical path method is to produce an **activity network** (or project network) to model the situation.

- The nodes (vertices) in this network represent **events**. Each event is the completion of one or more activities.
- The arcs in this network represent **activities** and the weight on each arc represents the **duration** of the corresponding activity.
- The **source node** represents the *beginning of the project* and the **sink node** represents the *end of the project.*

Here is a typical activity network:

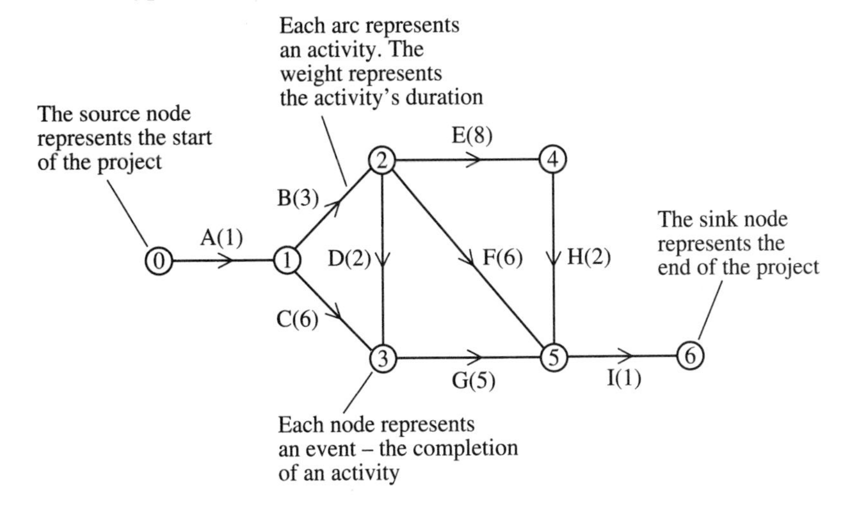

Notice that the events represented by the nodes are numbered. The lengths and orientation of the arcs are not significant.

- **It is conventional to assume that time flows from left to right when possible, but arrows are used to make the directionality clear.**

Consider node ④:

with the arc representing E *entering* and the arc representing H *leaving*. The numbers in brackets are the durations of these activities. This models the situation 'activity E must be completed before activity H can begin'. Event ④ denotes the completion of activity E.

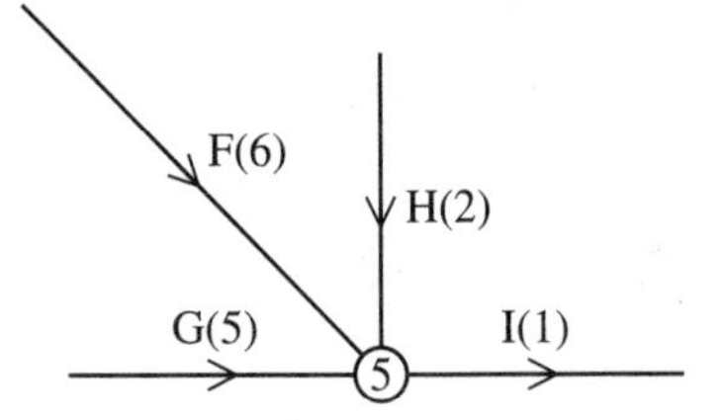

Node ⑤ represents a more complicated situation. This models the situation 'activity I cannot start until activities F, G and H are all complete'. Event ⑤ represents the completion of all of the activities F, G and H.

Consider now a typical *activity*, say, G:

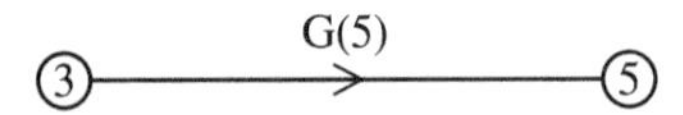

This activity has to be performed between the two events indicated by the numbers at the end of the arcs. The event at the beginning of the arc, in this case ③, is called the **tail event**; that at the end of the arc, in this case ⑤, is called the **head event**. Any activity can be specified by giving the numbers of its tail and head events. That is, arc (3, 5) represents G. In constructing a network the head event is always given a number greater than the corresponding tail event.

Drawing an activity network

The network on page 103 models this precedence table:

Activity	Depends on
A	—
B	A
C	A
D	B
E	B
F	B
G	C, D
H	E
I	F, G, H

The network was constructed like this:

1 The activity A has no predecessor and so may begin at any time. Label source node ⓪ and insert activity A.

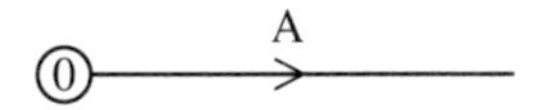

2 Activities B and C depend only on A. Therefore insert node ① indicating the completion of activity A and add activities B and C.

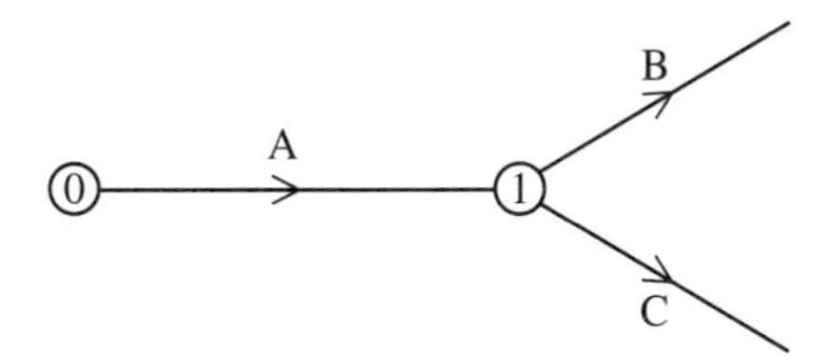

Arcs representing B and C are any two arcs leaving ①; their orientation is irrelevant but the convention is that they go from left to right when possible.

3 Activities D, E and F depend only on B. Therefore insert node ② indicating the completion of activity B and add activities D, E and F.

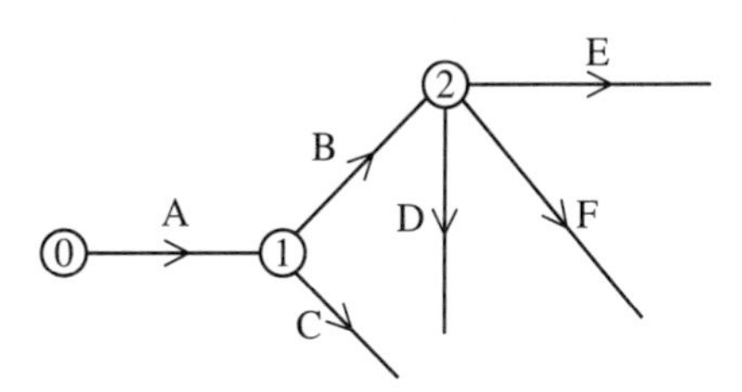

4 Activity G depends on C and D. Therefore insert node ③ indicating the completion of both activity C and activity D and add activity G.

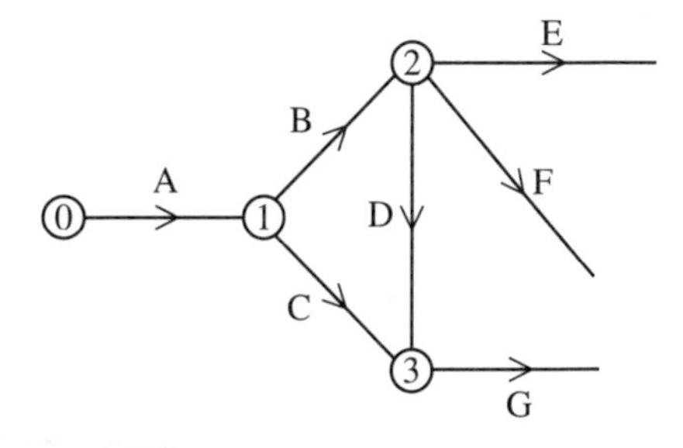

5 Activity H depends on E. Therefore insert node ④ indicating the completion of activity E and add activity H.

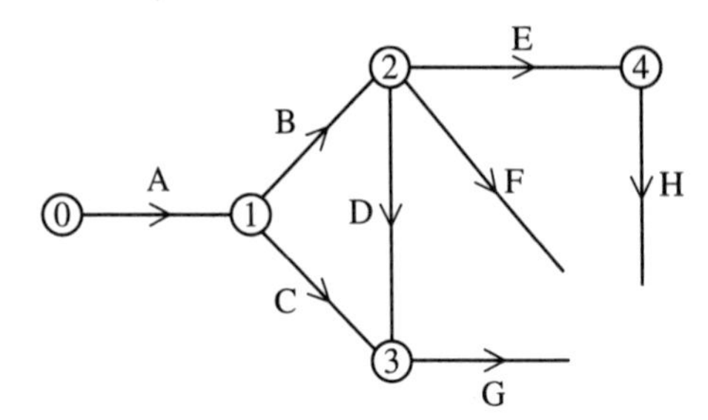

6 Finally, activity I depends on F, G and H. Therefore insert node ⑤ indicating the completion of activities F, G and H and add activity I.

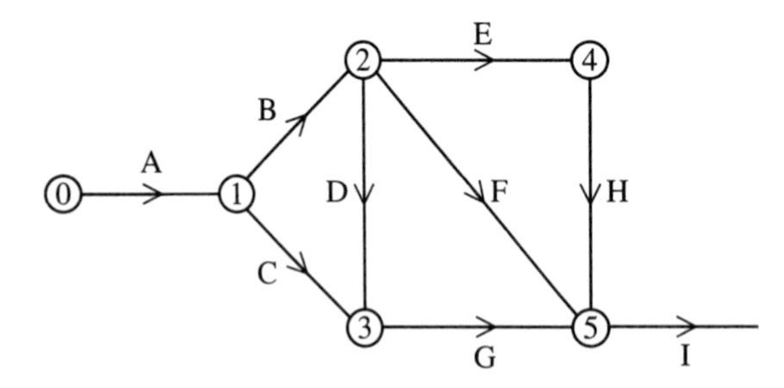

7 The project is complete when all activities are completed. So add a sink node ⑥.

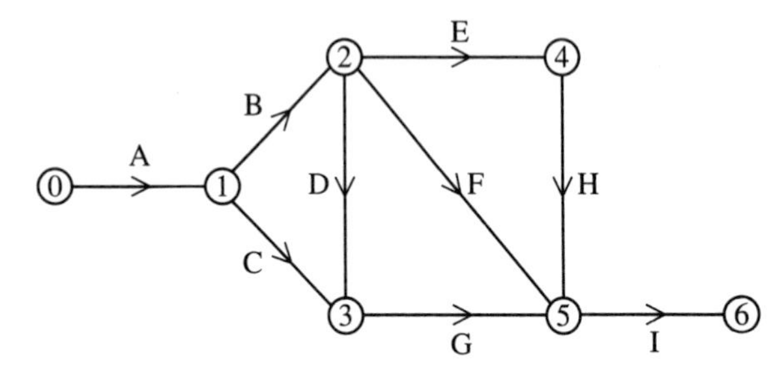

Your first attempt to draw a network may not look as neat as this. If possible you should try to produce a network without any crossing arcs. It is useful to look ahead down the 'depends on' column to see which arcs are likely to meet at a node.

In this example D was drawn downwards towards C for this reason. Further, it was noticed that F, G and H occurred together and so these arcs were drawn so that node ⑤ was easily added. Again a neat diagram results if all the arcs are straight lines, but this is not necessary.

Using dummies

In the above examples it was possible to represent the dependence table without too much difficulty. However, situations can occur when, in order to represent the dependence table, it is necessary to use a **dummy activity** or logical restraint. Such a situation occurs when two chains have a common event but they are themselves wholly or partly independent of each other.

- **A dummy activity is usually shown as a dotted line. *Its direction is important* but it has zero duration.**

Here is an example where a dummy activity is necessary:

Activity	Depends on
C	A, B
D	B

This situation can be represented by:

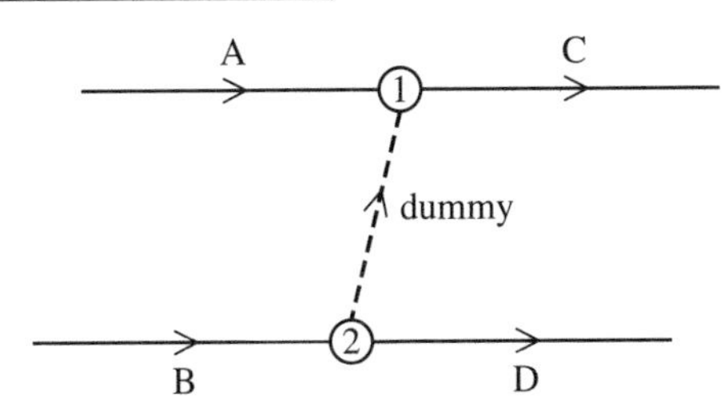

It is not possible to represent this situation without a dummy. Notice carefully the direction of the dummy.

It is a convention in the critical path method that we allow *at most one activity between any two events*. In order to meet this convention a dummy activity may be necessary. For example, consider

Activity	Depends on
A	—
B	A
C	A
D	B, C

This may be represented by

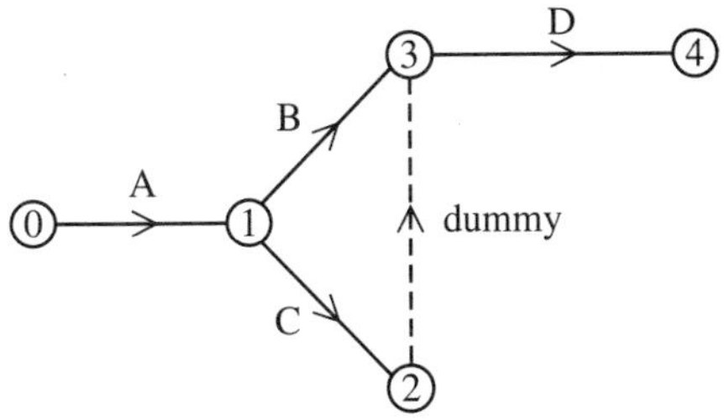

Example 2

Draw the activity network for the precedence table.

Activity	Depends on
A	—
B	—
C	—
D	A
E	A
F	C
G	B, C, D
H	G
I	E, G
J	F

1 Activities A, B, C

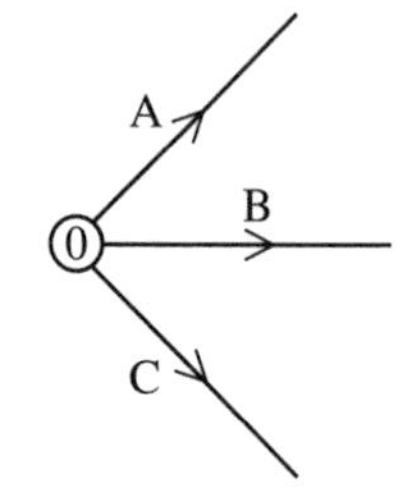

2 Activities D, E

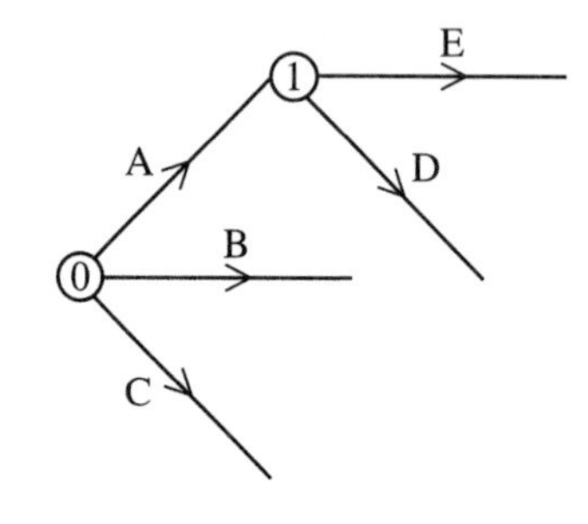

3 Activity F

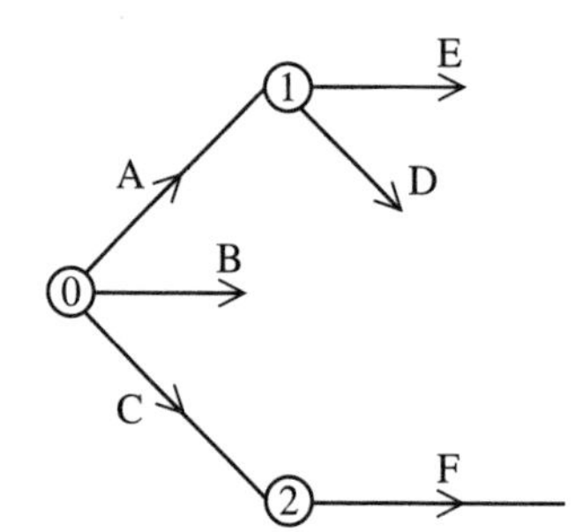

4 Activity G – you need a dummy here to represent this dependence:

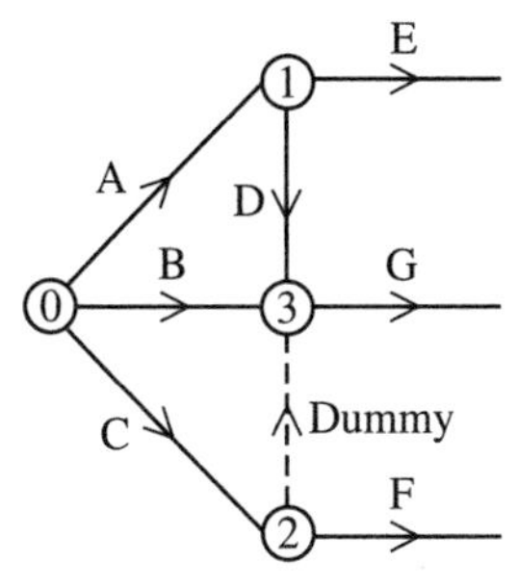

Notice the direction of the dummy as F does not depend on B.

5 Activity H

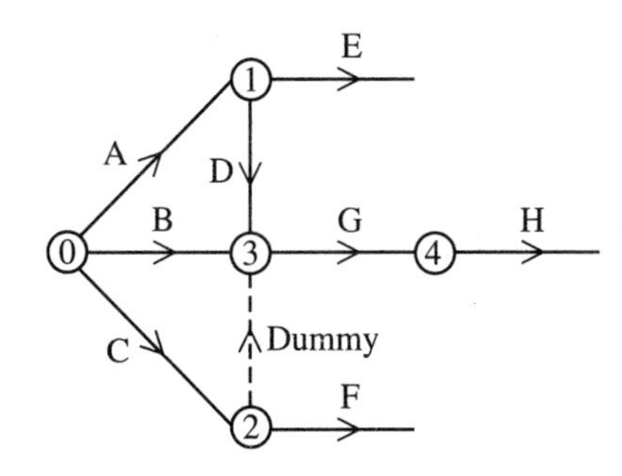

6 Activity I – as I depends on E and G, but H depends only on G, you need a dummy to represent this.

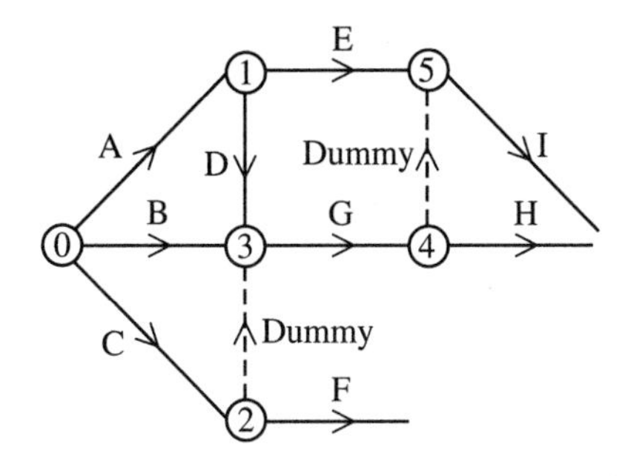

7 Activity J

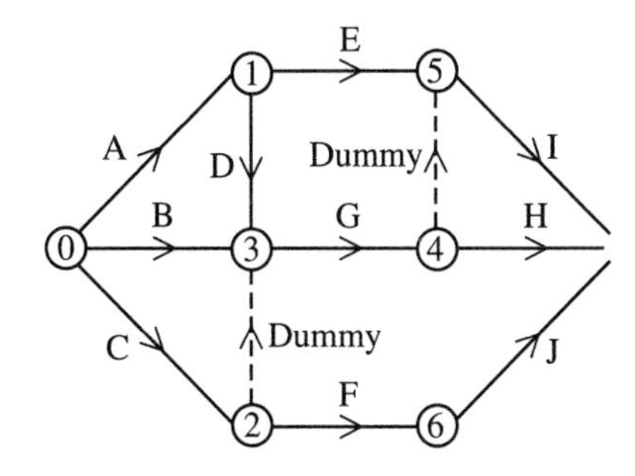

8 End of project when H, I and J are complete.

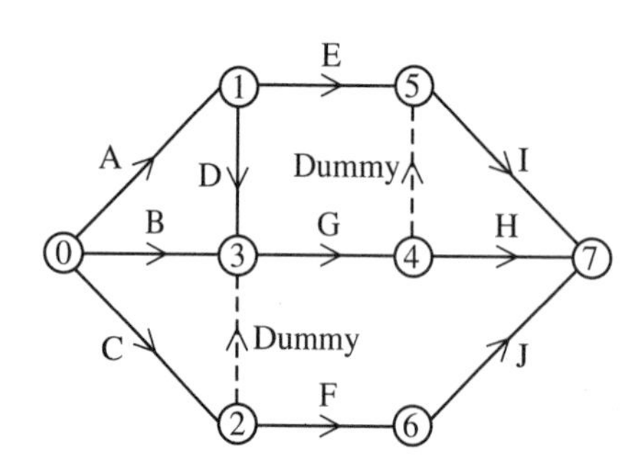

It is possible to draw different diagrams to represent the same dependencies.

Exercise 4B

1 Draw an activity network to represent the dependence table opposite, given that the project is complete when all activities are complete.

Activity	Depends on
A	—
B	—
C	—
D	A
E	A
F	B
G	C
H	D
I	E, F, G

2 Draw an activity network to represent the dependence table opposite, given that the project is complete when all activities are complete.

Activity	Depends on
A	—
B	—
C	B
D	B
E	A
F	D, E
G	D, E
H	C, F

3 Draw an activity network to represent the dependence table below, given that the project is complete when all activities are complete.

Activity	Depends on
A	—
B	—
C	—
D	A
E	B
F	B, C
G	B, D
H	G
I	E, F, H

4 The following information has been obtained concerning the activities A, B, C, ..., N involved in a project:

Activity A precedes B, C, D, E and F
Activity J follows B
Activity K follows C
Activity G follows E
Activity H follows D and G and precedes L
Activity I follows D and G
Activities F and I precede M
Activity L follows J and K
Activity N follows L and M.

(a) Draw up a precedence table for these activities.
(b) Draw an activity network to represent your precedence table.

4.3 Analysing the project (the critical path algorithm)

The **critical path algorithm** identifies the longest path or paths from the source node to the sink node. This is the **critical path** (or paths). The algorithm proceeds by finding two times associated with each event: the **earliest event time** and the **latest event time**.

Earliest event time e_i (forward scan)

- **The earliest time e_i for node i is the earliest time that you can arrive at event i with all the incoming activities completed. We work from source node to sink node, that is forwards, or from the left to right in the diagram.**

Initially you set the earliest time for the source node as zero. We will illustrate the algorithm by carrying out the calculations for the network on page 103, which for ease of reference we repeat here.

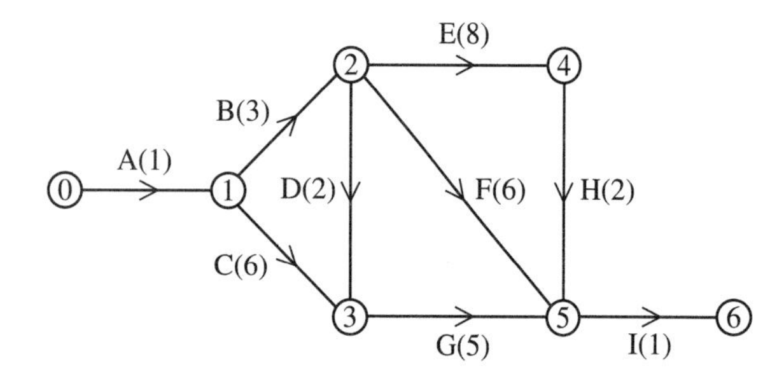

Node 0 $e_0 = 0$

Node 1 There is only one arc (0, 1) *leading into* node ① and so:

$$\begin{aligned} e_1 &= e_0 + (\text{duration of A}) \\ &= 0 + 1 \\ &= 1 \end{aligned}$$

Node 2 There is only one arc (1, 2) leading into node ② and so:

$$\begin{aligned} e_2 &= e_1 + (\text{duration of B}) \\ &= 1 + 3 \\ &= 4 \end{aligned}$$

Node 3 There are *two* arcs (1, 3) and (2, 3) leading into node ③ and so, since both activities C and D must be completed before event ③ can take place:

$$\begin{aligned} e_3 &= \text{larger of } [e_1 + (\text{duration of C}), e_2 + (\text{duration of D})] \\ &= \text{larger of } [1 + 6, 4 + 2] \\ &= \text{larger of } [7, 6] \\ &= 7 \end{aligned}$$

Node 4 There is only one arc (2, 4) leading into node ④ and so:

$$\begin{aligned} e_4 &= e_2 + (\text{duration of E}) \\ &= 4 + 8 \\ &= 12 \end{aligned}$$

Node 5 There are *three* arcs (2, 5), (3, 5) and (4, 5) leading into node ⑤ and so activities F, G and H must all be completed before event ⑤ can take place.

$$\begin{aligned} e_5 &= \text{largest of } [e_2 + (\text{duration of F}), \\ &\qquad e_3 + (\text{duration of G}), \\ &\qquad e_4 + (\text{duration of H}),] \\ &= \text{largest of } [4+6, 7+5, 12+2] \\ &= \text{largest of } [10, 12, 14] = 14 \end{aligned}$$

Node 6 There is only one arc (5, 6) leading into node ⑥ and so:

$$\begin{aligned} e_6 &= e_5 + (\text{duration of I}) \\ &= 14 + 1 = 15 \end{aligned}$$

- **Thus, in general:**

  $$e_i = \underset{k}{\textbf{maximum}}\,[e_k + \textbf{duration } (k, i)]$$

  where the maximum is taken over all arcs (k, i) leading into vertex i.

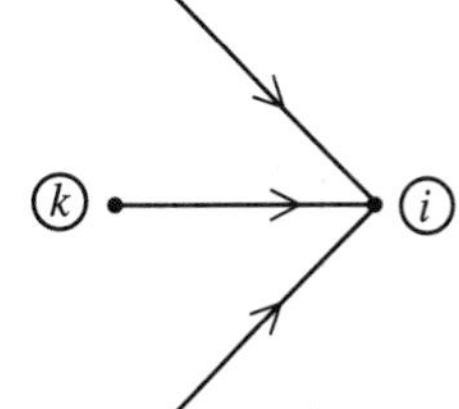

$\underset{k}{\text{maximum}}[e_k + \text{duration } (k, i)]$ means find the values of $[e_k + \text{duration } (k, i)]$ for all possible values of k, keeping i fixed, and then take the largest of these values.

The earliest time calculated for the sink node is the total amount of time required to complete the project. It is the length of the longest (critical) path (or paths).

For our project, therefore, the length of the critical path is 15.

Latest event time l_i (backward scan)

- **The latest time l_i for node i is the latest time you can leave event i without extending the length of the critical path. We work from sink node to source node, that is backwards, or from right to left in the diagram.**

Clearly for the sink node (finish event) the latest and earliest terms are the same so you have $l_6 = e_6 =$ the total project time $= 15$.

Node 5 Since there is only one arc (5, 6) *leaving* node ⑤:

$$\begin{aligned} l_5 &= l_6 - (\text{duration of I}) \\ &= 15 - 1 = 14 \end{aligned}$$

Node 4 Since there is only one arc (4, 5) leaving node ④:

$$\begin{aligned} l_4 &= l_5 - (\text{duration of H}) \\ &= 14 - 2 = 12 \end{aligned}$$

Node 3 Since there is only one arc (3, 5) leaving node ③:

$$\begin{aligned} l_3 &= l_5 - (\text{duration of G}) \\ &= 14 - 5 = 9 \end{aligned}$$

Node 2 There are *three* arcs (2, 3), (2, 4) and (2, 5) leaving node ② and so:

$$\begin{aligned} l_2 &= \text{smallest of } [l_3 - (\text{duration of D}), \\ &\qquad l_4 - (\text{duration of E}), \\ &\qquad l_5 - (\text{duration of F})] \\ &= \text{smallest of } [9-2, 12-8, 14-6] \\ &= \text{smallest of } [7, 4, 8] = 4 \end{aligned}$$

Node 1 There are *two* arcs (1, 2) and (1, 3) leaving node ① and so:

$$\begin{aligned} l_1 &= \text{smaller of } [l_2 - (\text{duration of B}), \\ &\qquad l_3 - (\text{duration of C})] \\ &= \text{smaller of } [4-3, 9-6] \\ &= \text{smaller of } [1, 3] = 1 \end{aligned}$$

Node 0 Since there is only one arc (0, 1) leaving node ⓪

$$\begin{aligned} l_0 &= l_1 - (\text{duration of A}) \\ &= 1 - 1 = 0 \end{aligned}$$

- **In general: $l_i = \underset{j}{\text{minimum}}\left[l_j - \text{duration } (i,j)\right]$**

  where the minimum is taken over all arcs (i, j) leaving the vertex i.

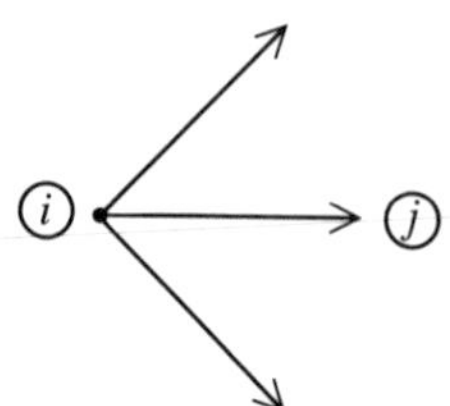

$\underset{j}{\text{minimum}}\left[l_j - \text{duration } (i,j)\right]$ means find the values of $\left[l_j - \text{duration } (i,j)\right]$ for all possible values of j, keeping i fixed, and then take the smallest of these values.

The information we have obtained so far is summarised in the table below:

Event, i	0	1	2	3	4	5	6
Earliest time, e_i	0	1	4	7	12	14	15
Latest time, l_i	0	1	4	9	12	14	15

This information may also be added to the network, as it is obtained, by placing at each node a box with e_i and l_i in the positions indicated: $\boxed{e_i \mid l_i}$.

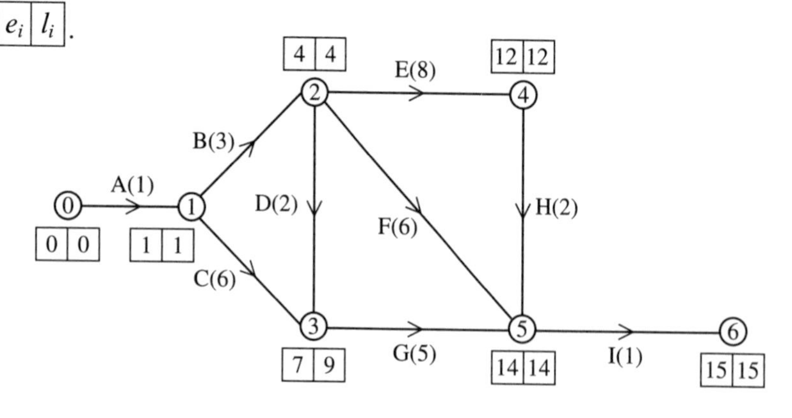

Exercise 4C

1

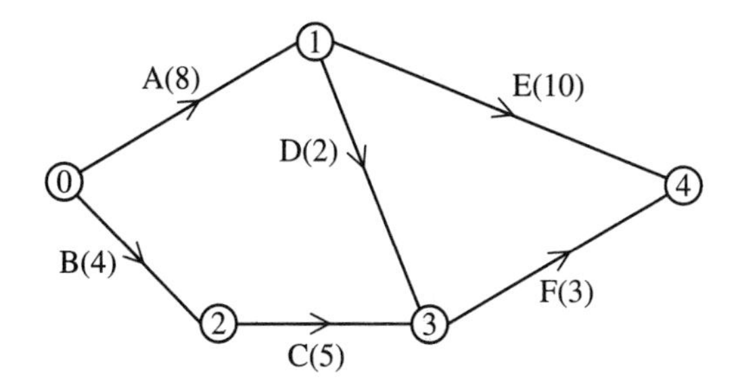

The above network models a project. Complete the table below by finding all the earliest times e_i and the latest times l_i.

Event, i	0	1	2	3	4
Earliest time, e_i Latest time, l_i	0	8	4	 15	 18

2 Complete the boxes $\boxed{e_i \mid l_i}$ for each event in the following network model.

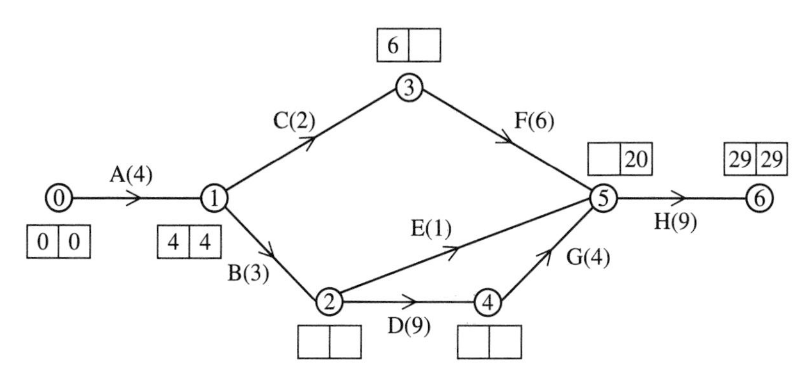

3 For the network model shown below find the earliest time e_i and the latest time l_i for each event.

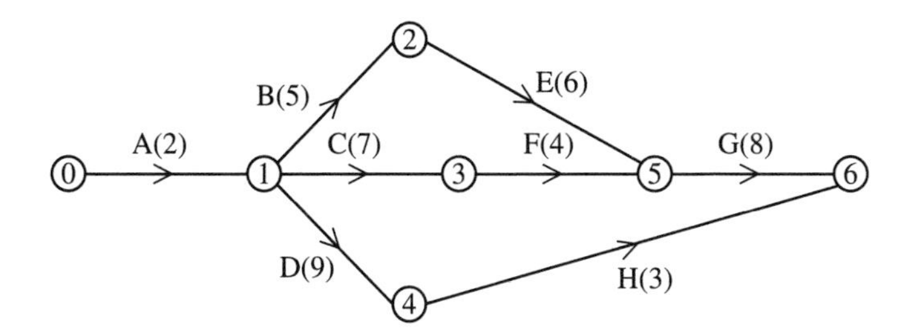

4.4 Critical events and critical activities

On a critical path(s) there can be no leeway. Any delay delays the completion of the entire project.

- **For all events on a critical path $e_i = l_i$.**

Such events are known as **critical events.**

In our example, events 0, 1, 2, 4, 5 and 6 are critical; only 3 is non-critical.

Any delay of a critical activity will increase the length of the critical path and so for any activity (i, j) on a critical path

$$[\text{duration of } (i,j)] = l_j - e_i$$

- **Activities for which $[l_j - e_i - \text{duration of } (i,j)] = 0$ are called critical activities.**

You should not assume that an activity joining critical events is a critical activity.

For any event the difference $s_i = l_i - e_i$ is called the **slack**. In general $s_i \geqslant 0$ and is **equal to zero for critical events**. It is the allowable delay at event i.

For any non-critical activity the quantity

$$F(i,j) = [l_j - e_i - \text{duration of } (i,j)]$$

is called the **total float of activity (i,j)**. The total float $F(i,j)$ of a critical activity is of course zero.

The total float represents the maximum possible delay that can be incurred in the processing of this activity without increasing the critical path length, providing there are no delays elsewhere. In our case we have

	A	B	C	D	E	F	G	H	I
Activity (i, j)	(0, 1)	(1, 2)	(1, 3)	(2, 3)	(2, 4)	(2, 5)	(3, 5)	(4, 5)	(5, 6)
Total float	0	0	2	3	0	4	2	0	0

Hence the critical activities are

(0, 1),	(1, 2),	(2, 4),	(4, 5),	(5, 6)
A	B	E	H	I

So A B E H I is the critical path of length 15. Notice that (2, 5) is *not* a critical activity although events 2 and 5 are critical events.

Exercise 4D

1 For the project in question 1 of Exercise 4C
(a) obtain the slack s_i for each event i and so identify the critical events
(b) obtain the total float $F(i,j)$ for each activity (i,j) and so identify the critical activities
(c) hence state the critical path and give its length.

2 For the project in question 2 of Exercise 4C obtain the information requested above in question 1.

3 For the project in question 3 of Exercise 4C obtain the information requested in question 1.

4.5 Time analysis of a network

It is useful in analysing a project to calculate four times associated with each *activity*: the earliest start time, the earliest finish time, the latest finish time and the latest start time. Let a typical activity be represented by the arc (i,j) and let the duration of this activity be a_{ij}.

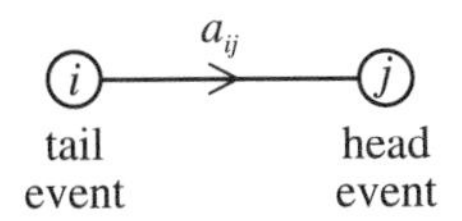

(i) Earliest start time – this is the *earliest possible time* at which activity (i,j) can start. By definition this is the earliest time for the tail event i and so is e_i.
(ii) Earliest finish time – this is the *earliest possible time* at which activity (i,j) can finish. It is obtained by adding the duration a_{ij} of the activity (i,j) to the earliest start time and so is

$$e_i + a_{ij}$$

(iii) Latest finish time – this is the *latest time* at which activity (i,j) can *finish* if the project is to be completed in the project time. It is therefore the latest time for the head event j and so is l_j.
(iv) Latest start time – this is the *latest time* at which activity (i,j) can *start* if the project is to be completed on time. This is therefore obtained by subtracting the duration a_{ij} of activity (i,j) from the latest finish time and so is

$$l_j - a_{ij}$$

For example, for our project consider activity (2, 5), that is activity F.

(i) Earliest start time $= e_2 = 4$
(ii) Earliest finish time $= e_2 + a_{25} = 4 + 6 = 10$
(iii) Latest finish time $= l_5 = 14$
(iv) Latest start time $= l_5 - a_{25} = 14 - 6 = 8$

Notice the order in which these times are obtained. If you calculate them in the order shown you will avoid errors.

The relationship between these times is often displayed in a diagram.

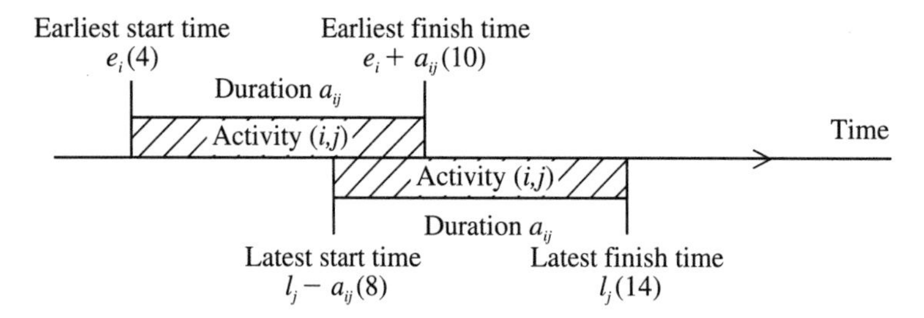

The figures in brackets are the values for the above example.

All the information obtained for the activities can be conveniently summarised in a table:

		Start		Finish		
Activity	Duration	Earliest	Latest	Earliest	Latest	Float
A (0, 1)	1	0	0	1	1	0
B (1, 2)	3	1	1	4	4	0
C (1, 3)	6	1	3	7	9	2
D (2, 3)	2	4	7	6	9	3
E (2, 4)	8	4	4	12	12	0
F (2, 5)	6	4	8	10	14	4
G (3, 5)	5	7	9	12	14	2
H (4, 5)	2	12	12	14	14	0
I (5, 6)	1	14	14	15	15	0
		(i)	(iv)	(ii)	(iii)	(v)

Notice that alternative definitions for the total float of activity (i, j), that is $(l_j - e_i - a_{ij})$ are:

(a) (latest finish time) – (earliest finish time)
(b) (latest start time) – (earliest start time).

The float (v) may therefore easily be calculated once columns (i), (ii), (iii) and (iv) have been obtained, in that order. Calculating the float using both (a) and (b) does provide a check on the other calculations.

Meaning of the total float

Let us look again at the activity F and illustrate the meaning of its entries in the table on the previous page.

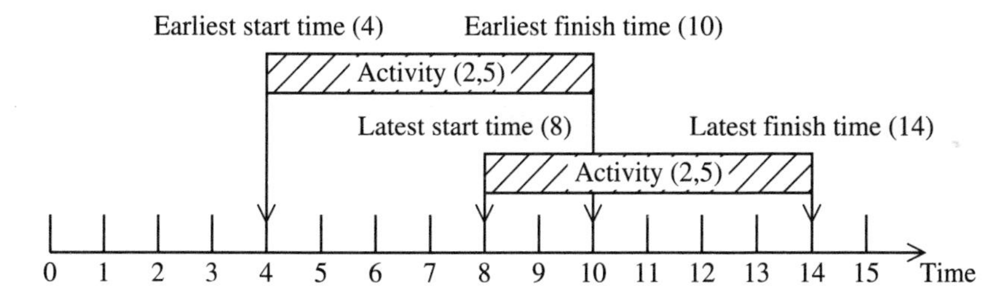

The activity F may start at any time between the earliest start time 4 and the latest start time 8 and it may finish at any time between the earliest finish time 10 and the latest finish time 14. The activity F may therefore float between the boundaries 4 and 14. The difference between the length of this interval ($14 - 4 = 10$) and the duration of the activity (6) is the total float 4. As we shall see in the next section there may be particular reasons for not starting this activity as early as possible.

We may illustrate the information in the table in a graphical way.

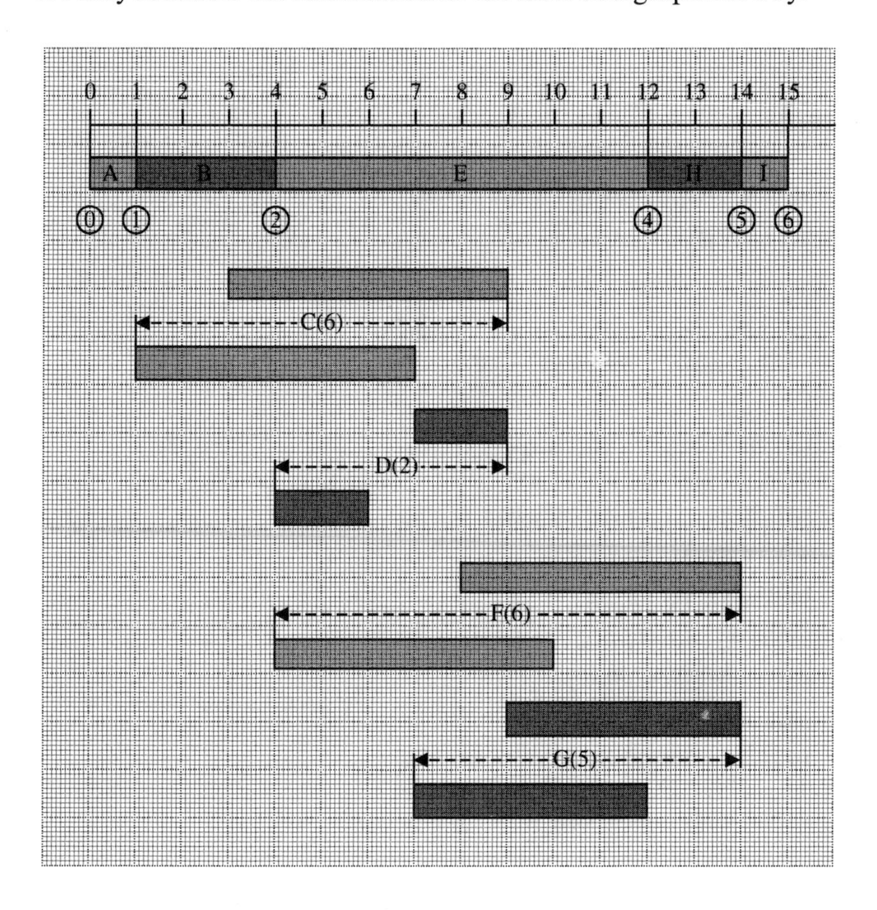

The critical activities are drawn across the top. They have zero float. For each other activity the 'boundaries' are indicated by |←----→|. The shaded bars indicate the activity as early and as late as possible. The critical events are indicated at the junctions of the critical activities.

Exercise 4E

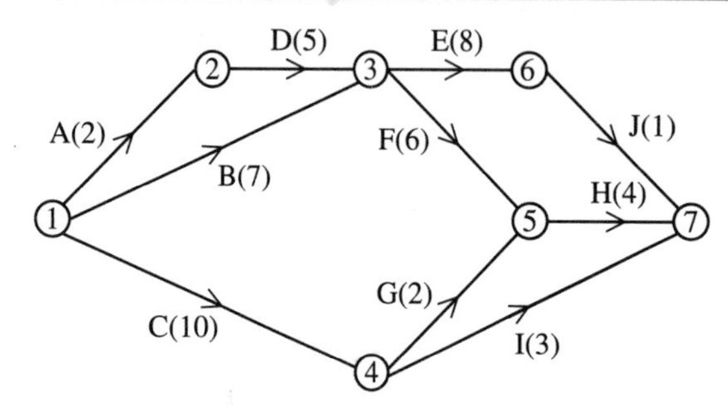

Analyse the above network and produce a table, giving for each activity the earliest and latest start times, the earliest and latest finish times and the total float. Illustrate the information in the table in a graphical manner.

4.6 Using the float

In this section we will consider two ways in which the float may be used.

Scheduling

For the present purpose we will assume that each activity of the project analysed in the previous section requires *one worker*. We now wish to consider such questions as:

(i) How many workers are required to complete the project in the critical time?
(ii) If only a limited number of workers is available what is the minimum completion time for the project?

The following operating rules will be assumed:

(i) No worker may remain idle if there is an activity that can be started.
(ii) Once a worker starts an activity he must continue with that activity until it is finished.

The objective is, of course, to complete the project with as few workers as possible or to complete the project, with the available workers, in a minimum time. There is no algorithm which guarantees an optimal solution.

The procedure we shall adopt is:

(a) When a worker completes an activity consider all the activities which have not been started but which can now be started.
(b) Assign the worker to the activity whose latest start time is the smallest – this is in a sense the 'most critical' activity.
(c) If there are no activities which can be started at this time the worker will have to wait until an activity can be assigned.

With reference to our example we will try and answer three questions:

(i) Can the project be completed by two workers in the project time? (For the present purpose we will refer to this as 15 days.)

The critical activities A, B, E, H and I can all be done by one worker in the project time of 15 days.

The non-critical activities are C, D, F and G. The sum of their durations is $(6 + 2 + 6 + 5)$ days $= 19$ days. Hence it is clear that not all these activities can be completed by a single worker in the project time.

(ii) If only two workers are available what is the minimum time required to complete the project?

We will use the procedure described above and the table on page 118, which for ease of reference we repeat here.

		Start		Finish		
Activity	Duration	Earliest	Latest	Earliest	Latest	Float
A (0, 1)	1	0	0	1	1	0
B (1, 2)	3	1	1	4	4	0
C (1, 3)	6	1	3	7	9	2
D (2, 3)	2	4	7	6	9	3
E (2, 4)	8	4	4	12	12	0
F (2, 5)	6	4	8	10	14	4
G (3, 5)	5	7	9	12	14	2
H (4, 5)	2	12	12	14	14	0
I (5, 6)	1	14	14	15	15	0
		(i)	(iv)	(ii)	(iii)	(v)

Begin by allocating worker 1 to activity A. Worker 2 cannot be assigned until A(1) is complete. So at $t = 1$ allocate worker 1 to activity B(3) and worker 2 to activity C(6). You have then, using a time line,

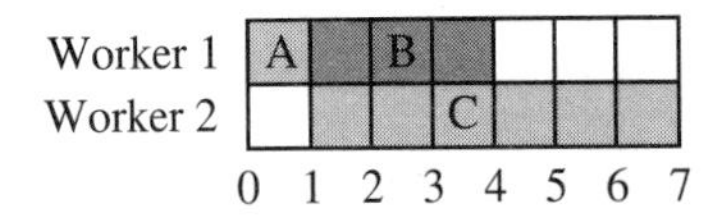

When $t = 4$ worker 1 becomes available. The activity remaining which has 'the smallest of the latest start times' is E. So allocate worker 1 to E(8) and obtain

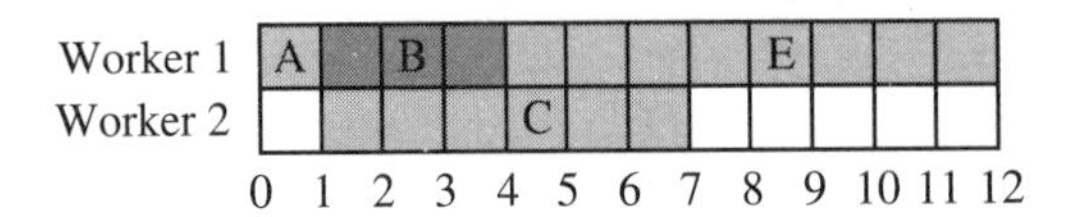

When $t = 7$ worker 2 becomes available. The activity remaining which has 'the smallest of the latest start times' is D. So allocate worker 2 to D. This activity will be finished at $t = 9$, as duration of D is 2 days. From the table you can see that he can then start activity F(6). You have then:

Worker 1 | A | B | E
Worker 2 | C | D | F

0 1 2 3 4 5 6 7 8 9 10 11 12 13 14 15

When $t = 12$ worker 1 becomes available. Using the criterion you can allocate worker 1 to activity G(5). When $t = 15$ worker 2 becomes available and using the criterion will then start activity H. You have then:

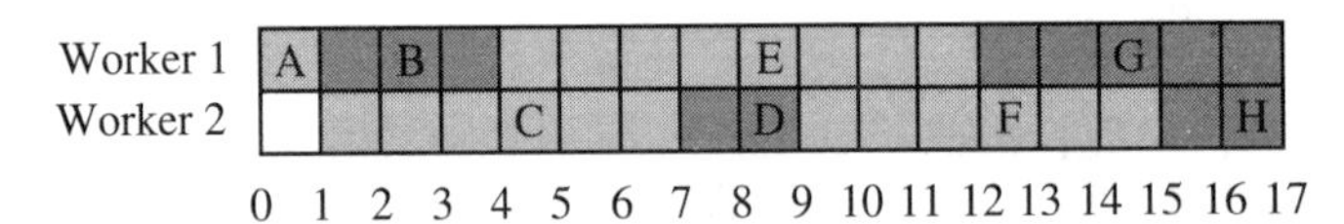

The final activity I(1) may be done by either worker 1 or worker 2 as both G and H will be finished at $t = 17$. In either case the minimum time for the project is 18 days when only two workers are available.

(iii) Can the project be completed by three workers in the project time of 15 days?

Consider first the critical activities. Activity B depends only on A, activity E depends only on B and activity H depends only on E. But activity I depends also on non-critical activities. It is therefore clear that critical activities A, B, E and H can be completed by worker 1 in 14 days.

Of the remaining activities, activity C has the smallest of the latest start times. Therefore, allocate worker 2 to activity C starting as early as possible, that is at $t = 1$.

The next activity to consider, according to (b) on page 121, is D. Allocate worker 3 to D starting as early as possible, that is at $t = 4$.

We have then

Worker 1 A B E H
Worker 2 C
Worker 3 D

0 1 2 3 4 5 6 7 8 9 10 11 12 13 14 15

Activity F may be done by worker 3 starting at $t = 6$ and finishing at $t = 12$. Activity G, which depends on C and D, can be done by worker 2 starting at $t = 7$ and finishing at $t = 12$. The final activity I can be done by worker 1 starting at $t = 14$ when H is complete.

So the project can be completed by 3 workers in the project time of 15 days.

Exercise 4F

In the project in Exercise 4E each activity requires one worker. Determine the minimum numbers of workers required if the project is to be completed in the project time. Draw a diagram to show how the workers should be allocated.

Resource levelling

There are, of course, situations when various activities require different numbers of workers. A common objective in such a situation is to minimise the maximum daily manpower that is required, subject of course to not extending the length of the critical path. Suppose that the manpower requirements for the activities in our problem are:

Activity	Workers needed
A	4
B	2
C	2
D	3
E	2
F	4
G	3
H	2
I	4

If all the activities start as *early as possible* then from the diagram on page 119 you can produce the following diagram, where the number of workers is given in the box. The total number of workers required each day may then be easily obtained and is given below the diagram.

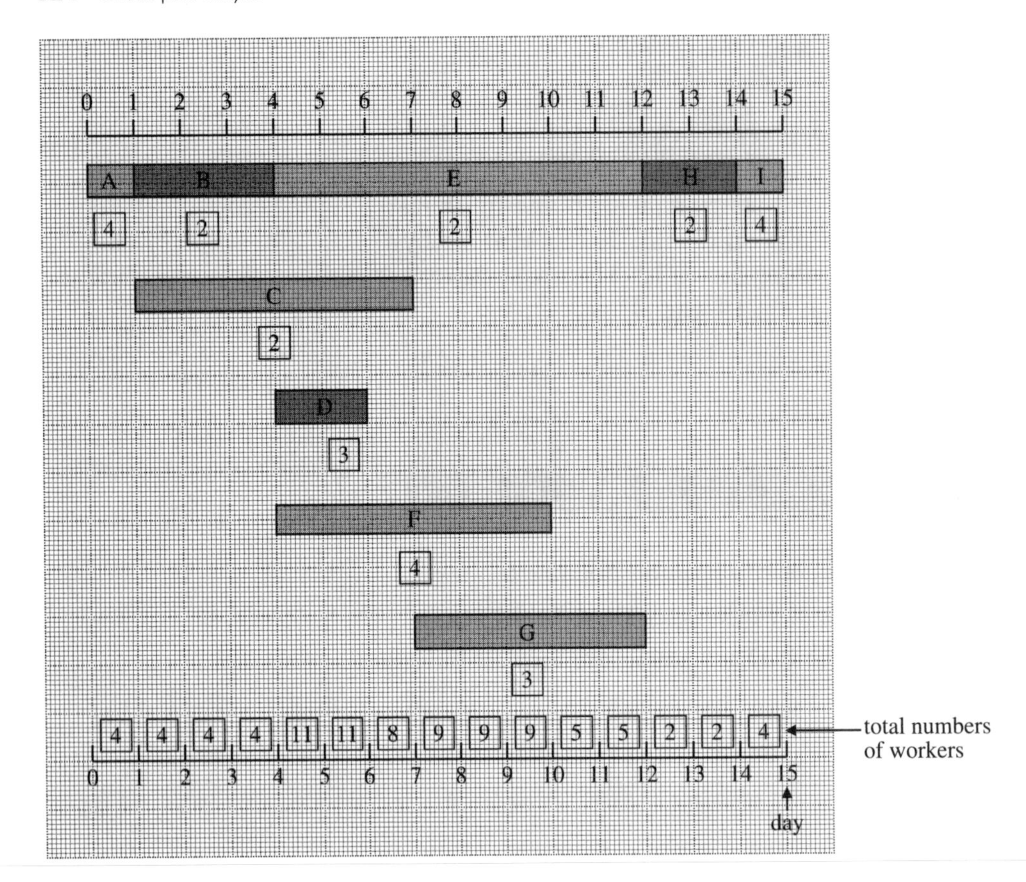

This information may be displayed as a graph of resource against time as below.

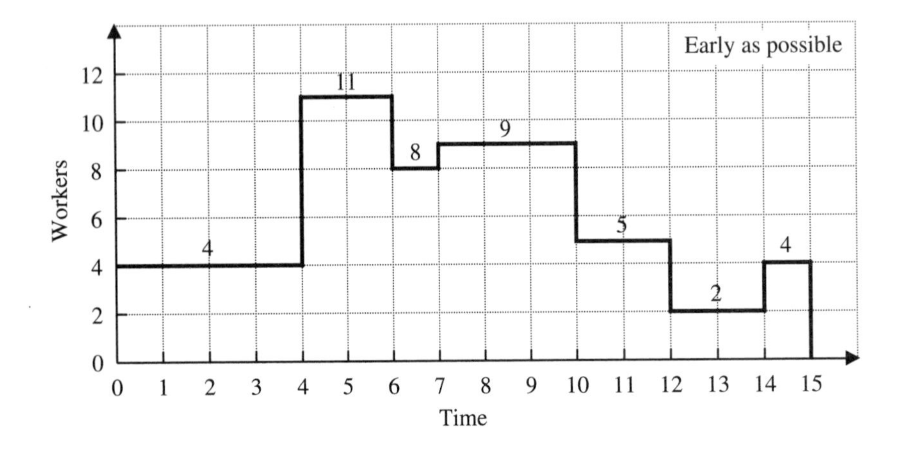

We may obtain a similar graph of resource against time when all activities terminate as late as possible. The diagram on page 119 gives you:

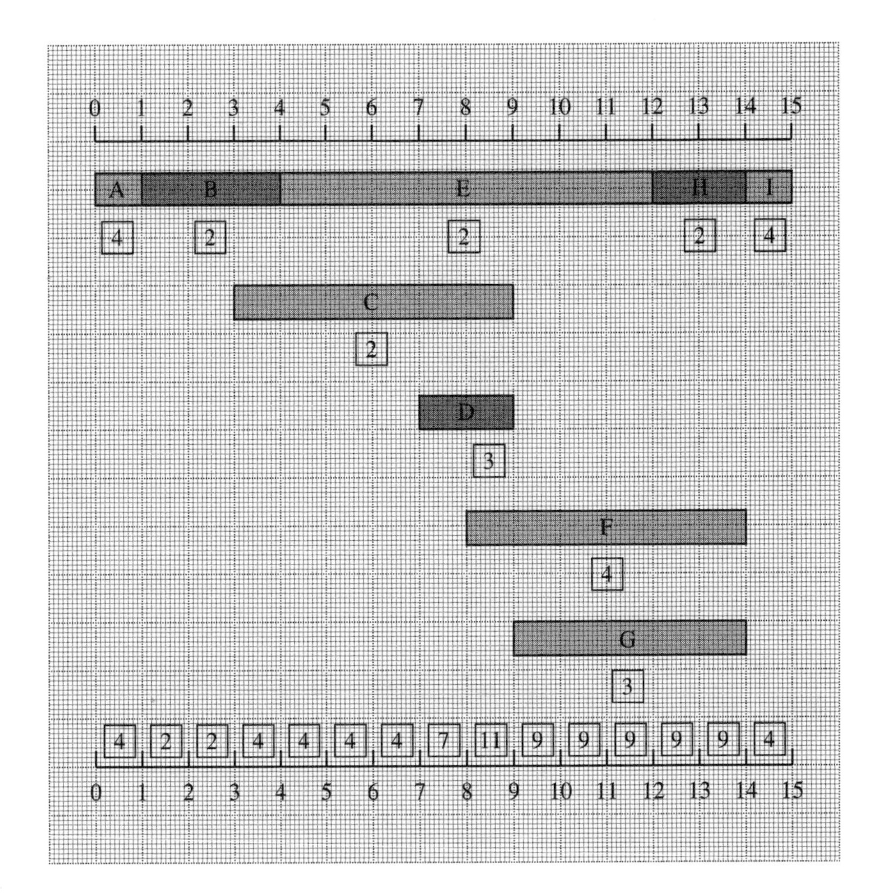

The graph of resource against time for this situation is given below:

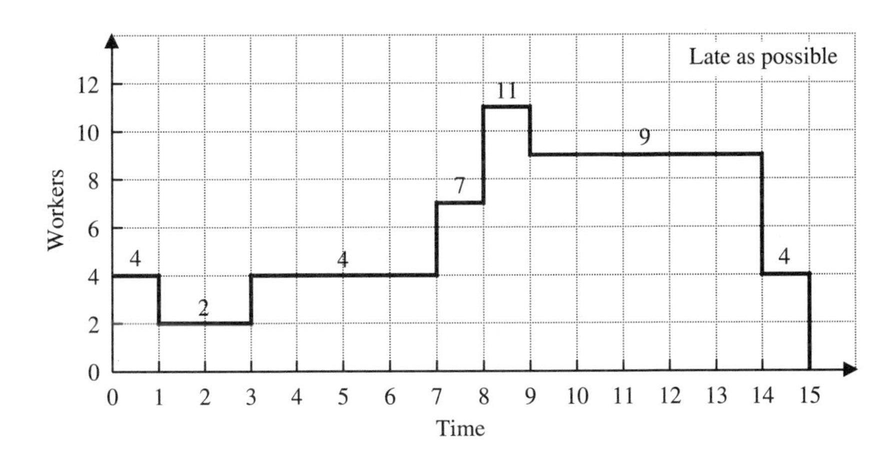

Now consider how by using the floats of the non-critical activities you can minimise the maximum daily manpower that is required. Since the second graph has a single peak of 11 workers we will consider this. (You could just as well begin with the first graph.)

Contributing to this peak in the period (8, 9) we have E[2], C[2], D[3] and F[4]. The activity E is critical and so has zero float and cannot be rescheduled. If you refer to the network representing this project (page 103) you can see that activity C may start as soon as A is finished. You may therefore use the float of 2 on activity C to reschedule C as early as possible. The effect of this is:

(a) To add 2 workers in (1, 2) and (2, 3) making these requirements 4 in each case.
(b) To subtract 2 workers in (7, 8) and (8, 9) making these requirements 5 and 9 respectively.

This produces the graph of resources shown below:

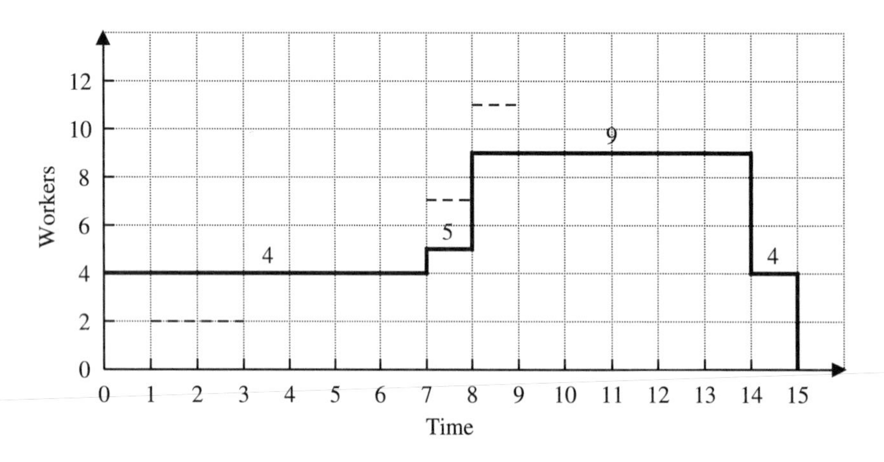

It is suggested that you consider the effect of rescheduling the activity D as early as possible.

Notice that, however you schedule the non-critical activities, in the interval (9, 10) you must always have activities E[2], F[4], and G[3] taking place and therefore you cannot reduce the maximum number of workers required below 9.

Example 3

The activity network for an industrial process is shown below with each arc labelled with the time, in days, required for the corresponding activity. S is the source node and T is the sink node.

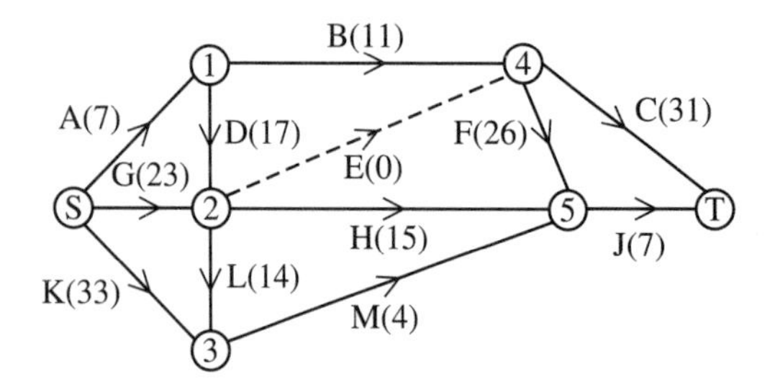

(a) Find the earliest time and the latest time for each event and the total float for each activity. Hence determine the length of the critical path, the critical events and the critical activities.
(b) What are the critical activities and the minimum project time if the dummy E is removed?

(a) The calculation of the earliest event times is summarised below:

$$e_S = 0,\ e_1 = 7$$
$$e_2 = \max\ (7 + 17, 23) = \max\ (24, 23) = 24$$
$$e_3 = \max\ (33, 24 + 14) = \max\ (33, 38) = 38$$
$$e_4 = \max\ (7 + 11, 24 + 0) = \max\ (18, 24) = 24$$
$$e_5 = \max\ (24 + 15, 38 + 4, 24 + 26) = \max\ (39, 42, 50) = 50$$
$$e_T = \max\ (24 + 31, 50 + 7) = \max\ (55, 57) = 57$$

The calculation of the latest event times for the events is summarised below:

$$l_T = 57, \quad l_5 = 50$$
$$l_4 = \min\ (57 - 31, 50 - 26) = \min\ (26, 24) = 24$$
$$l_3 = 50 - 4 = 46$$
$$l_2 = \min\ (24 - 0, 50 - 15, 46 - 14) = \min\ (24, 35, 32) = 24$$
$$l_1 = \min\ (24 - 11, 24 - 17) = \min\ (13, 7) = 7$$
$$l_S = \min\ (46 - 33, 24 - 23, 7 - 7) = \min\ (13, 1, 0) = 0$$

The total floats are

A (S, 1) $= 7 - 0 - 7 = 0$
B (1, 4) $= 24 - 7 - 11 = 6$
C (4, T) $= 57 - 24 - 31 = 2$
D (1, 2) $= 24 - 7 - 17 = 0$
E (2, 4) $= 24 - 24 - 0 = 0$
F (4, 5) $= 50 - 24 - 26 = 0$
G (S, 2) $= 24 - 0 - 23 = 1$
H (2, 5) $= 50 - 24 - 15 = 11$
J (5, T) $= 57 - 50 - 7 = 0$
K (S, 3) $= 46 - 0 - 33 = 13$
L (2, 3) $= 46 - 24 - 14 = 8$
M (3, 5) $= 50 - 38 - 4 = 8$

The length of the critical path is obtained from e_T and so is 57 days.

The critical events are those for which $e_i = l_i$ and so are Ⓢ, ①, ②, ④, ⑤, Ⓣ.

The critical activities are those for which the total float is zero and so are activities

A, D, E, F and J

(b) When the dummy E is removed you need to consider the changes that are necessary in the above calculation.

Let us recalculate the earliest event times:

$$\begin{aligned}
e_S &= 0 \text{ (unchanged)} \\
e_1 &= 7 \text{ (unchanged)} \\
e_2 &= 24 \text{ (unchanged)} \\
e_3 &= 38 \text{ (unchanged)} \\
e_4 &= 18 \text{ (changed)} \\
e_5 &= \max\,(24 + 15, 38 + 4, 18 + 26) \\
&= \max\,(39, 42, 44) = 44 \\
e_T &= \max\,(44 + 7, 18 + 31) = 51
\end{aligned}$$

The length of the critical path is now 51 days. You can obtain the critical path in this case by using an alternative method. For each event, in addition to recording the earliest time e_i you can record the vertex $p(i)$ from which the maximum is achieved. It is then possible to determine the critical path by using these and working backwards from Ⓣ.

In this case:

$e_S = 0$

$e_1 = 7$ $\quad$ $p(1) =$ Ⓢ

$e_2 = 24$ $\quad$ $p(2) =$ ①

$e_3 = 38$ $\quad$ $p(3) =$ ②

$e_4 = 18$ $\quad$ $p(4) =$ ①

$e_5 = 44$ $\quad$ $p(5) =$ ④

$e_T = 51$ $\quad$ $p(T) =$ ⑤

Since $p(T) =$ ⑤ you go to ⑤
and $p(5) =$ ④ so you go to ④
and $p(4) =$ ① so you go to ①
and $p(1) =$ Ⓢ

Then from the activity network the critical path is (S, 1) (1, 4) (4, 5) (5, T); that is, A B F J.

Exercise 4G

1

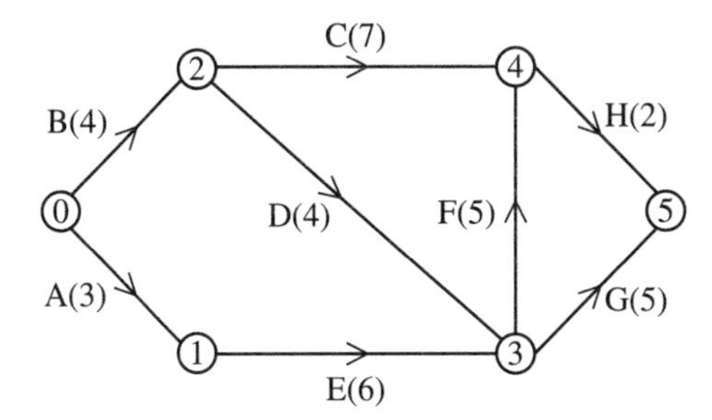

For the activity network shown calculate the earliest time and the latest time for each event and the total float for each activity. Hence determine the length of the critical path, the critical activities and the critical events.

2

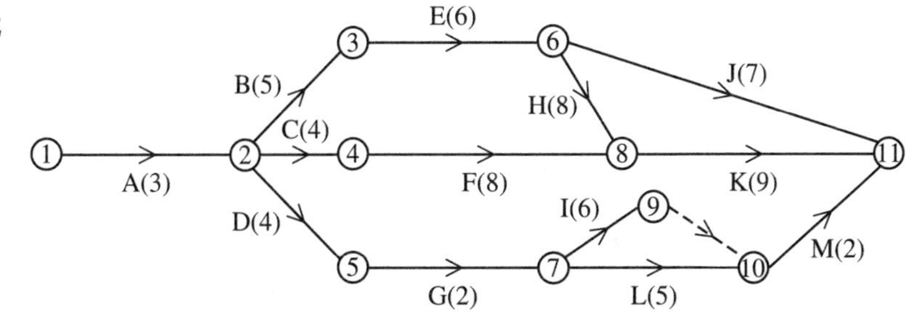

For the activity network shown above calculate the earliest and latest time for each event and record them in the form $\boxed{e_i \mid l_i}$ on a copy of the network. Indicate also the critical path by marking critical activities by a double dash, that is: —||— .

3 As a result of the breakdown of a piece of equipment the activity F in the previous example takes 16 days.

(a) Carry out further calculations to decide if the length of the project is changed.

(b) If the length of the project is changed, determine the new project time and the critical path.

4

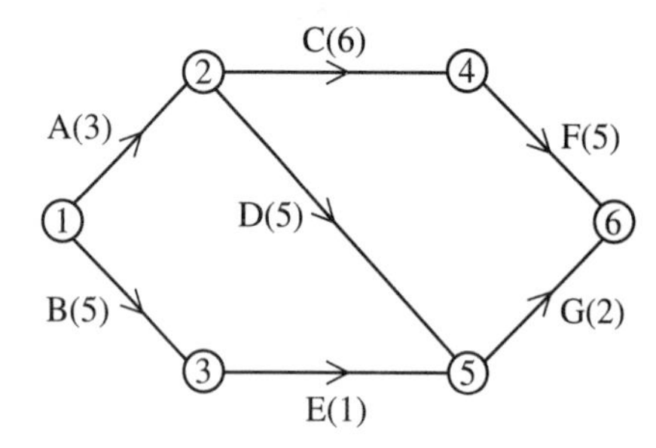

For the above activity network

(a) determine the earliest time and latest time for each event, and hence the project time and critical path

(b) produce a table including for each activity the earliest and latest start times and finish times and the total float.

Given that each activity requires one worker

(c) determine the minimum number of workers required if the project is to be completed in the project time.

5 Consider the project modelled in question 4 with the following manpower requirements for the activities:

Activity	Workers needed
A	5
B	4
C	3
D	1
E	2
F	4
G	4

Obtain a graph of resource against time

(a) when all activities take place as early as possible

(b) when all activities take place as late as possible.

(c) Starting from your graph in (b) use the float on activity D to show that the project may be completed with a maximum of 8 workers.

SUMMARY OF KEY POINTS

In an activity network:

1 The nodes represent events.

2 The arcs represent activities.

3 The weights represent the duration of the corresponding activity.

4 The source node represents the beginning of the project.

5 The sink node represents the end of the project.

6 The earliest time e_i for node i is the earliest time we can arrive at event i.

7 The latest time l_i for node i is the latest time we can leave event i without extending the length of the critical path.

8 The critical path is the longest path through the network from the source node to the sink node.

9 Events i for which $e_i = l_i$ are critical events.

10 Activities (i, j) for which $l_j - e_i - [\text{duration } (i, j)] = 0$ are critical activities.

11 For activity (i, j) of duration a_{ij}:

the earliest start time is e_i
the earliest finish time is $e_i + a_{ij}$
the latest finish time is l_j
the latest start time is $l_j - a_{ij}$
the total float on activity (i, j) is $l_j - e_i - a_{ij}$

Flows in networks 5

5.1 Flow networks

This chapter deals with networks representing **transmission systems**. These are systems of flows – of liquids, gases, other materials, information, etc. So the edges of the networks often (but not always) represent pipes, and then the vertices represent the intersections of the pipes.

The network below represents an electricity distribution system. A and B are power stations. J, K, L and M are substations. X, Y and Z are cities. The edge weights give the capacities of the transmission lines in gigawatts – the amount of electricity they can carry.

A gigawatt is 1 000 000 000 watts – enough to keep a million one kilowatt fires running

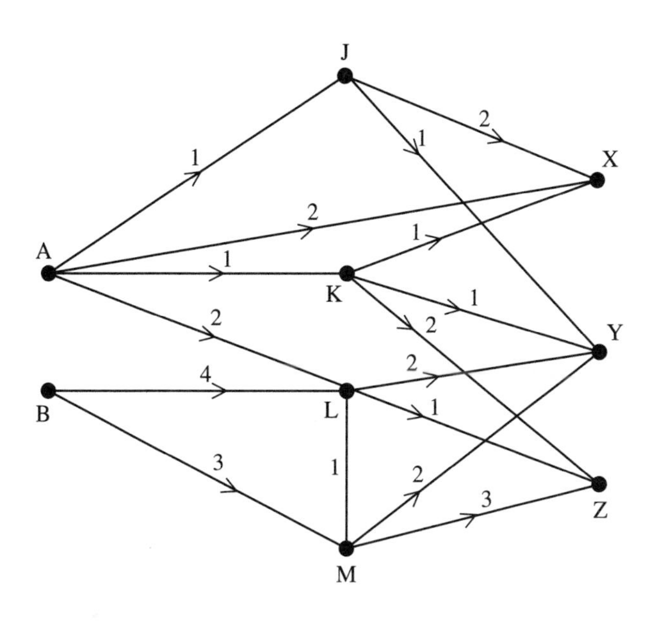

Suppose that, in addition, you are told the maximum output of the power stations (the amount of electricity they can produce) and the

peak requirements of the cities. Then you can also use weighted edges to represent these maximum outputs and requirements. Suppose that, for instance, the maximum output of A is 6 gigawatts and the maximum output of B is 8 gigawatts. City X has a peak requirement for 4 gigawatts, Y for 6 gigawatts, and Z for 3 gigawatts. We introduce a 'source' vertex, S, and a 'sink' vertex, T, and extra edges as shown in the diagram:

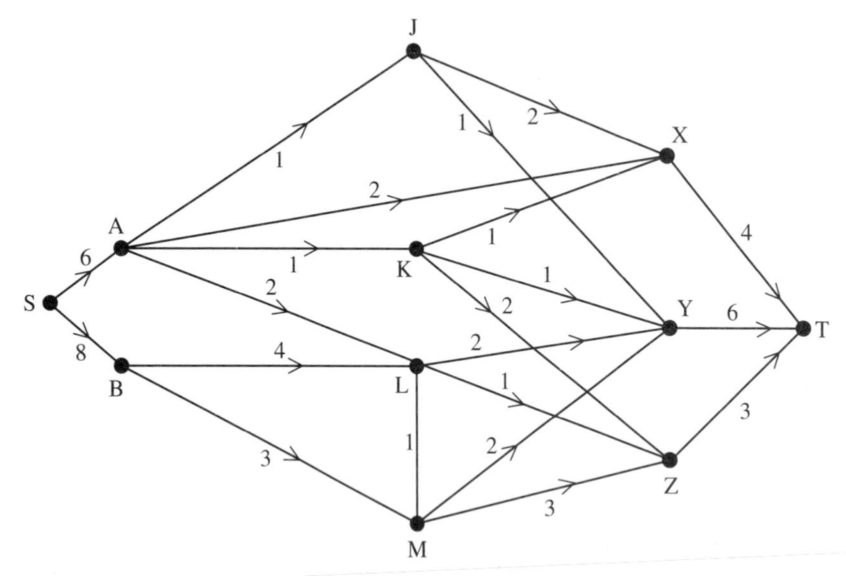

Adding a single source lets you model the *maximum outputs* with weighted edges. Adding a single sink lets you model the *requirements* with weighted edges.

This is a **transmission or flow network**. Most flow networks will have a single source and a single sink.

5.2 Capacities and flows

In a flow network there are two numbers associated with each edge. These numbers represent its **capacity** and the actual **flow**. The flow along an edge must be less than or equal to the capacity of that edge.

> **Warning — a common error**
>
> Mistakes are often made by confusing capacity and flow. In your diagrams put your flows in circles, e.g. ⑤. One way to remember this is to think of the number in the circle as being inside a pipe. Be sure that, in what follows, you distinguish between statements relating to capacities and statements relating to flows.

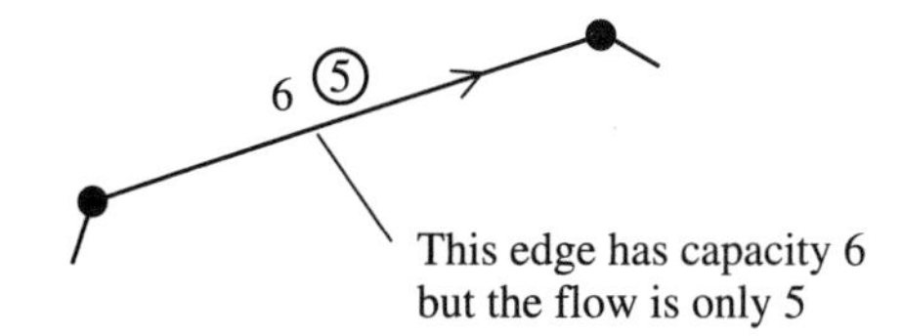

If a vertex is neither the source nor the sink of the network then all that flows into it must also flow out. A set of flows in edges which satisfy this condition, and in which the individual flows are within the capacities of the edges, is called **feasible**. The total flow through the network is then given by the sum of the flows out of the source, or into the sink, and these are equal.

Notice that circular flow patterns are not allowed:

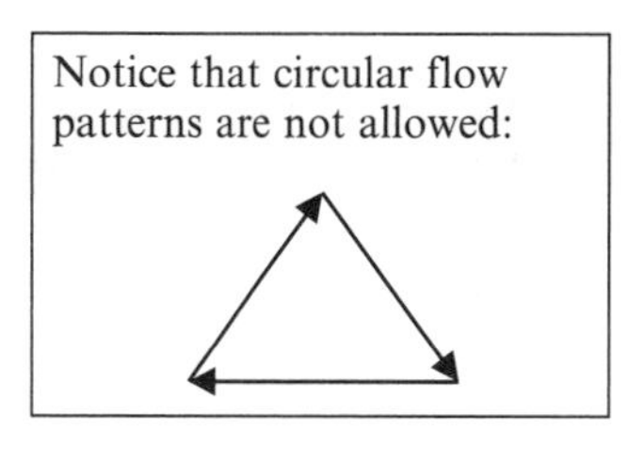

Example 1

In the diagram the unringed numbers represent the edge capacities. Consider the set of flows shown in the rings, and show that it is feasible by showing that it satisfies all of the necessary constraints.

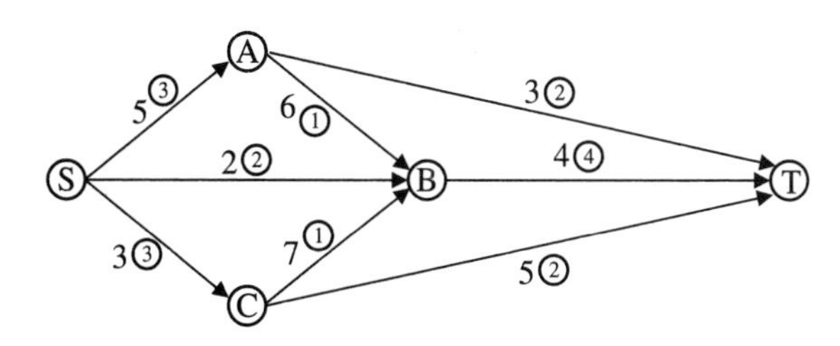

First check that the capacity constraint on each edge is not exceeded by the flow:

Capacity constraint on SA: $3 \leqslant 5$ ✓
Capacity constraint on SB: $2 \leqslant 2$ ✓
Capacity constraint on SC: $3 \leqslant 3$ ✓
Capacity constraint on AB: $1 \leqslant 6$ ✓
Capacity constraint on CB: $1 \leqslant 7$ ✓
Capacity constraint on AT: $2 \leqslant 3$ ✓
Capacity constraint on BT: $4 \leqslant 4$ ✓
Capacity constraint on CT: $2 \leqslant 5$ ✓

Then check that the flow into each vertex matches the flow out of it:

Vertex A:	In = ③	Out = ① + ② = ③
Vertex B:	In = ① + ② + ① = ④	Out = ④
Vertex C:	In = ③	Out = ① + ② = ③

Finally check that the total flow out matches the flow in:

Source: Flow out of S $= 3 + 2 + 3 = 8$
Sink: Flow into T $= 2 + 4 + 2 = 8$

In the power station example on page 133 the transmission line represented by LM allows flows in either direction. In example 1 the capacities are all directed. You need to be able to deal with both directed and undirected capacities.

Exercise 5A

1 The capacities and flows of a transmission network are shown below.

Supply point	A	B	C
Availability	10	20	15

Connection	AD	BD	BE	CE	CF	DG	EG	EH	FH
Capacity	10	15	10	10	10	20	10	9	10
Flow	8	10	10	4	6	18	5	9	6

Delivery point	G	H
Requirements	25	20

(a) Draw a network, with a single source and a single sink, to represent this information.

(b) Check that the set of flows is feasible.

2

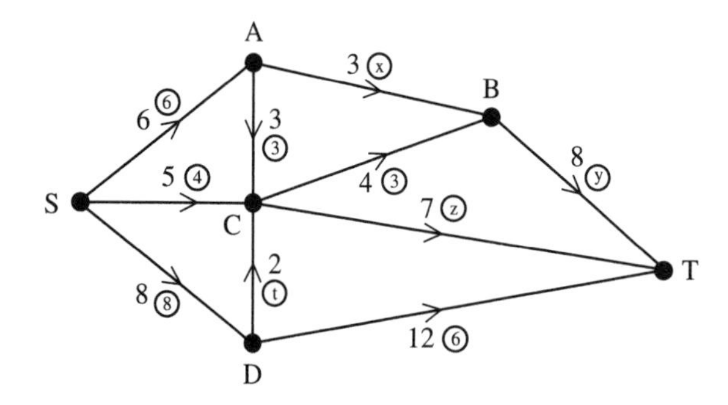

Given that the above network is feasible, find the values of:

(a) x

(b) y

(c) z

(d) t

3

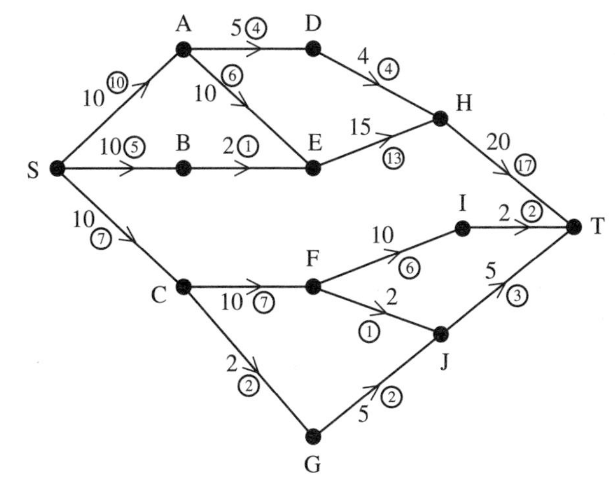

A number of pipes have been omitted from this drawing of a pipe transmission network. Given that the flows are feasible, draw the complete network, including the missing pipes. Mark on the flows in those pipes.

5.3 The maximum flow–minimum cut theorem

What is the maximum flow you can push through a network? One way of finding out is to cut the network into two parts by splitting the vertices into two sets, one containing the source and one containing the sink. To find out how much flow you might be able to push from source to sink across the cut, list all edges with flows from a vertex in the source set to a vertex in the sink set. You cannot pass more from source to sink than you can get across the cut.

- **The sum of the capacities from source set to sink set is known as the capacity of the cut.**

Example 2

The figure shows three cuts across a network.

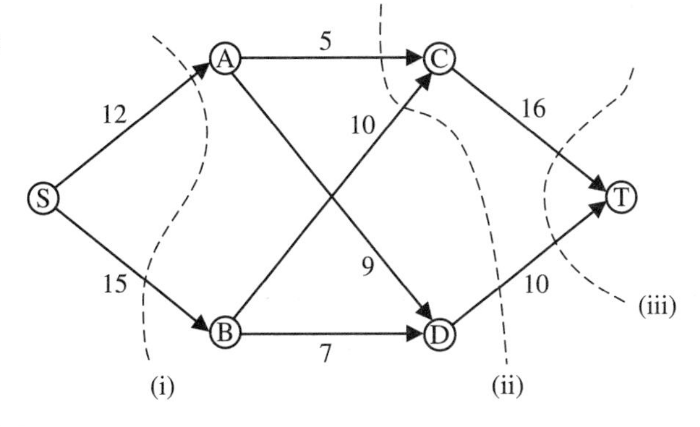

You can record the cut between S and A, B, C, D, T using set notation and the cut symbol | like this:

	Source Set	Sink Set	
Cut (i):	{S}	{A, B, C, D, T}	Capacity = 27
Cut (ii):	{S, A, B, D}	{C, T}	Capacity = 25
Cut (iii):	{S, A, B, C, D}	{T}	Capacity = 26

Suppose now that instead of having a capacity of 10 from B to C, the network were to have a capacity of 10 from C to B:

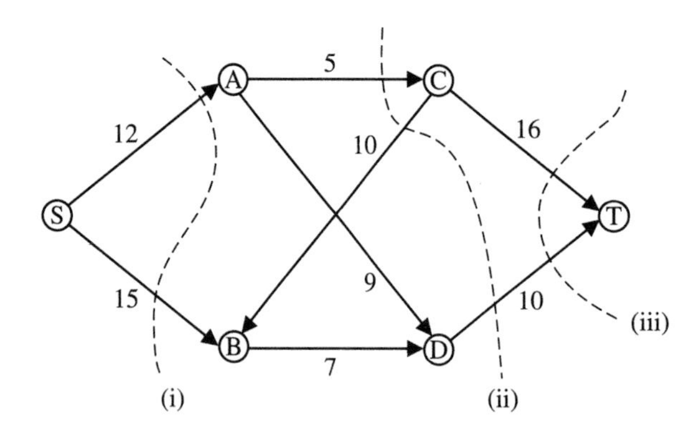

The capacities of cuts (i) and (iii) would be the same as before, but the capacity of cut (ii) is now only 15. This is because the capacity of the cut is defined to be *the sum of all the capacities from source set to sink set*, and in the revised transmission network there are only two such capacities, AC (5) and DT (10).

In a complex transmission network finding the capacity of a cut by looking at the diagram can be confusing – it isn't always clear whether an edge crossing the cut is going from source set to sink set or vice versa. You can draw up a table like this for cut (ii) to make it clear:

	Capacity
SC	—
ST	—
AC	5
AT	—
BC	0
BT	—
DC	—
DT	10
Total capacity	15

Remember that we are dealing here with edge *capacities*, and *not* with flows.

All of the possible combinations XY have been listed, where X is from the source set and Y is from the sink set. For some of these there is no edge (marked '—'), and for BC in the revised network there is an edge, but no capacity from source to sink. This approach will work equally well when the network includes edges with undirected capacities, or even when the capacities are different in opposite directions.

- **The max flow–min cut theorem states that the *maximum flow* that can be established through a flow network is equal to the *capacity of the minimum cut*, i.e. the cut with smallest capacity.**

How is the max flow–min cut theorem useful?

At first sight the max flow–min cut theorem does not appear to be a great deal of use:

- It seems obvious.
- It doesn't offer a way of finding the maximum flow other than by complete enumeration of all cuts – and there can be lots of them!
- Even if we find the minimum cut, and hence the maximum flow, it won't tell us how to construct a flow.

However, it does have one great virtue, not in terms of achieving the maximum flow, but in confirming that a maximum flow has been found. Suppose that you have a feasible flow through a network, and suppose that you have identified a cut with capacity equal to that flow. Then, by the theorem, you know that your cut is a minimum cut, and that the flow is a maximum flow.

Exercise 5B

1

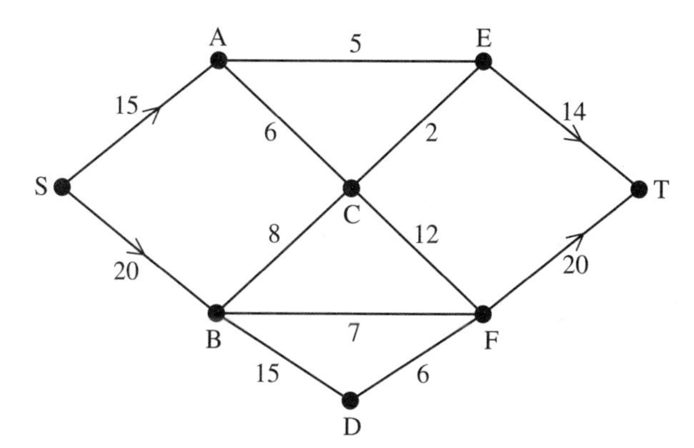

Give the capacities of the following cuts:

(a) {S} | {A, B, C, D, E, F, T}

(b) {S, A, B} | {C, D, E, F, T}

(c) {S, F} | {A, B, C, D, E, T}

(d) {S, A, B, D} | {C, E, F, T}

2

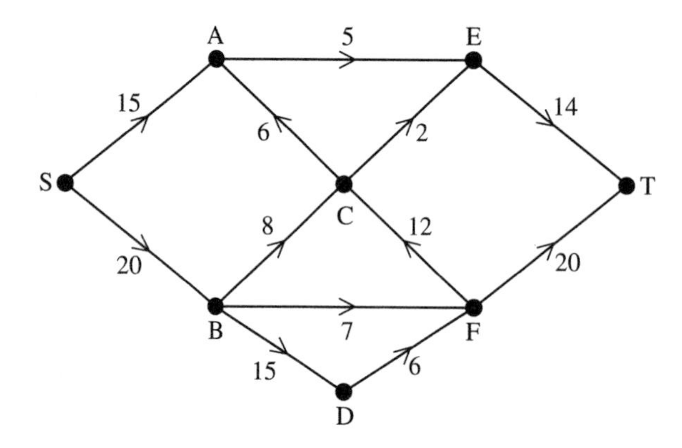

Give the capacities of the following cuts:

(a) {S} | {A, B, C, D, E, F, T}
(b) {S, A, B} | {C, D, E, F, T}
(c) {S, F} | {A, B, C, D, E, T}
(d) {S, A, B, D} | {C, E, F, T}

3 Explain why there are 512 different cuts for the electricity distribution network on page 133.

4

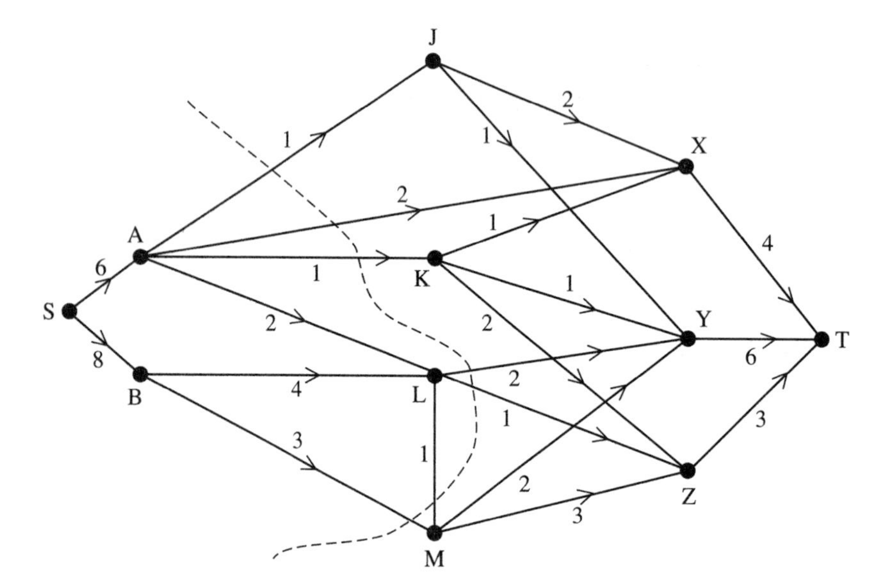

(a) By finding a flow pattern such that the total flow is equal to the capacity of the indicated cut, show that it is a minimum cut.
(b) What does this tell you about the total flow?

5.4 Finding the maximum flow using augmentation – a labelling algorithm

Finding the maximum flow from source to sink involves finding both its value and finding a set of flows along the individual edges of the network.

You can use an algorithm called the **labelling algorithm** to achieve both these objectives. The algorithm consists of two stages:

Stage 1 Obtain an initial flow by inspection.

Stage 2 Identify flow augmenting paths.
(There are several steps associated with stage 2. These will be listed later.)

At each step every edge of the network is *labelled* with three numbers – this is why the algorithm is called a labelling algorithm. The numbers are:

(i) The **flow** along the edge.

(ii) The **excess capacity** of the edge: that is, the amount by which flow along the edge may be increased.

(iii) The **back capacity:** that is, the amount by which flow along the edge may be reduced. For example, if the flow along a directed edge is 8 it can be reduced by up to 8.

For directed edges the back capacity is equal numerically to the flow. You will see later that this is not the case for undirected edges.

For example, suppose edge AB has capacity 12 from A to B, and a flow of 8. Then the arc will be labelled with a flow of 8, an excess capacity of $12 - 8 = 4$, and a back capacity of 8:

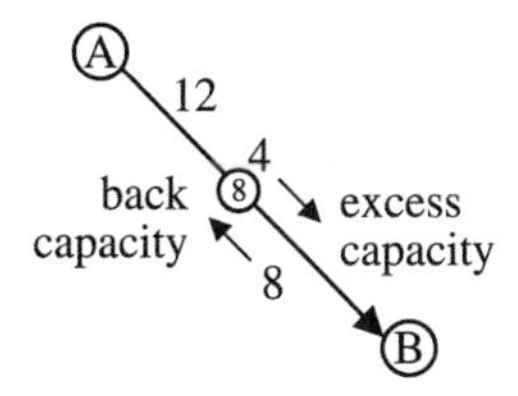

Notice that to avoid clutter the flow has now been marked in a circle on the edge itself.

If at some stage there is a path from the source to the sink along which each edge has unused capacity, then the flow along this path may be increased. Such a path is called a **flow augmenting path**.

augment here means increase

Example 3

In the directed network below, the numbers on the edges denote the capacities.

Find a feasible set of flows along edges which give a total flow of 14 through the network. By considering the cut {S, C, D} | {A, B, T} prove that 14 is the maximum flow.

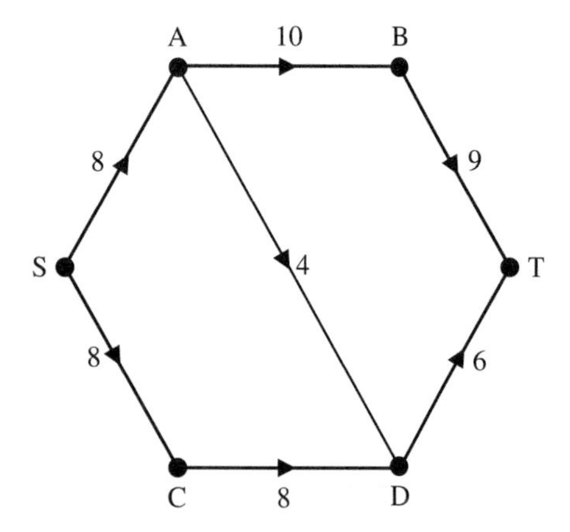

You should be able to see that the maximum flow through this network is 14. What we need is an algorithm which is guaranteed to establish the maximum flow in this and in much more complicated networks.

Start by listing all possible routes through the network from S to T. Although it seems to run counter to the directionality, you must include SCDABT in your list. You will see why in a moment:

Route	SABT	SADT	SCDT	SCDABT
Capacity	8	4	6	0

The capacity of a route is the smallest unused capacity among the edges of the route.

Next, augment (increase) the flow by sending as much as possible along one of the routes. Label each edge not only with the flow, but also with the remaining capacity of the edge.

First let us put a flow of 8 through SABT and label the edges used:

Then update the route capacities:

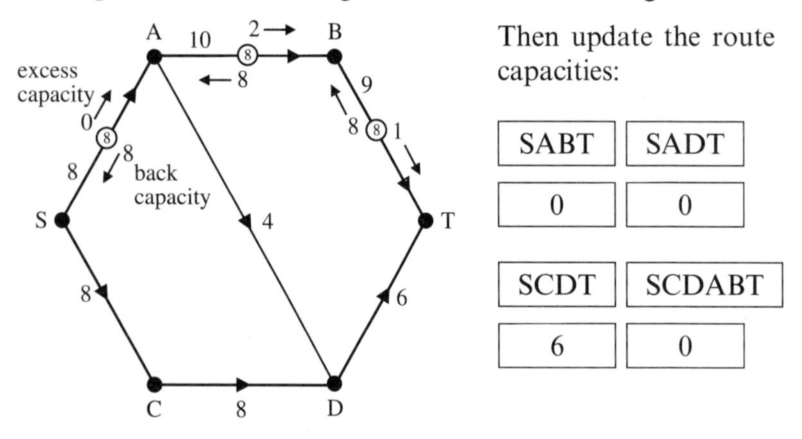

Continue systematically sending as much as possible down each route in turn until there are no further opportunities to increase the flow. In this simple application that happens after one further flow augmentation. Try putting a flow of 6 through SCDT:

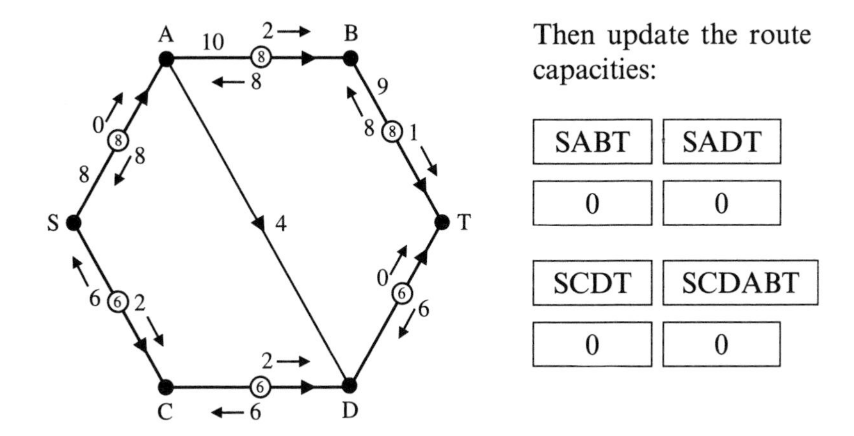

Then update the route capacities:

SABT	SADT
0	0

SCDT	SCDABT
0	0

We have now managed to augment the flow to 8 along one side and 6 on the other, giving a total of 14. This all seems very straightforward. In this simple network it hardly seems worth bothering with the labelling procedure. But of course there is no guarantee, particularly in a complex network, that you will be able to spot the easy way of doing things – if an easy way exists!

Suppose, for example that you had chosen SADT first, pushing a flow of 4 through that route. Notice what happens when you label the edges and update the route capacities:

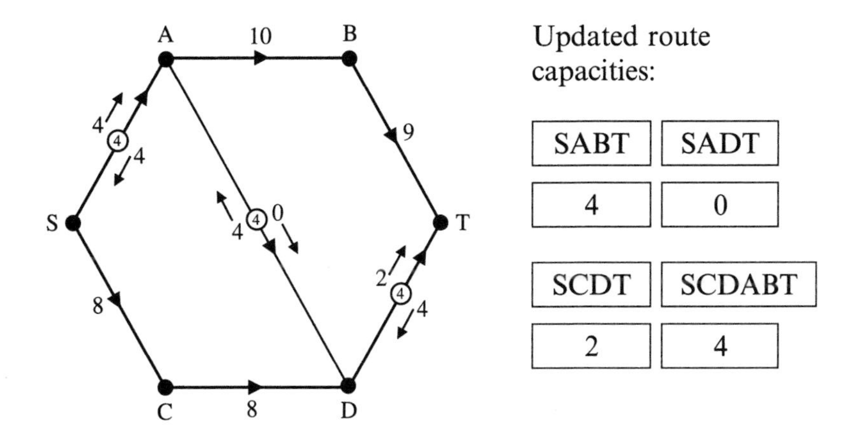

Updated route capacities:

SABT	SADT
4	0

SCDT	SCDABT
2	4

The interesting route is SCDABT. Although AD is a directed edge we can send flow in the reverse direction by cancelling out existing flow in the permitted direction. Let us see that in action by sending an augmenting flow of 4 along SCDABT:

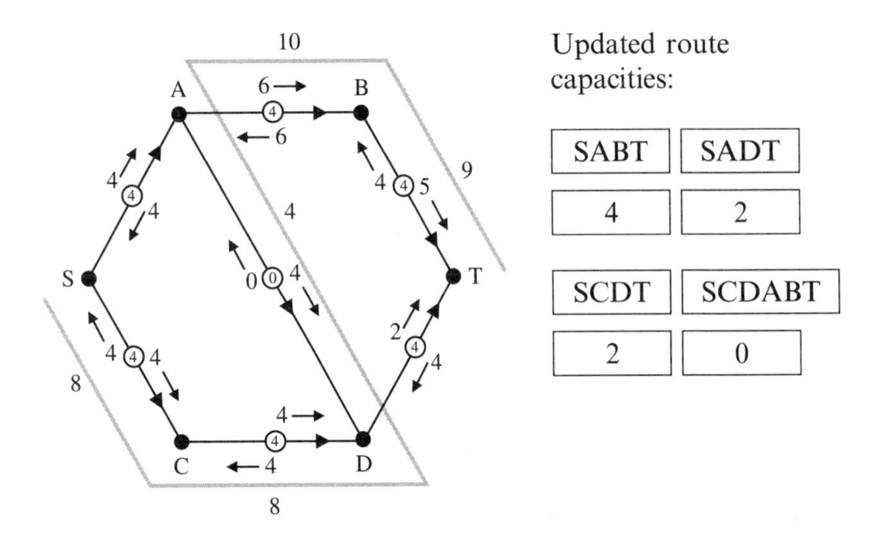

So we have progressed towards a solution by finding a feasible set of flows with an increased total flow.

Finish off the algorithm from this point.

Listing all the route capacities at each stage is tedious, but if you try to do without the list you must be careful to look out for paths with 'reverse' flows.

Example 4

Use labelling to find a set of flows which give the maximum total flow through the following transmission network:

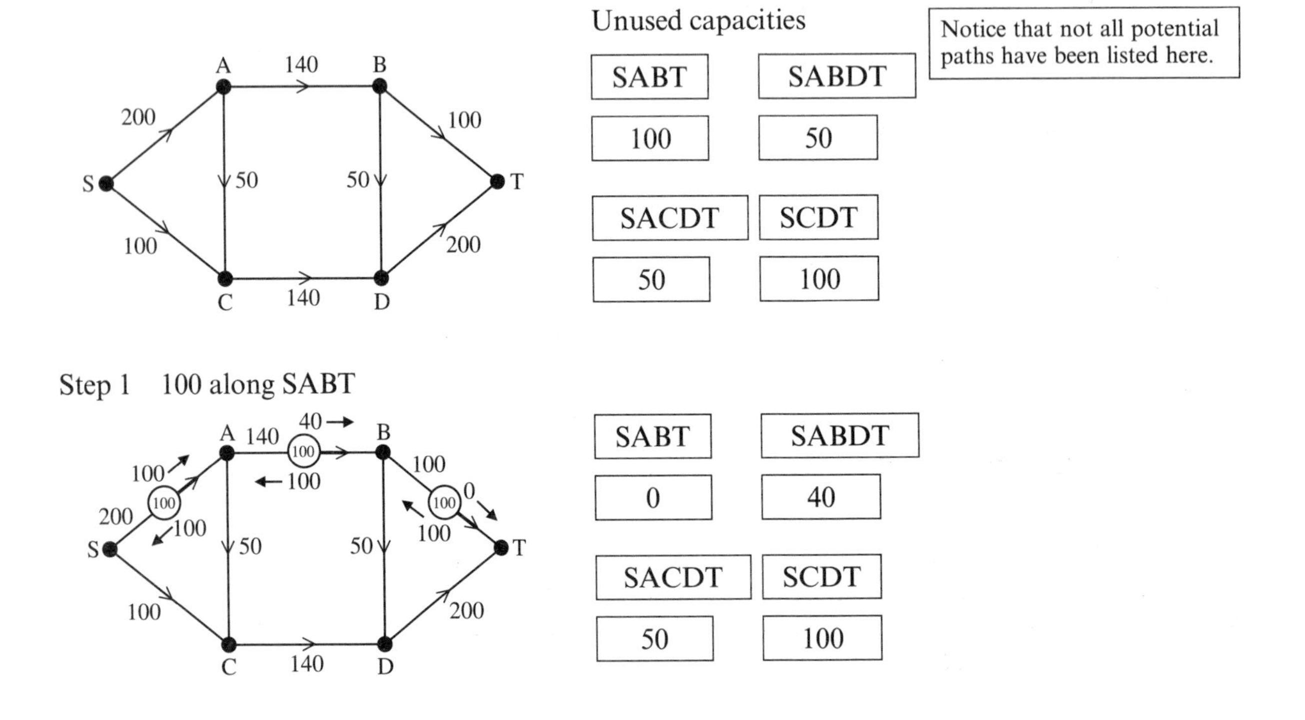

Step 2 40 along SABDT

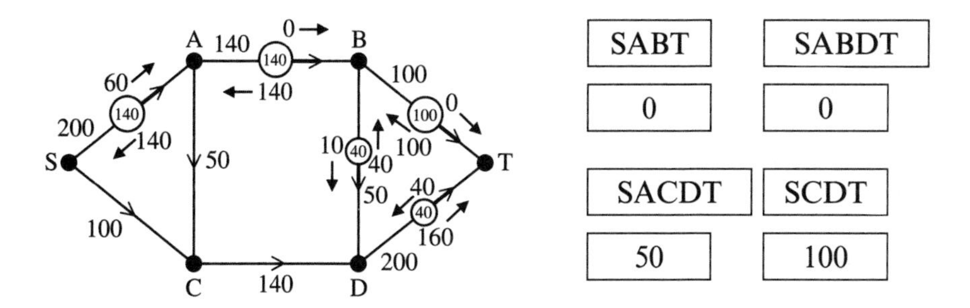

Step 3 50 along SACDT

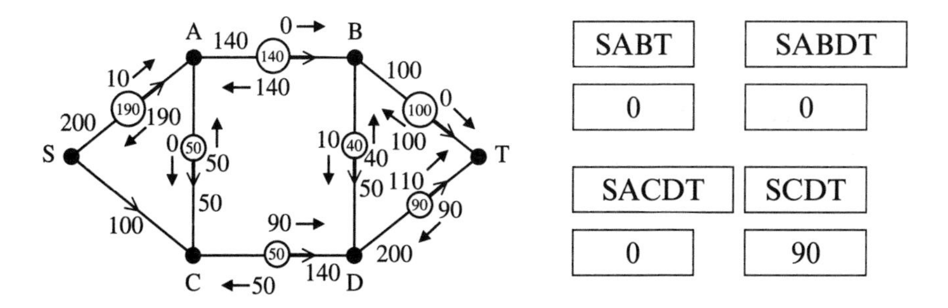

Step 4 90 along SCDT

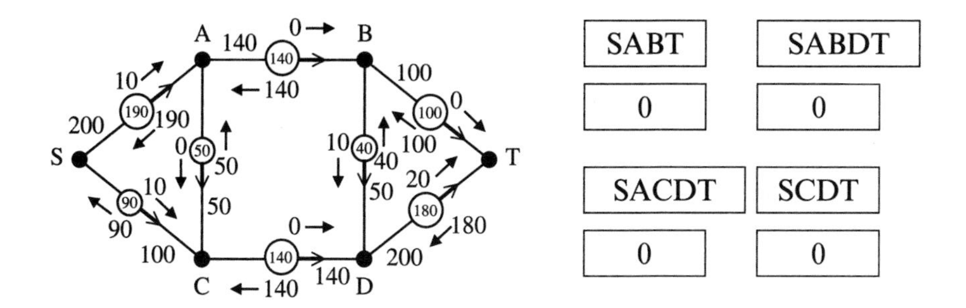

Note: Choosing augmenting paths in a different order may lead to a different pattern of flows, but the total flow through the network will always be 280.

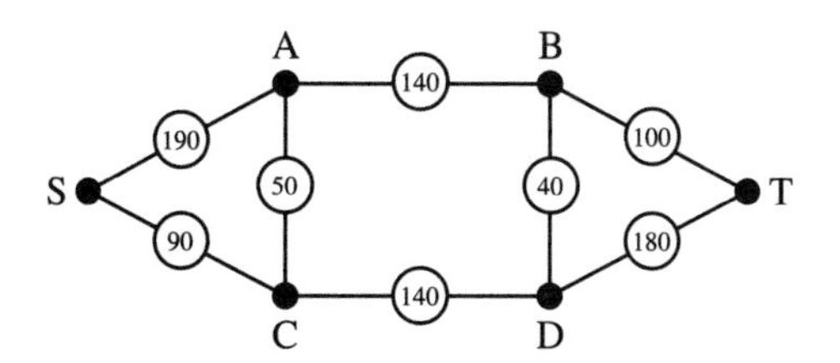

Exercise 5C

1 (a) Use labelling to find a set of flows which give the maximum total flow through the following transmission network:

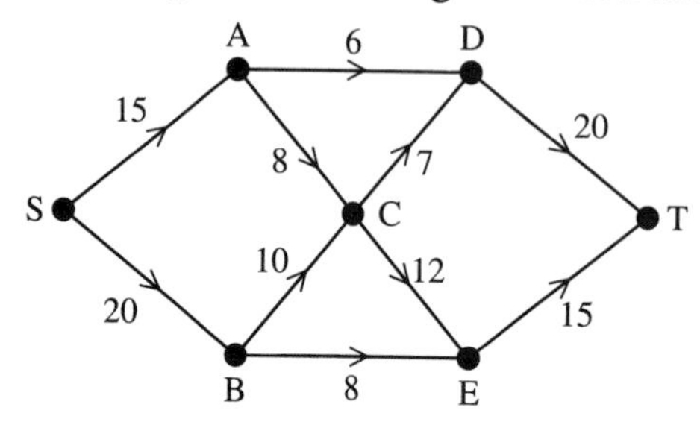

(b) Give a cut to prove that you have found the maximum flow.

2 Find a feasible set of flows along edges in the following networks which will give a total flow of the given value. In each case find a cut with this capacity.

(a)

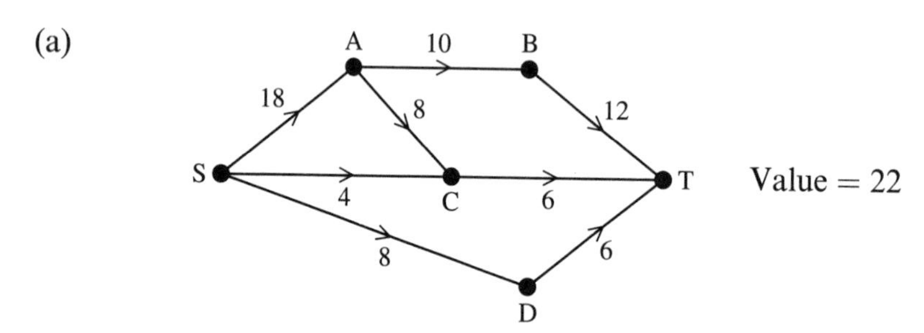

(b)

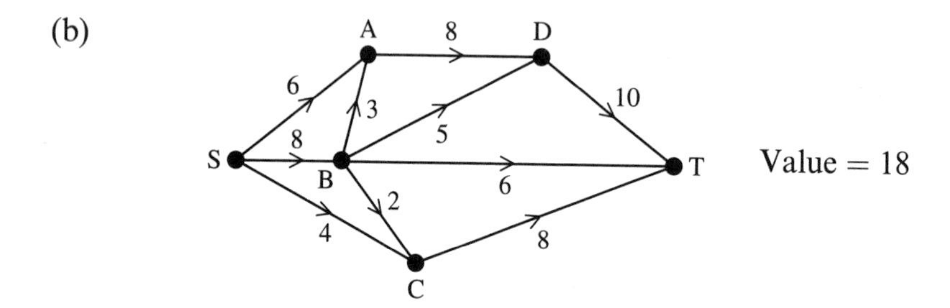

(c)

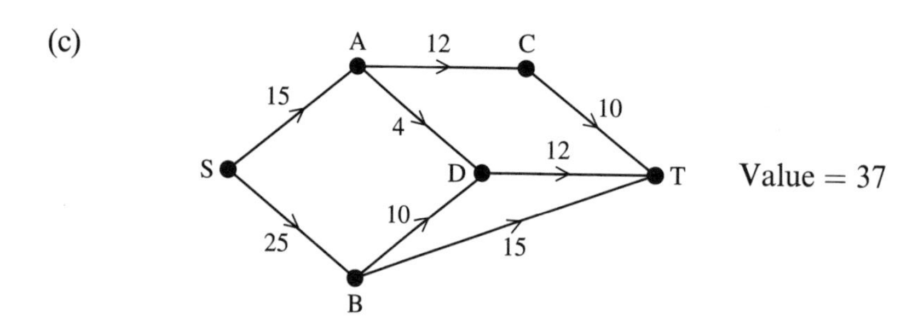

5.5 Flows in undirected edges

Example 5

Suppose that the edge AD in the network given in example 3 is now undirected:

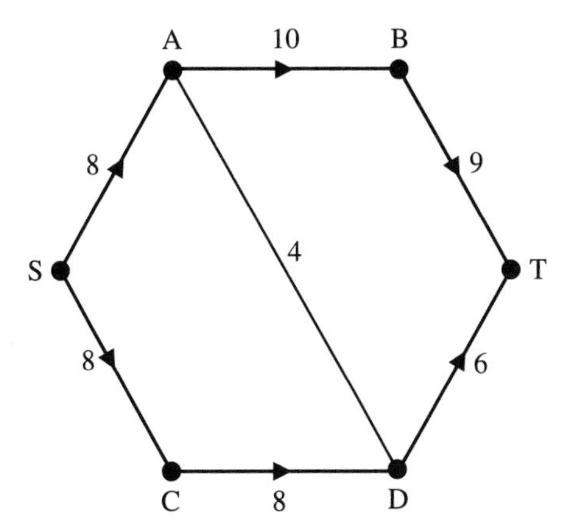

As before, we first find a flow augmenting route and push as much flow through it as possible. To demonstrate how to deal with undirected edges we shall choose to push 4 along SADT. This time the back capacity in AD is numerically equal to the flow **plus the capacity**, since there is the possibility of reducing the existing flow to zero and then pushing flow in the direction DA.

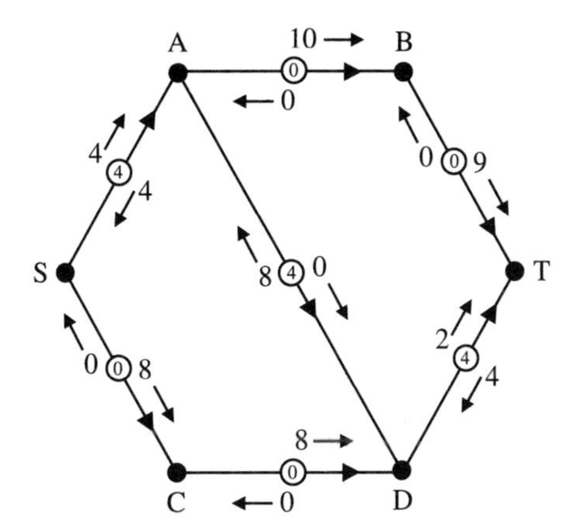

Notice that the back capacity on AD is 8, because we could consider moving from a flow of 4 along AD to a flow of 4 along DA. This is effectively the same as introducing a new flow of magnitude 8 along a route including DA.

- **For all edges, directed and undirected, the unused capacity in a given direction plus the flow in that direction is equal to the edge capacity.**

This applies in both directions, provided that you count a flow as negative when it is opposite to the direction being considered:

So for SA	unused capacity + flow $= 4 + 4 = 8 =$ edge capacity
For AS	unused capacity + flow $= 4 + (-4) = 0 =$ edge capacity
For AD	unused capacity + flow $= 0 + 4 = 4 =$ edge capacity
For DA	unused capacity + flow $= 8 + (-4) = 4 =$ edge capacity

Notice that for directed edges we can allow for there to be different capacities in different directions.

Now look for another flow augmenting path, augment the flow and update the labelling, and repeat until no more flow augmenting paths can be found. If at any stage we can find a cut with capacity equal to the current flow then the max flow–min cut theorem tells us that we need look no further.

- **Steps associated with the flow augmentation algorithm:**

  Step 1 **Find a flow augmenting path.**

  Step 2 **Augment the flow by a quantity equal to the minimum unused capacity of edges on the augmenting path.**

  Step 3 **Re-label used edges, both in the direction of the augmenting flow and in the reverse direction.**

  Step 4 **Repeat 1, 2 and 3 until there are no more flow augmenting paths. You can use the max flow–min cut theorem to check that there are no more flow augmenting paths.**

A sequence of augmentations for example 5 is:

using SABT

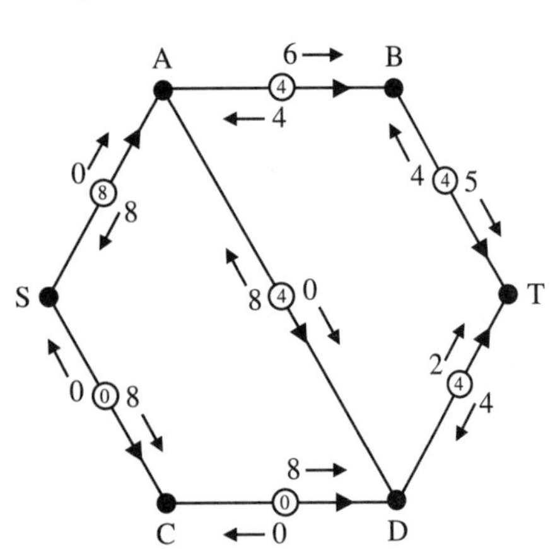

using SCDT

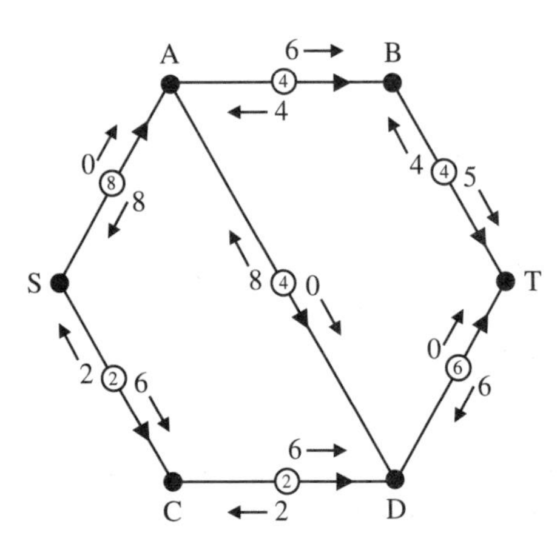

using SCDABT

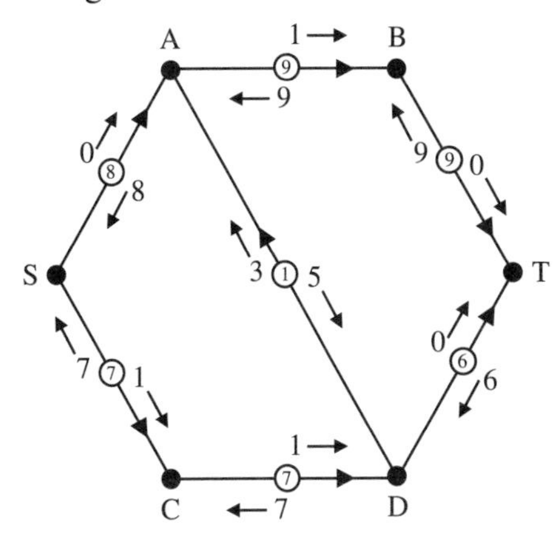

The optimal flow is $9 + 6 = 15$. This is optimal, as you can see by finding the capacity of the cut defined by the edges BT and DT.

The solution can be described in a table like this:

Edge	SA	SC	AB	DA	CD	BT	DT
Flow	8	7	9	1	7	9	6

Exercise 5D

Use the flow augmentation algorithm to find a set of flows which give the maximum total flow from S to T through the following transmission networks:

1

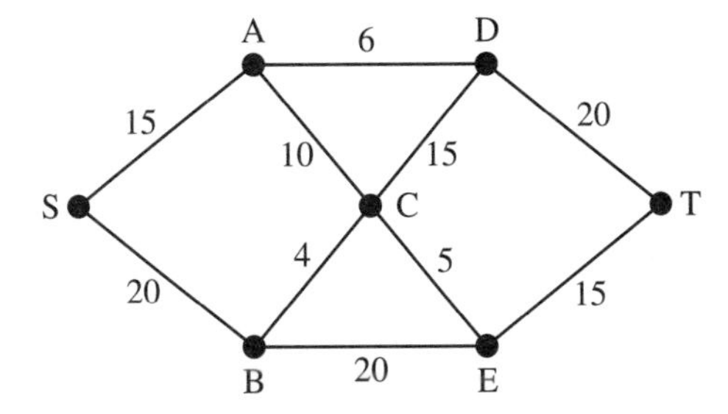

2

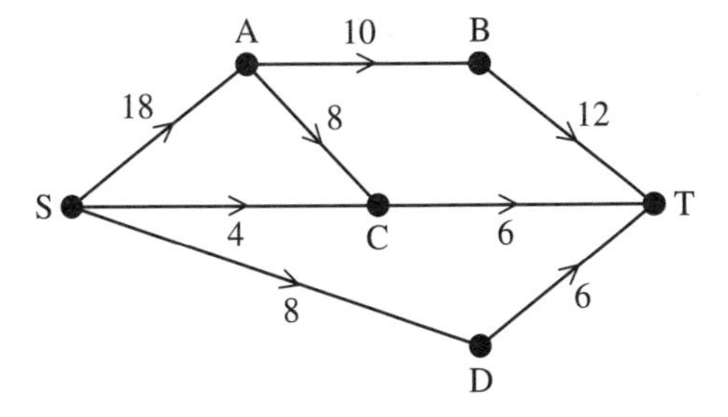

3

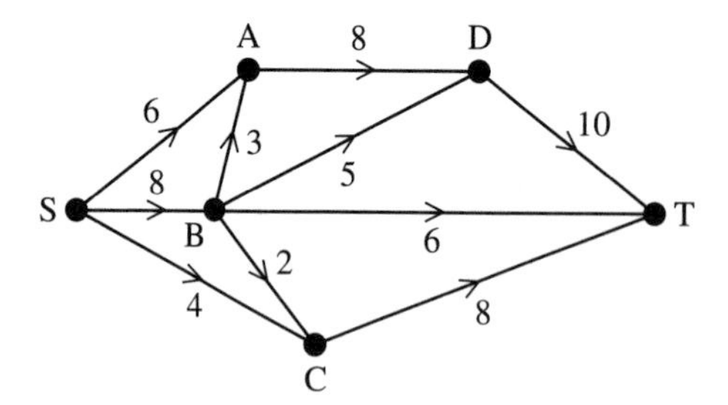

4

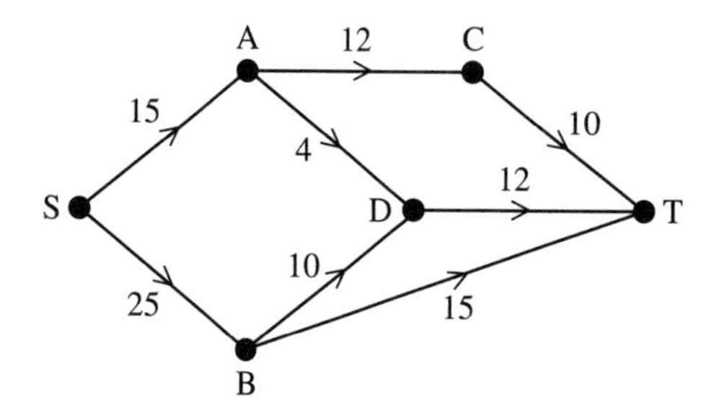

5.6 Spotting flow augmenting paths

For small networks you can spot flow augmenting paths from the labelled network diagram.

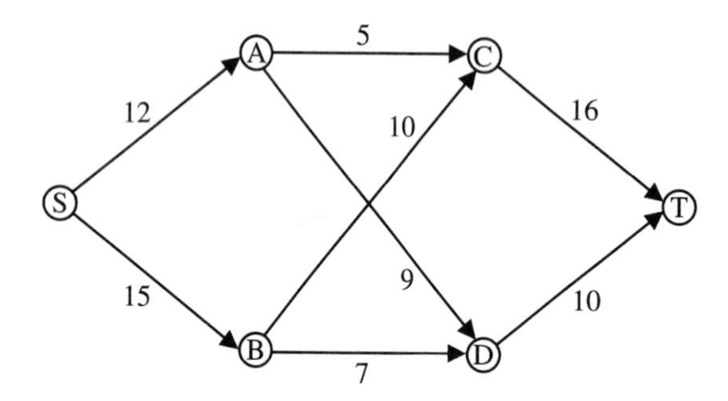

Stage 1 There are some fairly obvious initial flows in this network.

(i) Consider the path SACT. The edge with the smallest capacity on this path is AC. So it is possible to have a flow of 5 units along this path.

(ii) Consider the path SBDT. The edge with the smallest capacity on this path is BD. So it is possible to have a flow of 7 units along this path.

Using the labelling procedure this gives:

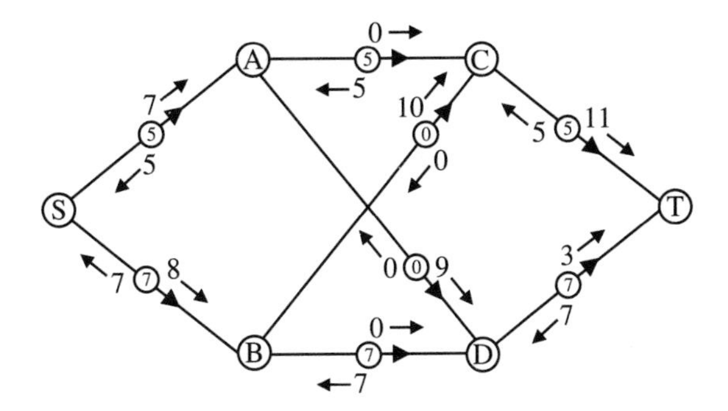

Here is how the labels were found for edge CT:

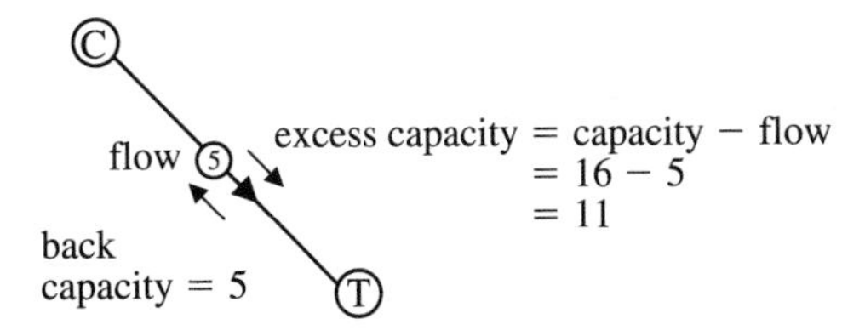

Stage 2 Flow augmenting paths are now required.

(i) Consider the path SADT. The edge with the smallest excess capacity on this path is DT. So it is possible to augment the flow by 3 units along this path.

(ii) Consider the path SBCT. The edge with the smallest excess capacity on this path is SB. So it is possible to augment the flow by 8 units along this path.

Using the labelling procedure this gives:

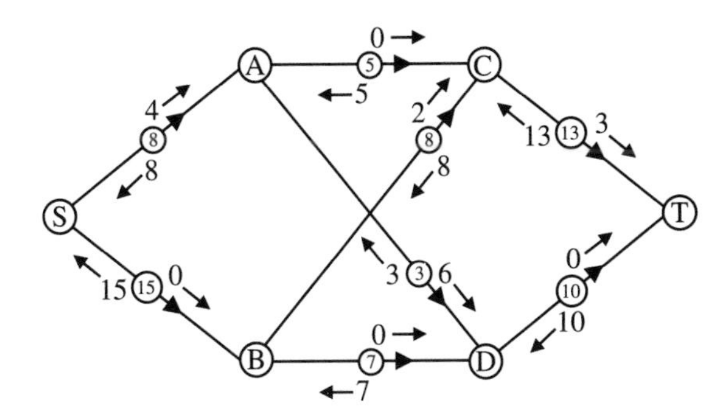

You may be tempted to think that there are no further flow augmenting paths, but there is one more.

Edge SB has zero excess capacity, so consider SA, which has excess capacity of 4 units. Edge AC has zero excess capacity, so consider AD, which has excess capacity of 6. Edge DT has zero excess capacity, but you can decrease the existing flow on BD by up to 7 units, as indicated by the back capacity of 7. A flow augmenting path can then be completed by using edges BC and CT, which have non-zero excess capacity.

So SADBCT is a flow augmenting path. The minimum excess capacity is that of BC, namely 2 units.

Using the labelling procedure this gives:

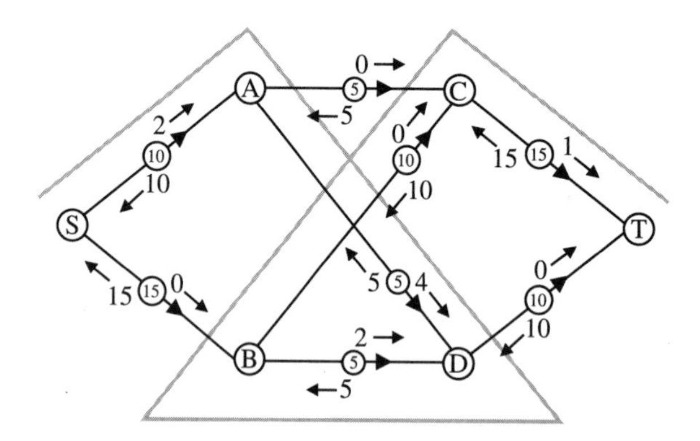

It seems as if this might give a maximal total flow. You can use the max flow–min cut theorem to check. If the flow is maximal, then you will be able to find a cut from among those edges with zero excess capacity or back capacity. In this case edges AC, BC and DT have zero excess capacity. Together they define the cut {S, A, B, D} | {C, T}, which has capacity $5 + 10 + 10 = 25$. Our flow pattern gives a total flow of $10 + 15 = 25$ (looking at the flows out of S), so the max flow–min cut theorem proves that we have a maximum total flow.

The final flow pattern is:

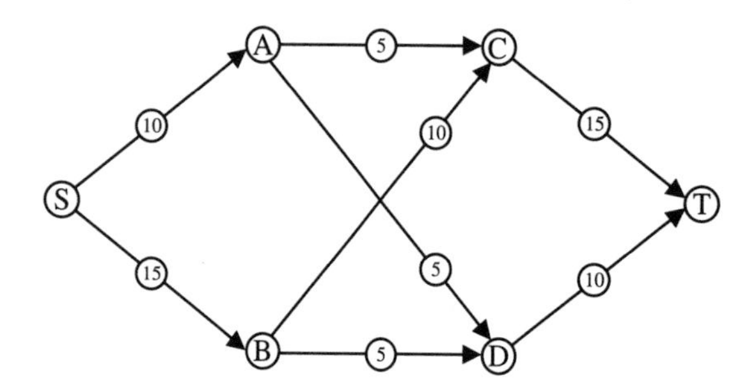

Exercise 5E

1 In this directed network the maximum flow allowed along each edge is given by the unringed numbers. A particular flow, marked by the ringed number on each edge, is established from source S to sink T.

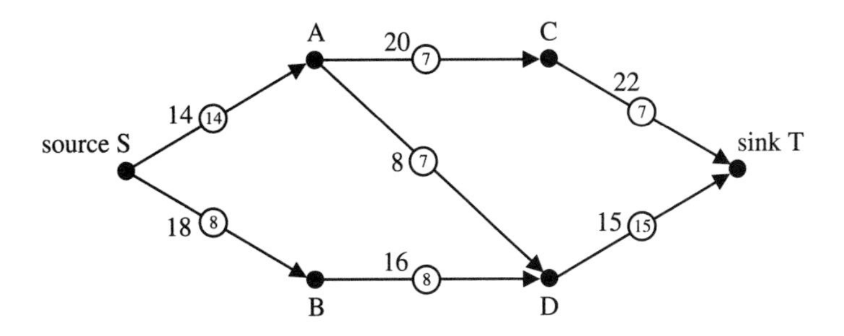

(a) Find a minimum cut.

(b) Explain why the minimum cut shows that the existing flow is not a maximum flow.

(c) On a copy of the network show flows along the edges which give the maximum flow from S to T.

2 For the network in question 1 suppose that the capacities remain the same but that the flow along the edge AD may now be in either direction:

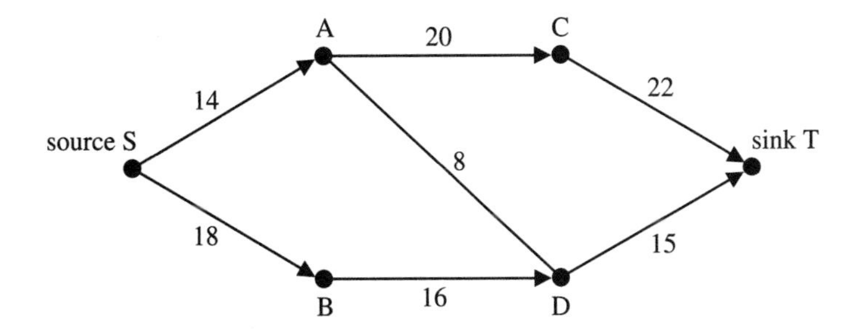

(a) Explain why the maximum flow from S to T is now greater than in question 1.

(b) State the value of the new maximum flow.

(c) On a copy of the network show the flows along the edges which give that maximum flow from S to T.

3 The diagram shows a directed flow network with two numbers on each edge, the capacity of the edge and, circled, the flow currently passing through the edge.

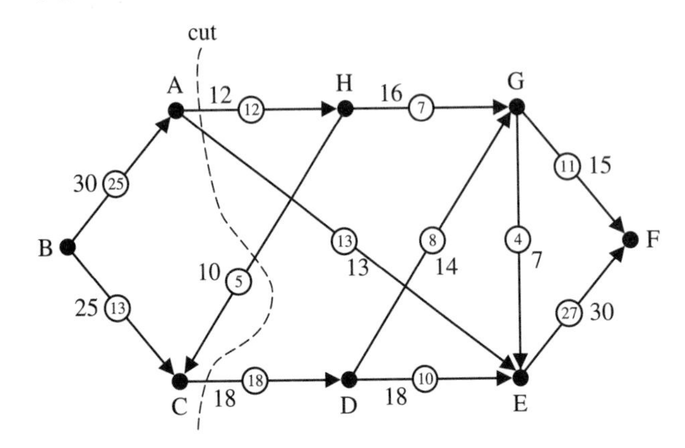

(a) Identify the source and the sink.

(b) What is meant by a *cut*? What is the capacity of the cut indicated?

(c) Find and indicate one flow augmenting path, and use it to augment the flow as far as is possible.

(d) Is the flow now maximal? If not use flow augmentation to find the maximal flow. Draw a diagram showing your maximal flow.

(e) Explain how you can tell whether the flow in part (d) is maximal.

4 The matrix represents the capacities of roads (in thousands of cars per hour) in a road network connecting 7 towns.

	A	B	C	D	E	F	G
A	–	2	2	–	–	1	–
B	2	–	–	5	1	–	–
C	2	–	–	1	–	3	–
D	–	5	1	–	–	1	1
E	–	1	–	–	–	–	1
F	1	–	3	1	–	–	1
G	–	–	–	1	1	1	–

road capacities

(a) Draw the road network.

(b) Find the maximum hourly flow of cars from B to F, showing how this may be achieved. Prove that this is a maximum.

5 A travel agent is asked by a youth organisation to try to organise coach transportation between London and Manchester for a musical extravaganza. He obtains from the coach operator the daily maximum capacity of coaches between various cities (in terms of hundreds of people).
These are shown in the table.

FROM \ TO	London	Leicester	Nottingham	Birmingham	Oxford	Manchester
London	–	–	–	30	15	–
Leicester	–	–	8	–	–	5
Nottingham	–	–	–	–	–	25
Birmingham	–	10	4	–	–	20
Oxford	–	–	10	15	–	–

(a) Draw a network showing the capacities.
(b) Find the maximum number of people that can be transported by coach from London to Manchester in a day.
(c) On a given day there is a strike at Nottingham coach station so that no coaches may use it.
Find the maximum number of people that can be transported from London to Manchester on that day.

6 A manufacturing company has two factories, F_1 and F_2, and wishes to transport its products to three warehouses, W_1, W_2 and W_3. The capacities of the various possible routes are shown in the diagram.

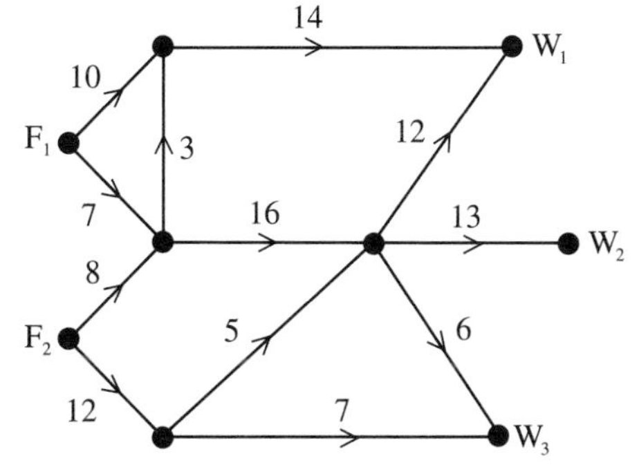

(a) By adding a source, S, and a sink, T, find the maximum flow to find the maximum number of products that can be transported.
(b) Interpret your flow pattern giving
(i) the number of products leaving F_1 and F_2
(ii) the number of products reaching W_1, W_2 and W_3.

7 The diagram shows a gas distribution network consisting of three supply points, A, B and C, three intermediate pumping stations, P, Q and R, two delivery points, X and Y, and connecting pipes. The figures on the arcs are measures of the amounts of gas which may be passed through each pipe per day. The figures by A, B and C are measures of the daily availability of gas at the supply points.

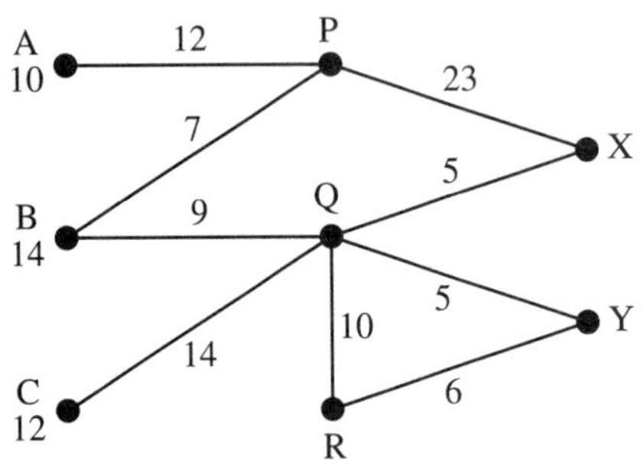

(a) Copy the network, introducing a single source with links to A, B and C, the capacities on the links showing the supply availabilities.

(b) Introduce a single sink linked to X and Y. Give large capacities to the links.

(c) Find the maximal daily flow through your network, making a list of the flows through each pipe.

(d) Find a suitable cut to prove that your flow in part (c) is maximal.

(e) Interpret the flows in the new links, from the source and to the sink.

(f) A new pipeline is constructed with capacity 6 units per day, connecting P and Q.
Use a labelling procedure to augment the flow, and thus find the new maximal flow.

(g) R is now to become a delivery point. Show how to adapt the approach of part (b) to find the maximal daily flow of gas in total that can be delivered to R, X and Y.

8 A transmission network has n vertices, each being connected directly to every other vertex. One vertex is the unique source for flows, another is the unique sink.
How many routes are there from source to sink?
What does your answer tell you about the complexity of the labelling algorithm for flow augmentation?

SUMMARY OF KEY POINTS

1 In a flow (or transmission) network:
  (a) there is a single source and a single sink
  (b) each edge has a capacity and a flow associated with it
  (c) the *source* vertex has no flows into it
  (d) the *sink* vertex has no flows out of it
  (e) for other vertices the total flow out = the total flow in

2 A cut is a division (partition) of the vertices of a flow network into two sets, one containing the source and the other containing the sink.

3 The capacity of a cut is the sum of the capacities of all edges which connect a vertex in the source set to a vertex in the sink set. If an edge is directed then it contributes to the cut capacity only if it is directed from the source set to the sink set.

4 The max flow–min cut theorem states that the maximum total flow that can be established through a flow network is equal to the capacity of the minimum cut, i.e. the cut with the smallest capacity.

5 If a cut is found with capacity equal to the existing flow, then that cut is a minimum cut and the flow is maximal.

6 In the flow augmentation algorithm edges are labelled with their flows, and with their unused capacities in both directions.

7 For all edges, directed and undirected, the unused capacity in a given direction plus the flow in that direction is equal to the edge capacity in that direction. (In this a flow is counted as negative when it is in the direction opposite to that being considered.)

8 The flow augmentation algorithm has two stages:
  (i) Obtain an initial flow by inspection.
  (ii) Identify and use flow augmenting paths.

9 Steps associated with flow augmentation:
  (i) Find a flow augmenting path.
  (ii) Augment the flow by the unused capacity of the edge on the augmenting path which has the minimum unused capacity. The flow along each edge of the augmenting path is increased by this amount.
  (iii) Re-label edges in the augmenting path, both in the direction of the augmentation and in the reverse direction.
  (iv) Repeat (i), (ii) and (iii) until there are no more flow augmenting paths.

Linear programming 6

Decision making is a process that has to be carried out in many areas of life. Usually there is a particular aim in making one decision rather than another. Two particular aims which are often considered in commerce are maximising profit and minimising costs.

During, and after, the Second World War a group of American mathematicians developed some mathematical methods to help with decision making. They sought to produce mathematical models of situations in which all the requirements, constraints and objectives were expressed as algebraic equations. They then developed methods for obtaining the **optimal solution** – the maximum or minimum value of a required function.

In this chapter you will study problems for which all the algebraic expressions are linear, that is, of the form

$$(\text{a number})\,x + (\text{a number})\,y + (\text{a number})\,z$$

For example,

$$\text{Profit} = 4x + 3y + 2z$$

is a linear equation in x, y and z and

$$16x + 18y + 9z \leqslant 25$$

is a linear inequation or inequality in x, y and z. This area of mathematics is called **linear programming**.

Linear programming methods are some of the most widely used methods employed to solve management and economic problems. They have been applied to a variety of contexts, some of which will be discussed later in this chapter, with enormous savings in money and resources.

6.1 Formulating linear programming problems

The first step in formulating a linear programming problem is to determine which quantities you need to know to solve the problem. These are called the **decision variables**.

The second step is to decide what are the **constraints** in the problem. For example, there may be a limit on resources, or a maximum or minimum value a decision variable may take, or there could be a relationship between two decision variables.

The third step is to determine the objective to be achieved. This is the quantity to be maximised or minimised, that is optimised. The function of the decision variables which is to be optimised is called the **objective function**.

The examples which follow illustrate the varied nature of problems which can be modelled by a linear programming, or LP, model. We will not at this stage attempt to solve these problems but instead concentrate on producing the objective function and the constraints, writing these in terms of the decision variables. As an aid to this it is often useful to summarise all the given information in the form of a table as is illustrated in example 1.

Example 1

A manufacturer makes two kinds of chairs, A and B, each of which has to be processed in two departments I and II. Chair A has to be processed in department I for 3 hours and in department II for 2 hours. Chair B has to be processed in department I for 3 hours and in department II for 4 hours.

The time available in department I, in a given month is 120 hours and the time available in department II, in the same month, is 150 hours.

Chair A has a selling price of £10 and chair B has a selling price of £12.

The manufacturer wishes to maximise his income. How many of each chair should be made in order to achieve this objective? You may assume that all chairs made can be sold.

The information given in the problem can be summarised by constructing the following table.

Type of chair	Time in Dept I (hours)	Time in Dept II (hours)	Selling price
A	3	2	10
B	3	4	12
Total time available	120	150	

Step 1 Which quantities do you need to know to solve the problem – the decision variables?
It is clear that in this case you wish to know how many type A chairs are to be made and how many type B chairs are to be made.
The decision variables are therefore:

x = number of type A chairs made
y = number of type B chairs made.

Step 2 What are the constraints?
Consider what happens in department I; that is, concentrate on the column in the table labelled department I.

Since: the production of 1 type A chair uses 3 hours
then: the production of x type A chairs uses $3x$ hours.
Similarly: the production of 1 type B chair uses 3 hours
so: the production of y type B chairs uses $3y$ hours.

The total time used is therefore:

$$(3x + 3y) \text{ hours}$$

Since only 120 hours are available in department I, one constraint is:

$$(3x + 3y) \text{ hours} \leqslant 120 \text{ hours}$$

or:

$$3x + 3y \leqslant 120$$

Considering department II in a similar way produces the second constraint:

$$2x + 4y \leqslant 150$$

In addition to these two constraints you also require that x and y be non-negative, that is:

$$x \geqslant 0, \quad y \geqslant 0$$

These are called the **non-negativity constraints**.

Step 3 What is the objective?
The objective is to maximise the income. If x chairs of type A are made the income is £$10x$ and if y chairs of type B are made the income is £$12y$.
The total income is then £z = £$(10x + 12y)$.
The aim therefore is to maximise

$$z = 10x + 12y$$

The problem can be modelled by the linear programming problem:

Maximise $z = 10x + 12y$
subject to the constraints:

$$3x + 3y \leqslant 120$$

$$2x + 4y \leqslant 150$$

$$x \geqslant 0, \quad y \geqslant 0$$

Notice that all the constraints are dimensionless, that is they only involve numbers. Also z is a number. It is advisable, as in this case, to write the income as £z so that all the variables in the problem have purely numerical values.

Example 2

A book publisher is planning to produce a book in three different bindings: paperback, book club and library. Each book goes through a sewing process and a gluing process. The table gives the time required, in minutes for each process and for each of the bindings.

	Paperback	Book club	Library
Sewing (min)	2	2	3
Gluing (min)	4	6	10

The sewing process is available for 7 hours per day and the gluing process for 11 hours per day. The profits are 25p on a paperback edition, 40p on a book club edition and 60p on a library edition. How many books in each binding should be manufactured to maximise profits? (Assume that the publisher can sell as many of each type of book as she produces.)

For this problem it is a good idea to extend the table to include all the information given by adding the restrictions on time, and also the profits.

	Paperback	Book club	Library	Total time available
Sewing (min)	2	2	3	420
Gluing (min)	4	6	10	660
Profit (P)	25	40	60	

Notice that the time available has been converted to minutes to be consistent with the other times given.

Step 1 – decision variables

The decision variables are the numbers of books to be made in each binding.

Let x = number in paperback binding
y = number in book club binding
z = number in library binding.

Step 2 – constraints

The constraints are:

Sewing: $2x + 2y + 3z \leqslant 420$
Gluing: $4x + 6y + 10z \leqslant 660$
together with the non-negativity conditions
$x \geqslant 0, y \geqslant 0, z \geqslant 0$.

Step 3 – objective

The objective is to maximise the profit P pence. That is, to maximise $P = 25x + 40y + 60z$.

So the linear programming problem to be solved is:

Maximise $P = 25x + 40y + 60z$
subject to the constraints:

$$2x + 2y + 3z \leqslant 420$$
$$4x + 6y + 10z \leqslant 660$$
$$x \geqslant 0, y \geqslant 0, z \geqslant 0$$

Example 3

KJB Haulage receives an order to transport 1600 packages. They have large vans, which can take 200 packages each, and small vans which can take 80 packages each.

The cost of running each large van on the required journey is £40 and the cost of running each small van on the same journey is £20.

There is a limited budget for the job which requires that not more than £340 be spent. It is additionally required that the number of small vans used must not exceed the number of large vans used.

How many of each kind of van should be used if costs are to be kept to a minimum?

The table summarises the given information:

	Capacity	Cost (in £)
Large van	200	40
Small van	80	20
Limits	1600	340

Step 1 – decision variables

The decision required is how many of each type of van to use to fulfil the order. The decision variables are:

l = number of large vans used
s = number of small vans used

Step 2 – constraints

If all the packages are to be transported then one constraint is:

$$200l + 80s \geqslant 1600$$

You can simplify this inequality, as 40 is a factor of each of the numbers in it, giving:

$$5l + 2s \geqslant 40$$

The restriction on total cost means that another constraint is:

$$40l + 20s \leqslant 340$$

This inequality may also be simplified by dividing by 20, giving:

$$2l + s \leqslant 17$$

Also the number of small vans must not exceed the number of large, so a third constraint is:

$$s \leqslant l \quad \text{or} \quad s - l \leqslant 0$$

Finally you must have $l \geqslant 0$ and $s \geqslant 0$.

Step 3 – objective

The objective is to keep costs to a minimum, that is minimise the cost £C where $C = 40l + 20s$.

So the linear programming problem to be solved is:

Minimise $C = 40l + 20s$
subject to the constraints:

$$5l + 2s \geqslant 40$$
$$2l + s \leqslant 17$$
$$s - l \leqslant 0$$
$$l \geqslant 0, s \geqslant 0$$

Example 4

When a lecturer retires he will have a 'lump sum' of £30 000 to invest. He decides to invest in stock AA and bond BBB. Stock AA yields 7% per annum and bond BBB yields 5% per annum. He decides that no more than £20 000 shall be invested in either option. How much should he invest in each option in order to maximise his yield?

Step 1 – decision variables

Suppose he invests £x in stock AA
and £y in bond BBB

Step 2 – constraints

The total amount he has to invest is £30 000 so £x + £y = £30 000

or $$x + y = 30\,000$$

Since he decides not to invest more than £20 000 in either option

$$x \leqslant 20\,000$$

$$y \leqslant 20\,000$$

We must also have here $x \geqslant 0$, $y \geqslant 0$.

Step 3 – objective

The yield £Y which he wishes to maximise is

$$£Y = \tfrac{7}{100}£x + \tfrac{5}{100}£y$$

or $$Y = 0.07x + 0.05y$$

The linear programming problem to be solved is then:

Maximise $Y = 0.07x + 0.05y$
subject to the constraints:

$$x + y = 30\,000$$
$$x \leqslant 20\,000$$
$$y \leqslant 20\,000$$
$$x \geqslant 0,\ y \geqslant 0$$

Notice that one of the constraints is an equation, $x + y = 30\,000$, rather than an inequation.

Exercise 6A

The following problems will help you practise **formulating** linear programming problems that can be solved. You are not being asked to solve them at this stage.

1 Allwood PLC plans to make two kinds of table. For table A the cost of the materials is £20, the number of person-hours needed to complete it is 10 and the profit, when it is sold, is £15. For table B the cost of the materials is £12, the number of person-hours needed to complete it is 15 and the profit, when it is sold, is £17. The total money available for materials is £480 and the labour available is 330 person-hours. Find the maximum profit that can be made and the number of each type of table that should be made to produce it. Formulate this as a linear programming problem.

2 To ensure that her family has a healthy diet Mrs Brown decides that the family's daily intake of vitamins A, B and C should not fall below 25 units, 30 units and 15 units respectively. To provide these vitamins she relies on two fresh foods α and β. Food α provides 30 units of vitamin A, 20 units of vitamin B and 10 units of vitamin C per 100 g. Food β provides 10 units of vitamin A, 25 units of vitamin B and 40 units of vitamin C per 100 g. Food α costs 40p per 100 g and food β costs 30p per 100 g. How many grams of food α and food β should she purchase daily if the food bill is to be kept to a minimum? Formulate this as a linear programming problem.

3 A factory is to install two types of machines A and B. Type A requires 1 operator and occupies 3 m^2 of floor space, type B requires 2 operators and occupies 4 m^2 of floor space. The maximum number of operators available is 40 and the floor space available is 100 m^2. Given that the weekly profits on type A and type B machines are £75 and £120, find the number of each machine that should be bought to maximise the profit and calculate the profit. Formulate this problem as a linear programming problem.

4 A sports club wishes to hire a minibus for a trip. Children under 16 can travel for £5 and adults for £10. The minibus company have the following conditions.
The total number of passengers must not exceed 14. There must be not less than 10 passengers on each trip.
There must be at least as many half-fare passengers as full-fare passengers.
Determine the maximum amount the minibus company can receive for the hire. Formulate this as a linear programming problem.

5 The Midlands Furniture Company manufactures bookshelves in three sizes: small, medium and large. Small bookshelves require 4 m of board, medium 8 m of board and large 16 m of board. The assembly time required is 2 hours, 4 hours and 6 hours

respectively. Only 500 m of board are available and assembly time is restricted to 400 hours. All bookshelves produced can be sold, the profits being £4, £6 and £12 respectively. How many of each should be made if profit is to be maximised? Formulate this as a linear programming problem.

6 Chair Supplies makes three types of wooden chairs. Each type is manufactured in a 4-stage process. The company is able to obtain all the raw materials it needs. The available production capacity during the 60 hour production work week is as follows:

	Weekly capacity in number of chairs				
Process	Chair A		Chair B		Chair C
1	400	or	600	or	900
2	1800	or	400	or	300
3	200	or	900	or	600
4	600	or	400	or	450

It is assumed that there are 60 hours of labour available for each process. The profits on each of the three types of chair are £15, £20 and £25 respectively. Formulate this as a linear programming problem, given that the profit is to be maximised.

6.2 Graphical solutions for two-variable problems

Linear programming problems which involve only two decision variables, x and y, may be solved graphically; that is by drawing lines associated with the constraints, identifying a region of possible solutions and then locating a point in this region that has a particular property. To solve linear programming problems using graphical methods you need to be able to use the following basic mathematical ideas.

Sets of points defined by a linear inequality

Any equation of the form $ax + by = c$, where a, b and c are numbers, is called a **linear equation**. For example, $3x + 4y = 12$ is a linear equation. In the x, y plane this is the equation of a **straight line**. This line may be drawn by identifying any two points on it. For example, when $x = 0$, $4y = 12 \Rightarrow y = 3$. So, $x = 0$, $y = 3$, or (0,3) is on the line.

Similarly, when $y = 0$, $3x = 12 \Rightarrow x = 4$. So, $x = 4$, $y = 0$, or (4,0), is also on the line. Plotting both points allows you to draw the line:

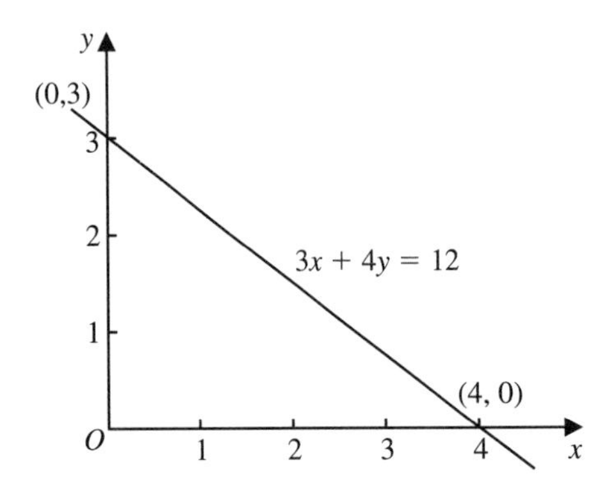

A special case of $ax + by = c$ is that in which $c = 0$, for example $2y - x = 0$. In such cases $x = 0$, $y = 0$, or (0, 0), is on the line. Taking $x = 1$ gives $y = \frac{1}{2}$. So a second point on the line is $x = 1$, $y = \frac{1}{2}$, or $(1, \frac{1}{2})$.

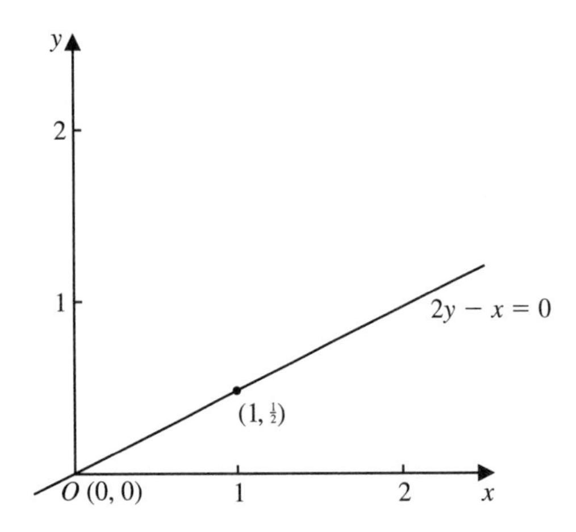

In linear programming problems we usually have the non-negativity conditions $x \geqslant 0$, $y \geqslant 0$, so we are only concerned with values of x and y *in the first quadrant*.

Any straight line divides the x,y-plane into two half-planes. If the equation of the line is $ax + by = c$ then on one side of the line you have $ax + by < c$ and on the other side of the line you have $ax + by > c$. For example,

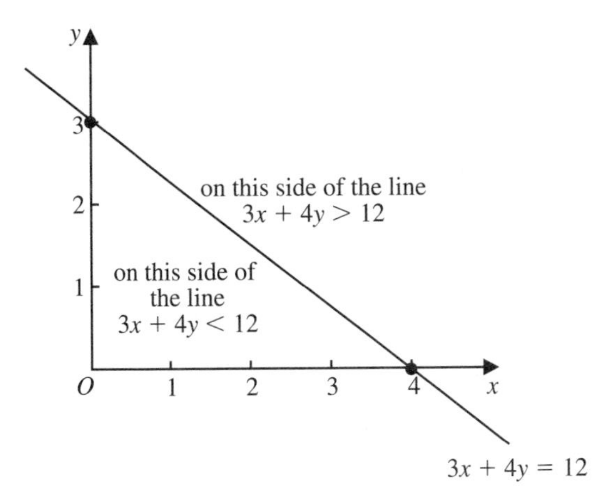

An inequality of the form $ax + by \leqslant c$ therefore defines a set of points (x, y) which *either lie on the line or else in the half-plane on one side of the line*. You can easily decide which side of the line is the required half-plane by inserting the point (0,0) into the inequality. If $x = 0$, $y = 0$ satisfies the inequality then (0,0), the origin, is in the required region.

Example 5

Show on a diagram the region for which

$$3x + 4y \leqslant 12$$

Draw the line $3x + 4y = 12$. Since $3(0) + 4(0) = 0$, which is less than 12, the origin lies in the required region, and the half-plane required contains the origin. So all points on the line or below it satisfy the inequality. We call this the **admissible set** defined by the inequality. You can show this by using arrows to indicate the required region and shading marks placed on the line to indicate the half-plane of inadmissible points.

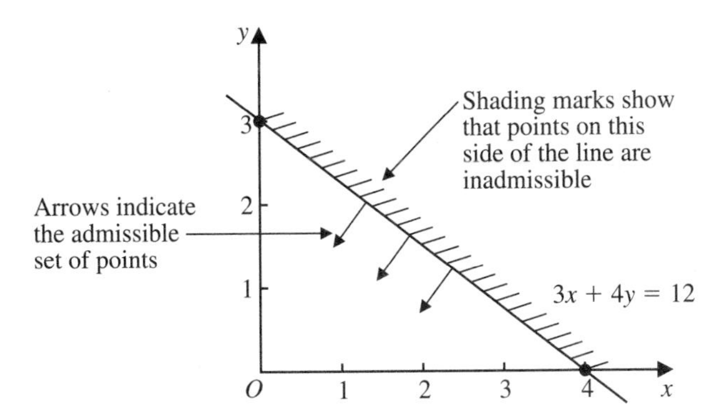

This method of indicating the admissible set is simple, and produces clear solutions.

Sets of points defined by a collection of inequalities

Each inequality in a linear programming problem will produce an admissible set. To find a solution to the problem you need to find the set of points which satisfy all the inequalities simultaneously. This is obtained graphically by drawing a diagram like the one on page 169 but showing *all* the inequalities. **The required region is then the one which does not contain any shading marks and into which all arrows point.**

Example 6

Indicate on a diagram the region for which

$$3x + 4y \leqslant 12$$
$$3x + 2y \leqslant 9$$
$$x \geqslant 0, y \geqslant 0$$

You can find the region for which the second inequality is satisfied by following the procedure outlined in example 5. The line with equation $3x + 2y = 9$ passes through the points with coordinates $(0,4\frac{1}{2})$ and $(3,0)$. The origin $(0,0)$ is in the admissible set since $3(0) + 2(0) = 0$.

You have then:

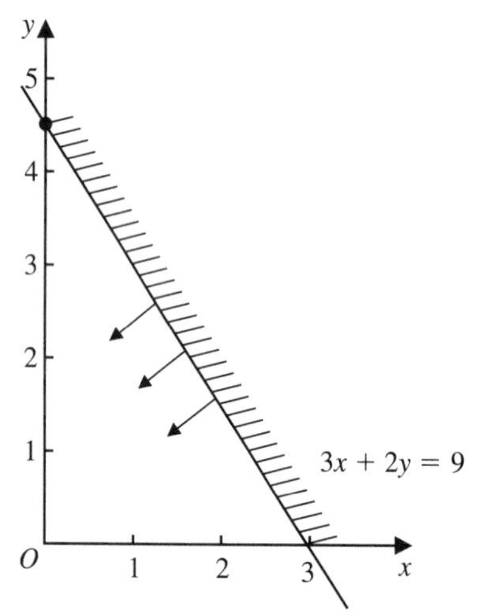

The inequality $x \geqslant 0$ gives:

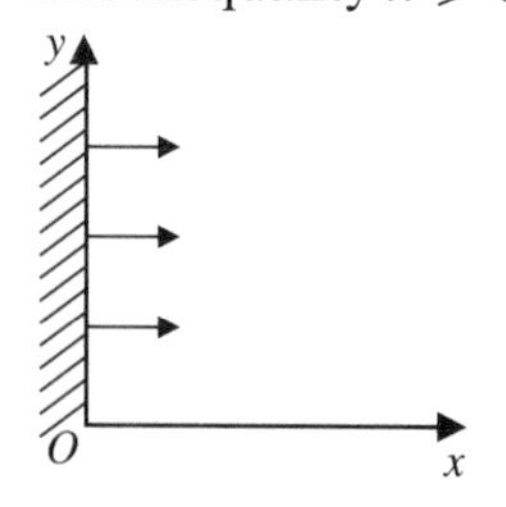

and the inequality $y \geqslant 0$ gives:

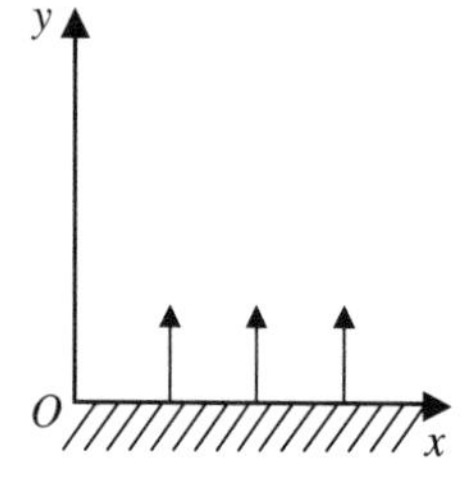

Combining these gives:

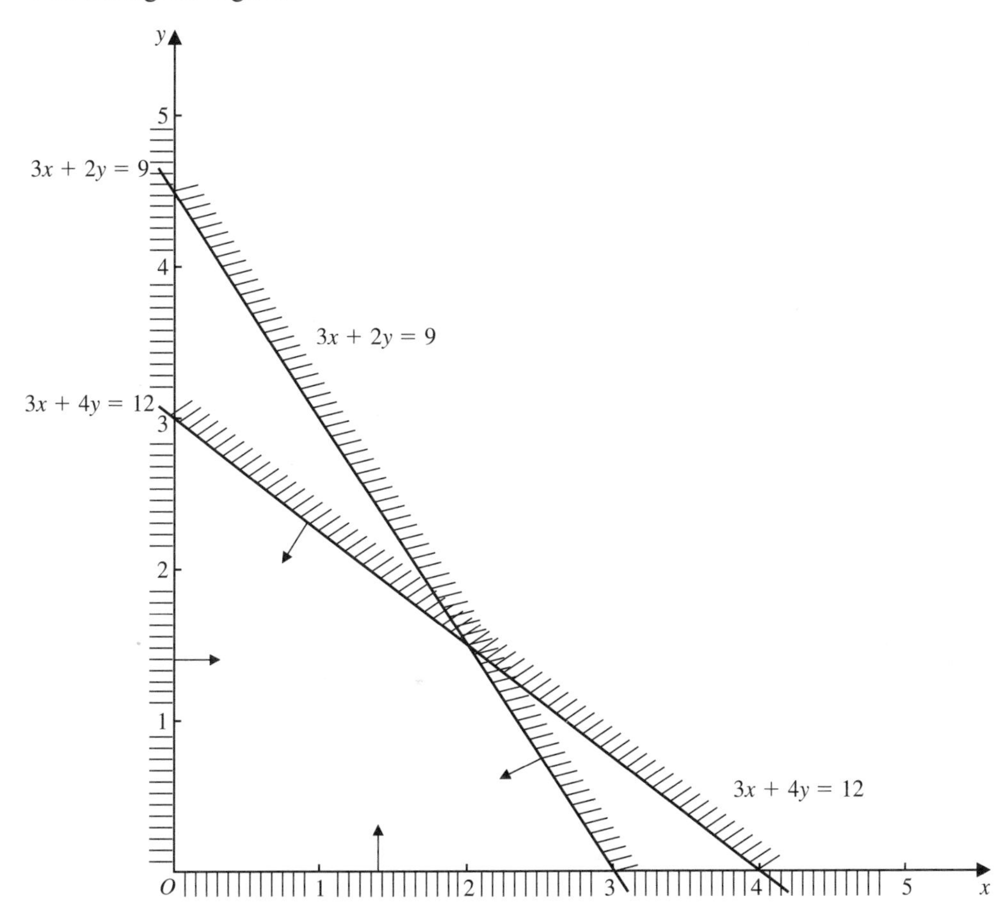

Exercise 6B

1 Indicate, on a diagram, the region for which

$$5x + 3y \leqslant 15$$
$$x \geqslant 0$$
$$y \geqslant 0$$

2 Indicate, on a diagram, the region for which

$$4x + 3y \leqslant 12$$
$$2x + 5y \leqslant 10$$
$$x \geqslant 0$$
$$y \geqslant 0$$

3 Indicate, on a diagram, the region for which

$$2x + y \leqslant 8$$
$$y \leqslant 7$$
$$x \leqslant 3$$
$$x \geqslant 0, y \geqslant 0$$

4 Indicate, on a diagram, the region for which

$$y + 2x \leqslant 12$$
$$x \geqslant 2$$
$$y \geqslant 4$$

5 Indicate, on a diagram, the region for which

$$3x + 2y \leqslant 12$$
$$3x + y \geqslant 6$$
$$x + y \geqslant 4$$

6.3 Feasible solutions of a linear programming problem

From the discussion in section 6.2 you can see that a point with coordinates (x,y) which lies in the intersection of all the admissible sets, defined by the constraints, represents a **feasible solution**.

■ **Any pair of values of x and y which satisfy all the constraints in a linear programming problem is called a feasible solution.**
The region which contains all such points is called the *feasible region*.

In example 6 all the constraints are satisfied in the unshaded region.

A linear programming problem is therefore solved by finding which member, or members, of the set of feasible solutions gives the optimal (maximum or minimum) value of the objective function.

6.4 Finding the optimal solution of a linear programming problem

Think about a linear programming problem in which you wish to maximise $z = \alpha x + \beta y$, where α and β are positive numbers. You have seen that $\alpha x + \beta y = P$ is the equation of a straight line. When P takes different values you get a family of parallel straight lines.

For example, the diagram below shows

$$x + 2y = P$$

for the cases $P = 2, 4, 6$.

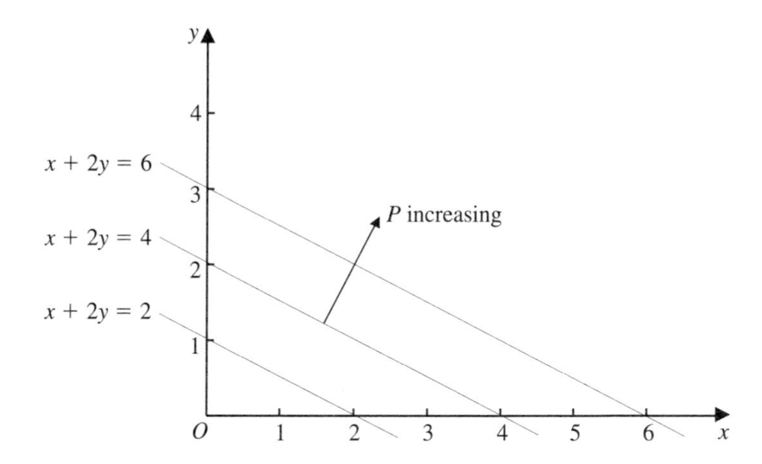

You can see that the points at which the line intercepts the x and y axes move further and further away from the origin O as P increases. To check each x intercept, find x when $y = 0$. To check each y intercept, find y when $x = 0$.

Now you can use the fact that the line with equation $\alpha x + \beta y = P$, α and β positive numbers, moves further from the origin as P increases to find the maximum value of P. The largest value of P will occur therefore at the point in the feasible region which is on the line *furthest* from the origin. You can find this point by sliding a ruler over the feasible region so that it is always parallel to the family of straight lines given by $\alpha x + \beta y =$ constant. This will enable you to identify the point which is *furthest* from the origin:

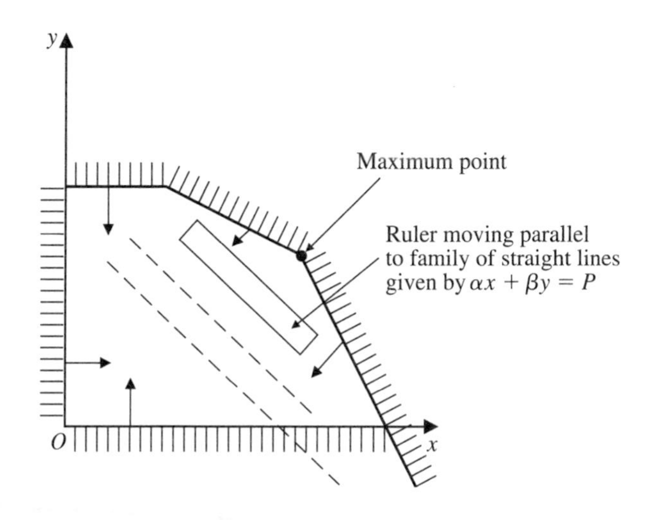

In the case of a *minimising* problem the point in the feasible region which is on the line *closest* to the origin is the one that is required:

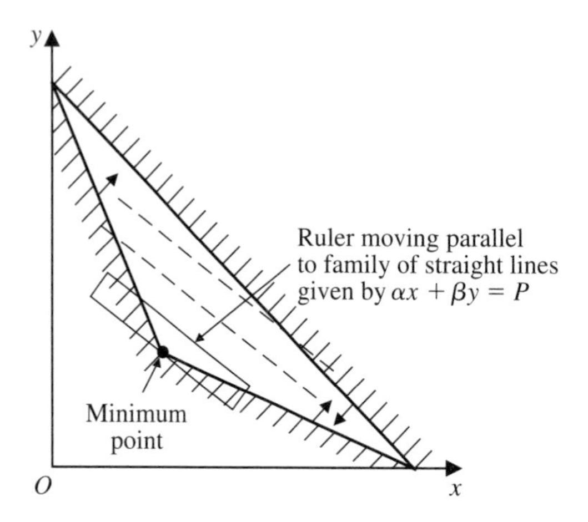

Suppose you have drawn the region, on a graph, in which the solution must lie. You can then start the process of maximising or minimising $\alpha x + \beta y$ by drawing an arbitrary member of the family of parallel lines, $\alpha x + \beta y = \gamma$, say.

Choose γ so that the intercepts on the x and y axes are integers – that is, γ is a number which has both α and β as factors. For example, if $\alpha = 3$ and $\beta = 4$ then draw $3x + 4y = 12$. Then when $y = 0$ the intercept is $x = 4$ and when $x = 0$ the intercept is $y = 3$. These ideas are illustrated in the following examples.

Example 7

Solve graphically the linear programming problem:

Maximise $\quad z = x + y$

subject to the constraints

$$3x + 4y \leqslant 12$$
$$3x + 2y \leqslant 9$$
$$x \geqslant 0, \; y \geqslant 0$$

The feasible region for this problem was obtained above (page 171) and is shown on the next page.

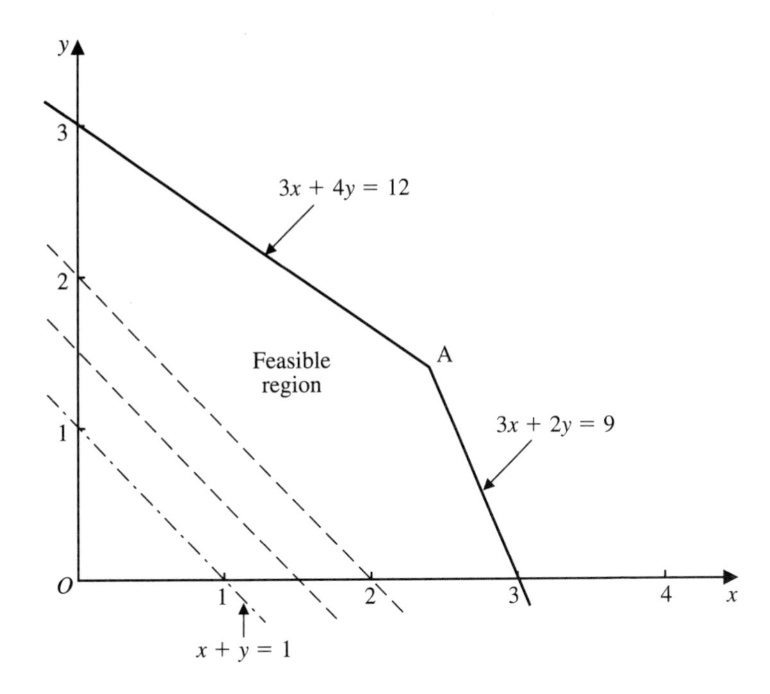

Start by drawing the line $x + y = 1$ (–·–· on graph above). This passes through $(0, 1)$ and $(1, 0)$. Other members of the family $x + y = c$ are shown like this – – –. The furthest you can move the ruler without losing contact with the feasible region is to the point A. The point A therefore gives the optimal value of z.

The coordinates of the point A are obtained from the fact that A is the point of intersection of the two lines with equations

$$3x + 4y = 12$$

and $$3x + 2y = 9$$

Subtracting: $$2y = 3 \Rightarrow y = 1\tfrac{1}{2}$$

Substituting into the second equation:

$$3x + 2\left(1\tfrac{1}{2}\right) = 9$$

So: $$3x = 9 - 3 = 6 \Rightarrow x = 2$$

Now substitute these values in the equation $z = x + y$.

When $x = 2$ and $y = 1\frac{1}{2}$ then $z = 2 + 1\frac{1}{2} = 3\frac{1}{2}$.

So the maximum value of z is $3\frac{1}{2}$ and occurs when $x = 2$ and $y = 1\frac{1}{2}$.

Example 8

Solve graphically the linear programming problem:

Minimise $z = 3x + 4y$

subject to the constraints:

$$x + 4y \geqslant 20$$
$$x + y \geqslant 8$$
$$x + 2y \leqslant 16$$
$$x \geqslant 0,\ y \geqslant 0$$

(a) The line $x + 4y = 20$ passes through the points (0, 5) and (20, 0). Since $0 + 4(0) = 0$, which is not greater than 20, the origin *does not lie* in the admissible region for the inequality $x + 4y \geqslant 20$.

(b) The line $x + y = 8$ passes through the points $(0, 8)$ and $(8, 0)$. Since $0 + 0 = 0$, which is not greater than 8, the origin *does not lie* in the admissible region for the inequality $x + y \geqslant 8$.

(c) The line $x + 2y = 16$ passes through the points $(0, 8)$ and $(16, 0)$. Since $0 + 2(0) = 0$, which is less than 16, the origin *does lie* in the admissible region for the inequality $x + 2y \leqslant 16$.

The diagram summarises the information in (a), (b) and (c) and shows the feasible region. The non-negativity conditions merely restrict you to the first quadrant.

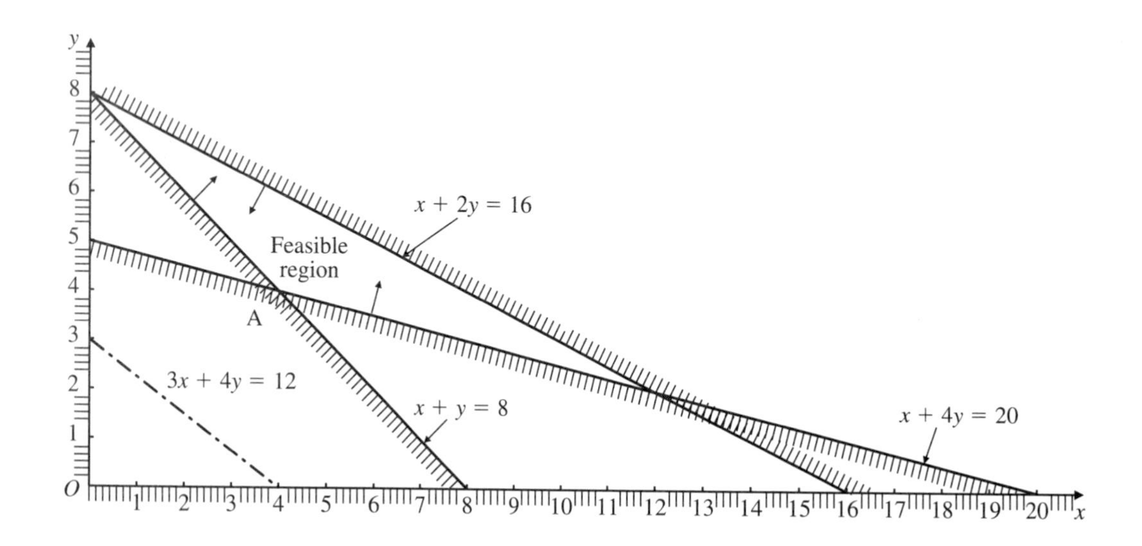

To locate the point where the objective function $z = 3x + 4y$ has its minimum value first draw a line of the form $3x + 4y = \text{constant}$.

The line $3x + 4y = 12$ is shown $- \cdot - \cdot$ on the diagram. (The number 12 was chosen because 3 and 4 are factors of 12 and so the points where it intersects the axes have integer values.) This line passes through the points $(0, 3)$ and $(4, 0)$.

If you move a ruler parallel to this line then you see that the point of the feasible region nearest to the origin, first reached, is the point A. As A is the point of intersection of the lines $x + y = 8$ and $x + 4y = 20$ you can obtain the coordinates of A by solving the simultaneous equations:

$$x + 4y = 20$$
$$x + y = 8$$

Subtracting: $3y = 12 \Rightarrow y = 4$

Substituting this value of y into the second equation gives

$$x + 4 = 8$$

so: $x = 8 - 4 = 4$

So, the minimum value of z is $3(4) + 4(4) = 28$ and this occurs when $x = 4$ and $y = 4$.

Example 9

Solve graphically the linear programming problem

Maximise $z = 16x + 24y$

subject to the constraints:

$$2x + 3y \leqslant 24$$
$$2x + y \leqslant 16$$
$$y \leqslant 6$$
$$x \geqslant 0, y \geqslant 0$$

(a) The line $2x + 3y = 24$ passes through the points $(0, 8)$ and $(12, 0)$. Since $2(0) + 3(0) = 0$ which is less than 24 the origin *does lie* in the admissible region for the inequality $2x + 3y \leqslant 24$.

(b) The line $2x + y = 16$ passes through the points $(0, 16)$ and $(8, 0)$. Since $2(0) + 0 = 0$ which is less than 16 the origin *does lie* in the admissible region for the inequality $2x + y \leqslant 16$.

(c) The admissible region for the inequality $y \leqslant 6$ is the region on and below the line $y = 6$.

The diagram summarises the information in (a), (b) and (c), together with the non-negativity conditions, and shows the feasible region.

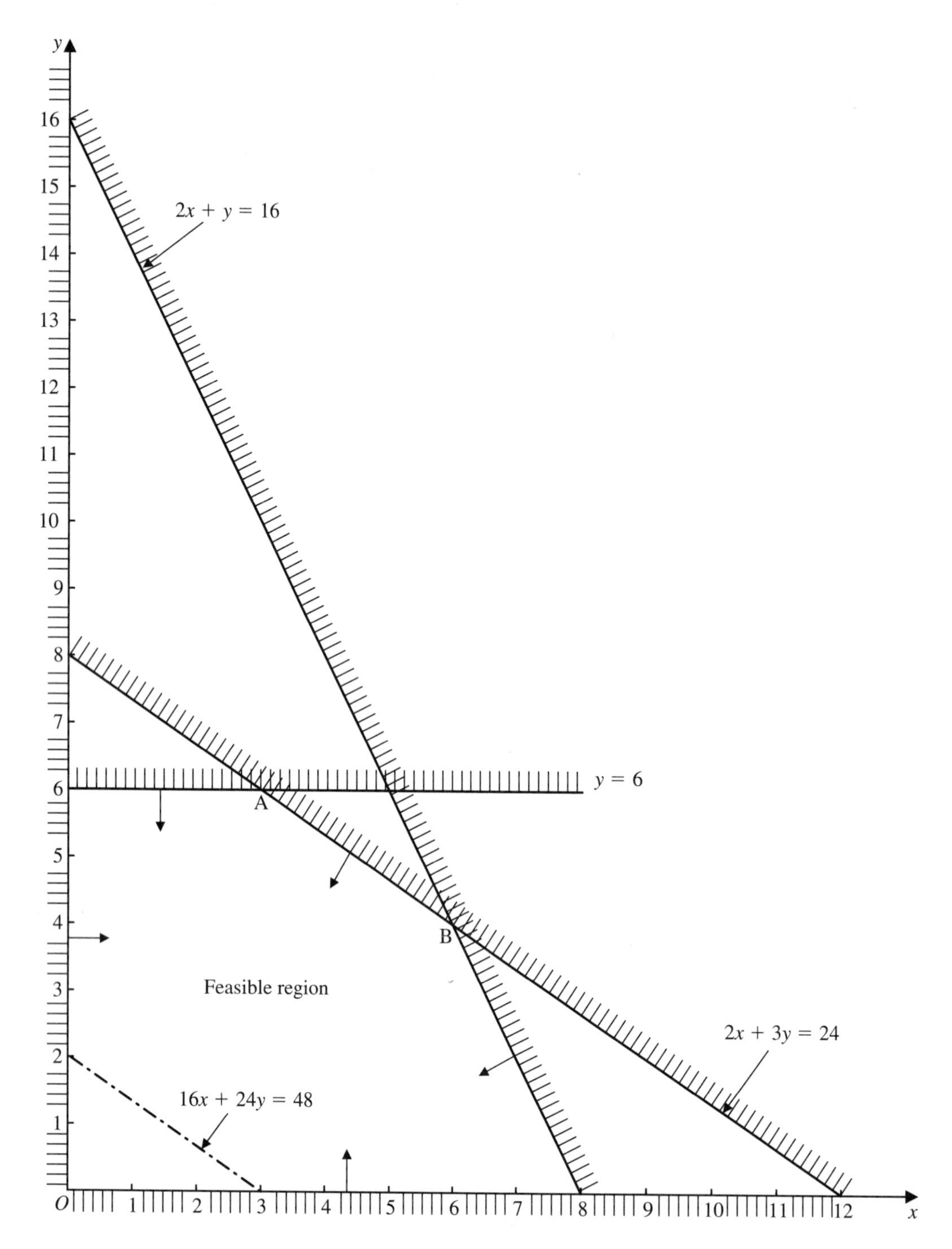

As in the previous examples you need first to draw a member of the family of parallel lines $16x + 24y =$ constant. In the diagram the line $16x + 24y = 48$ has been drawn. This line passes through the points (0, 2) and (3, 0). If you move the ruler parallel to this line as far as possible, so as still to contact the feasible region, you ultimately reach the boundary AB of the feasible region.

The coordinates of the point A are easily obtained. Its y-coordinate is 6 and its x-coordinate can be found by using the fact that A lies on the line with equation $2x + 3y = 24$. So:

$$2x + 18 = 24$$

$$2x = 24 - 18 = 6$$

$$\Rightarrow \quad x = 3$$

So A has coordinates (3,6).

The point B is the intersection of the lines

$$2x + y = 16$$

and $$2x + 3y = 24$$

Subtracting: $$2y = 8 \Rightarrow y = 4$$

Substituting in the first equation: $2x + 4 = 16$

$$2x = 16 - 4 = 12$$

so $$x = 6$$

So B has coordinates (6,4).

The value of z at A is $z_A = 16(3) + 24(6) = 192$ and the value of z at B is $z_B = 16(6) + 24(4) = 192$.

In fact the value of z is 192 for every point on the line AB. The maximum value of z for this problem occurs at every point of the line segment AB rather than for a single point as in the previous cases.

Example 10

Solve graphically example 9 when the objective function is changed to

(a) $z = 24x + 16y$ (b) $z = 12x + 24y$

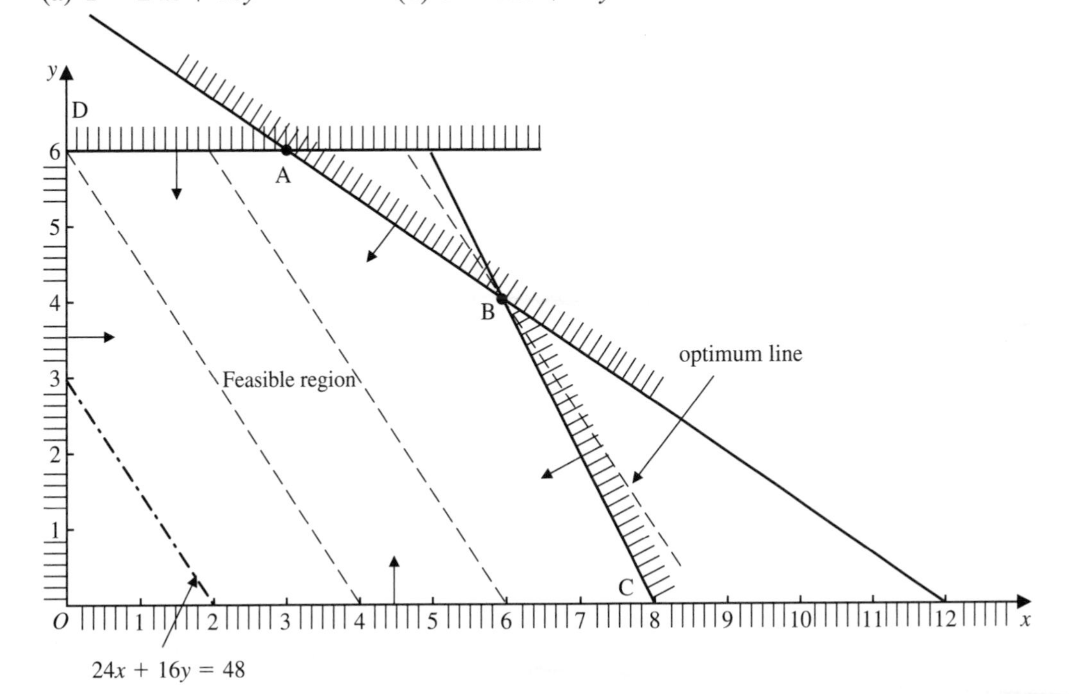

One member of the family of lines $24x + 16y = c$ is $24x + 16y = 48$ which passes through (0,3) and (2,0) shown $-\cdot-\cdot$. Other members of the family are shown $- - -$. The maximum value of z occurs at the point B(6,4) and $z_{\text{max}} = 24(6) + 16(4) = 208$. The optimum line passes through point B.

(b)

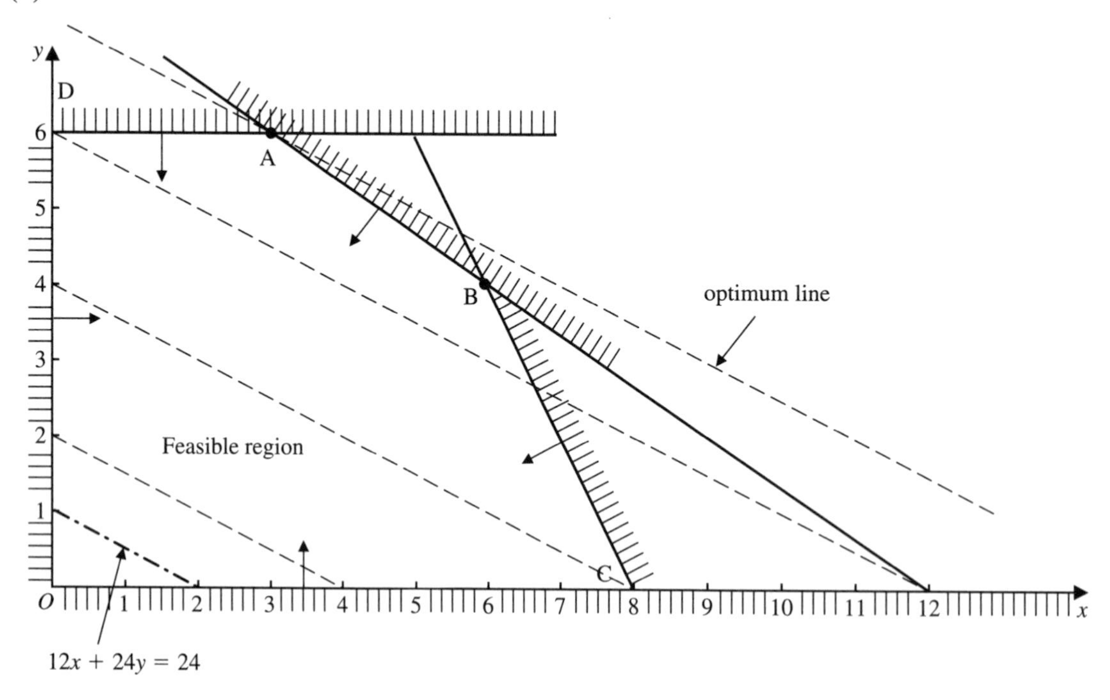

One member of the family of lines $12x + 24y = c$ is $12x + 24y = 24$ which passes through (0, 1) and (2, 0), shown $-\cdot-\cdot$. Other members of the family are shown $- - -$. The maximum value of z occurs at A (3,6) and $z_{\text{max}} = 12(3) + 24(6) = 180$.

The method used in these examples is usually called the **objective line method**.

Exercise 6C

Solve graphically the problems in questions 1–5. Draw the feasible region and use the objective line method.

1 Maximise $\quad z = 2x + y$

subject to the constraints:

$$x + y \leqslant 6$$
$$x \leqslant 5$$
$$y \leqslant 4, x \geqslant 0, y \geqslant 0$$

2 Maximise $z = x + 5y$
subject to the constraints:

$$4x + 3y \leqslant 12$$
$$2x + 5y \leqslant 10$$
$$x \geqslant 0$$
$$y \geqslant 0$$

3 Maximise $z = 3x + 2y$
subject to the constraints:

$$y + 2x \leqslant 12$$
$$x \geqslant 2$$
$$y \geqslant 4$$

4 Minimise $z = 4x + y$
subject to the constraints:

$$3x + y \geqslant 6$$
$$x + y \geqslant 4$$
$$x \leqslant 4,\ y \leqslant 6$$

5 Minimise $z = 2x + y$
subject to the constraints:

$$3x + y \geqslant 6$$
$$x + y \geqslant 4$$
$$x \leqslant 3$$
$$y \leqslant 4$$

6.5 Extreme points and optimality

In each of the above examples **the optimal value occurred at a corner (extreme point) of the feasible region**. In example 9 it occurred at two corners and at every point on the line joining them. This is not a coincidence. It can be proved mathematically that:

- **The optimal solution, if it exists, will occur at one or more of the extreme points (vertices) of the feasible region.**

This provides us with an alternative way of finding the optimal value of z and the values of x and y for which it occurs. So for a maximising problem:

1. Having determined the feasible region obtain the coordinates of the vertices of this region.

2. Evaluate the objective function at each of these vertices.
3. The maximum of the values found in step 2 gives the optimal value of z and the coordinates of the corresponding vertex give the values of x and y for which this occurs.

Example 11
Use the extreme point method to obtain the maximum value of

(a) $z = 24x + 16y$
(b) $z = 12x + 24y$

subject to the constraints:

$$2x + 3y \leqslant 24$$
$$2x + y \leqslant 16$$
$$y \leqslant 6$$
$$x \geqslant 0,\ y \geqslant 0$$

The feasible region has been determined in example 9 and is shown below. Label the vertices A, B, C and D as shown.

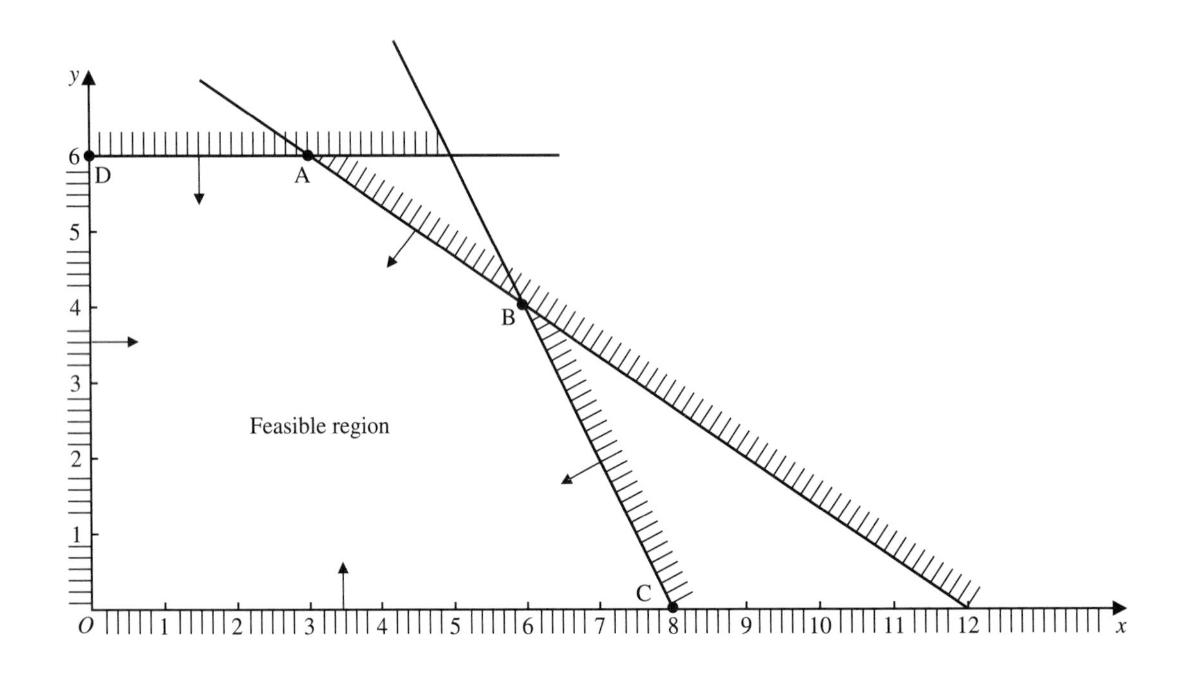

We have previously shown that A has coordinates (3,6) and B has coordinates (6,4).

It is clear from the construction of the feasible region that C has coordinates (8,0) and D has coordinates (0,6). The origin O has coordinates (0,0).

(a) Which of these vertices gives the maximum value of $z = 24x + 16y$? Substituting the coordinates of each point in turn in the equation gives:

$z_O = 0,\ z_A = 24(3) + 16(6) = 168$

$z_B = 24(6) + 16(4) = 208,\ z_C = 24(8) + 16(0) = 192$

$z_D = 24(0) + 16(6) = 96$

So the maximum value occurs at B where $x = 6,\ y = 4$ and $z = 208$ as found previously.

(b) Now try the coordinates of each vertex in turn in the equation $z = 12x + 24y$.

$z_O = 0,\ z_A = 12(3) + 24(6) = 180,\ z_B = 12(6) + 24(4) = 168,$

$z_C = 12(8) + 24(0) = 96,\ z_D = 12(0) + 24(6) = 144$

So the maximum value occurs at A where $x = 3,\ y = 6$ and $z = 180$ as found previously.

Exercise 6D

For the problems given in Exercise 6C:

(a) Determine the coordinates of the vertices of the feasible region.

(b) Evaluate the objective function at each of these vertices.

(c) Hence solve the linear programming problem and confirm your previous answers.

6.6 Integer valued solutions

All the examples considered so far have produced solutions for which the values of the decision variables have been integers. This is not true of all linear programming problems, in fact it is rare.

For some contexts solutions with non-integer values are acceptable. For example, in an investment problem if £x and £y are the investments in two options a solution of $x = \frac{4}{5}$ and $y = 1\frac{4}{5}$ is acceptable since £'s are subdivisible. However in other contexts non-integer solutions are not acceptable. In example 3 (page 163), involving large and small vans, the solution must clearly involve integer numbers of large and small vans.

In the next example we will show how to proceed when the additional constraint *decision variables must be integers* is included.

Example 12

A manufacturer makes two kinds of toys, A and B. The machine time and the craftsman's time required for each are shown in the table, together with the limitations on time per week and the profit on each.

	Machine time (h)	Craftsman's time (h)	Profit (£)
Toy A	3	2	10
Toy B	3	4	12
Time available (h)	40	50	

The manufacturer wishes to maximise profit. In this context the numbers of A and B made must be integers.

Let x be number of toy A made
y be number of toy B made.

Then you have:

Machine time:	$3x+3y \leqslant 40$	constraint (i)
Craftsman's time:	$2x+4y \leqslant 50$	constraint (ii)
Profit:	$z=10x+12y$	
Non-negativity condition:	$x \geqslant 0, y \geqslant 0$	constraint (iii)
Integer condition:	x and y must be integers	constraint (iv)

The problem is to maximise the profit $z = 10x + 12y$ subject to the constraints.

First draw the feasible region:

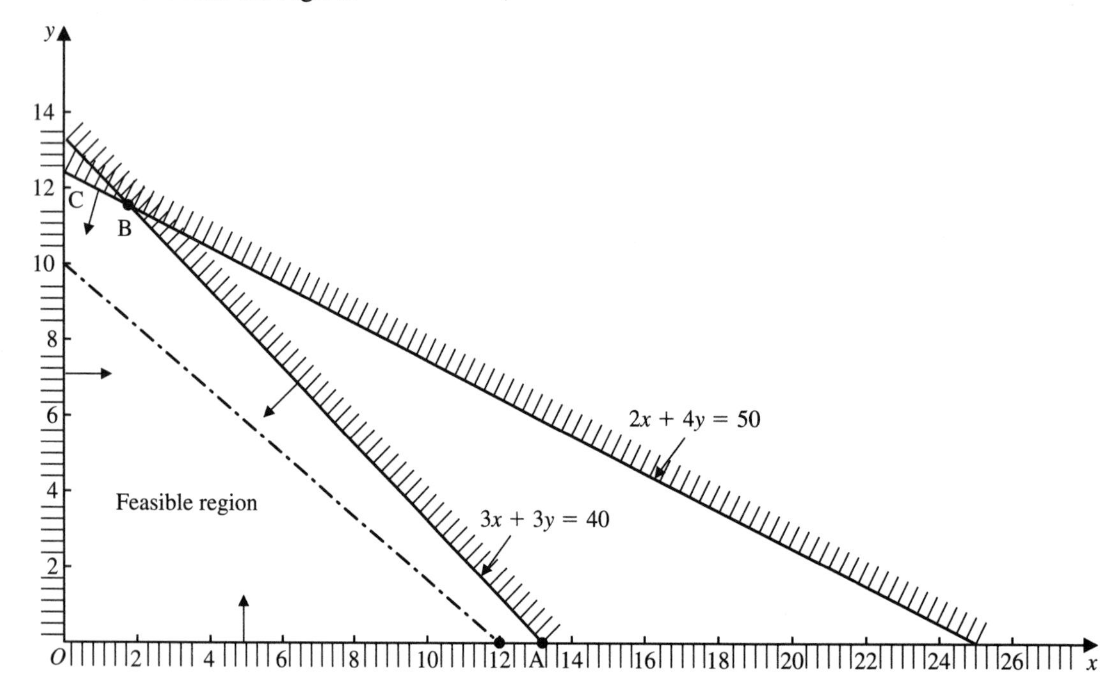

(i) $3x + 3y = 40$ passes through $(13\frac{1}{3},0)$, $(0,13\frac{1}{3})$. The origin lies in the admissible set for the inequality $3x + 3y \leqslant 40$.
(ii) $2x + 4y = 50$ passes through $(0,12\frac{1}{2})$, $(25,0)$. The origin lies in the admissible set for the inequality $2x + 4y \leqslant 50$.

The problem can now be solved using either the ruler method or the extreme point method.

Method 1 – using a ruler Draw an arbitrary member of the family of lines $10x + 12y = \text{constant}$.

The line $10x + 12y = 120$ is drawn in the diagram as –·–·. If you move a ruler parallel to this you will find the optimal value at the point B. The coordinates of B are obtained by solving the simultaneous equations

$$3x + 3y = 40$$
$$2x + 4y = 50$$

Multiplying the first equation by 2 gives $6x + 6y = 80$.

Multiplying the second equation by 3 gives $6x + 12y = 150$.

Subtracting: $6y = 70$

$$\Rightarrow \quad y = 11\tfrac{2}{3}$$

Substituting into the first equation:

$$3x + 3(11\tfrac{2}{3}) = 40$$
$$3x = 40 - 35 = 5$$
$$\Rightarrow \quad x = 1\tfrac{2}{3} \quad \text{and} \quad z = 10(1\tfrac{2}{3}) + 12(11\tfrac{2}{3})$$

At $B(1\frac{2}{3},11\frac{2}{3})$ the value of z is $156\frac{2}{3}$.

Method 2 – extreme points The coordinates of the vertices (extreme points) of the feasible region are:

B $(1\frac{2}{3}, 11\frac{2}{3})$ – from above
O (0, 0) – origin
A $(13\frac{1}{3}, 0)$ – from (i) above
C $(0, 12\frac{1}{2})$ – from (ii) above

The values of $z = 10x + 12y$ at B, O, A and C are:

$$z_B = 10(1\tfrac{2}{3}) + 12(11\tfrac{2}{3}) = 156\tfrac{2}{3}$$
$$z_O = 10(0) + 12(0) = 0$$
$$z_A = 10(13\tfrac{1}{3}) + 12(0) = 133\tfrac{1}{3}$$
$$z_C = 10(0) + 12(12\tfrac{1}{2}) = 150$$

So the maximum occurs at $B(1\frac{2}{3},11\frac{2}{3})$ and $z_{max} = 156\frac{2}{3}$.

Now let's consider how to apply the **integer constraint**. The solution obtained above is clearly not an acceptable one as $1\frac{2}{3}$ of a toy is not a credible answer.

You may be tempted just to take the integer part of the solutions but, as you will see below, this does not give the optimal solution. If the feasible region is small then it is possible to find z for each point (x, y) in the feasible region for which both x and y are integers. In our case there are many such points and this is therefore not a sensible approach.

Here is a systematic procedure which, for many situations, leads to the optimal value with integer values of the decision variables.

First find the points close to the optimal point which have integer coordinates and are in the feasible region. Then calculate z for each of these points and choose the maximum.

In this case, since $x = 1\frac{2}{3}$, $y = 11\frac{2}{3}$ at the optimal point, consider the lines $x = 1$ and $x = 2$ and the lines $y = 11$ and $y = 12$:

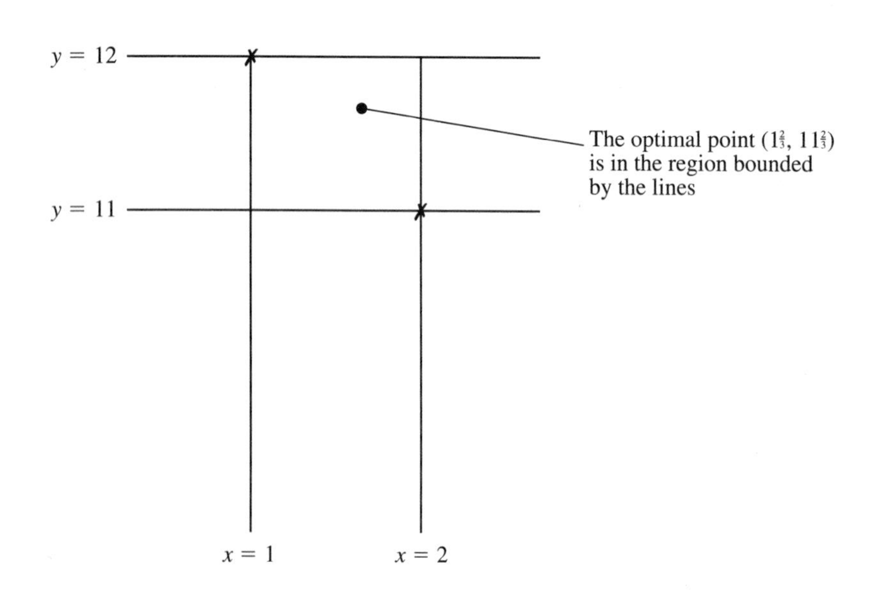

$x = 1$: When $x = 1$

Constraint (i) gives $3(1) + 3y \leqslant 40$

that is: $3y \leqslant 37$ or $y \leqslant 12\frac{1}{3}$

Constraint (ii) gives $2(1) + 4y \leqslant 50$

that is: $4y \leqslant 48$ or $y \leqslant 12$

For integral values of y, then, you require $y \leqslant 12$.

So $(1, 12)$ does lie in feasible region and

$$z = 10(1) + 12(12) = 154$$

$x = 2$: When $x = 2$

Constraint (i) gives $3(2) + 3y \leqslant 40$

that is: $3y \leqslant 34$ or $y \leqslant 11\frac{1}{3}$

Constraint (ii) gives $2(2) + 4y \leqslant 50$

That is: $4y \leqslant 46$ or $y \leqslant 11\frac{1}{2}$

For integral values of y, then, you require $y \leqslant 11$.

So (2, 11) does lie in feasible region and

$$z = 10(2) + 12(11) = 152$$

The point (1, 11) also satisfies constraint (i) and the value of z there is $z = 10(1) + 12(11) = 142$.

From these calculations it is clear that as your ruler moves out from the origin it first reaches the point (1,11), where z is 142, then the point (2,11) where z is 152 and then the point (1,12) where z attains its maximum value for integer solutions of 154. Notice that (1,11), obtained by rounding down each of the coordinates, *does not give the optimal solution.*

Exercise 6E

Solve graphically the problems in questions 1–3. Draw the feasible region and use (a) the objective line method (b) the extreme point method.

1 Maximise $z = 2x + 3y$

subject to the constraints:

$$3x + y \leqslant 30$$
$$x + 2y \leqslant 30$$
$$x \geqslant 0, y \geqslant 0$$

2 Minimise $w = 12u + 10v$

subject to the constraints:

$$2u + 3v \geqslant 16$$
$$4u + 2v \geqslant 24$$
$$u \geqslant 0, v \geqslant 0$$

3 Maximise $z = 16x + 12y$

subject to the constraints:

$$3x + 4y \leqslant 84$$
$$4x + 3y \leqslant 84$$
$$x \leqslant 15$$
$$x \geqslant 0, y \geqslant 0$$

4 The variables x and y are subject to the constraints

$$3x + 4y \leqslant 84$$
$$4x + 3y \leqslant 84$$
$$x \leqslant 15$$
$$x \geqslant 0,\ y \geqslant 0$$

as in question 3. Use graphical methods, both objective line and extreme point methods, to determine the maximum value of

(a) $z = 10x + 10y$

(b) $z = 30x + 15y$.

5 Maximise $z = x + y$

subject to the constraints:

$$3x + 4y \leqslant 12$$
$$2x + y \leqslant 4$$
$$x \geqslant 0,\ y \geqslant 0$$

(a) Solve this linear programming problem.

(b) Identify all the points in the feasible region which have integer coordinates. Which ones give the maximum value of z?

(c) Proceeding from your answer to part (a) find the solution to the linear programming problem when the integer constraint is added.

6 A builder has a plot of land available on which he can build houses. He can build either luxury houses or standard houses. He decides to built at least 5 luxury houses and at least 10 standard houses. Planning regulations prevent him from building more than 30 houses altogether. Each luxury house requires 300 m^2 of land and each standard house requires 150 m^2 of land. The total area of the plot is 6000 m^2. The builder makes a profit of £12 000 on each luxury house and a profit of £8000 on each standard house. The builder wishes to make the maximum possible profit.

(a) Formulate the above problem as a linear programming problem.

(b) Determine how many of each type of house he should build to make the maximum profit and state what this profit is.

7 Example 3 on page 163 is about the KJB Haulage company. Solve the linear programming problem formulated there to obtain the number of large and small vans to be used to minimise the cost.

8 Solve the linear programming problem formulated in example 4 (page 164) to determine how much the lecturer should invest in each option to maximise his yield.

6.7 The algebraic method for solving linear programming problems

The graphical method for solving linear programming problems discussed above obviously has limitations since it requires the drawing of the feasible region. This is only possible when you have just two decision variables. For more than two decision variables you need an algebraic method. Such a method has been developed which relies on the fact stated earlier that 'the optimal solution may be found by examining the value of the objective function at the extreme points of the feasible region'.

Slack variables

Before we discuss this method let's consider another way of looking at the graphical methods discussed above. Example 1 (page 160) produced the following linear programming problem:

Maximise $z = 10x + 12y$
subject to the constraints:

$$3x + 3y \leqslant 120 \quad \text{(i)}$$
$$2x + 4y \leqslant 150 \quad \text{(ii)}$$
$$x \geqslant 0,\ y \geqslant 0$$

You can easily construct the feasible region:

(i) the line $3x + 3y = 120$ passes through $(0, 40)$ and $(40, 0)$ and the origin does lie in the admissible set for the inequality $3x + 3y \leqslant 120$.

(ii) the line $2x + 4y = 150$ passes through $\left(0, 37\frac{1}{2}\right)$ and $(75, 0)$ and the origin does lie in the admissible set for the inequality $2x + 4y \leqslant 150$.

Here is the feasible region:

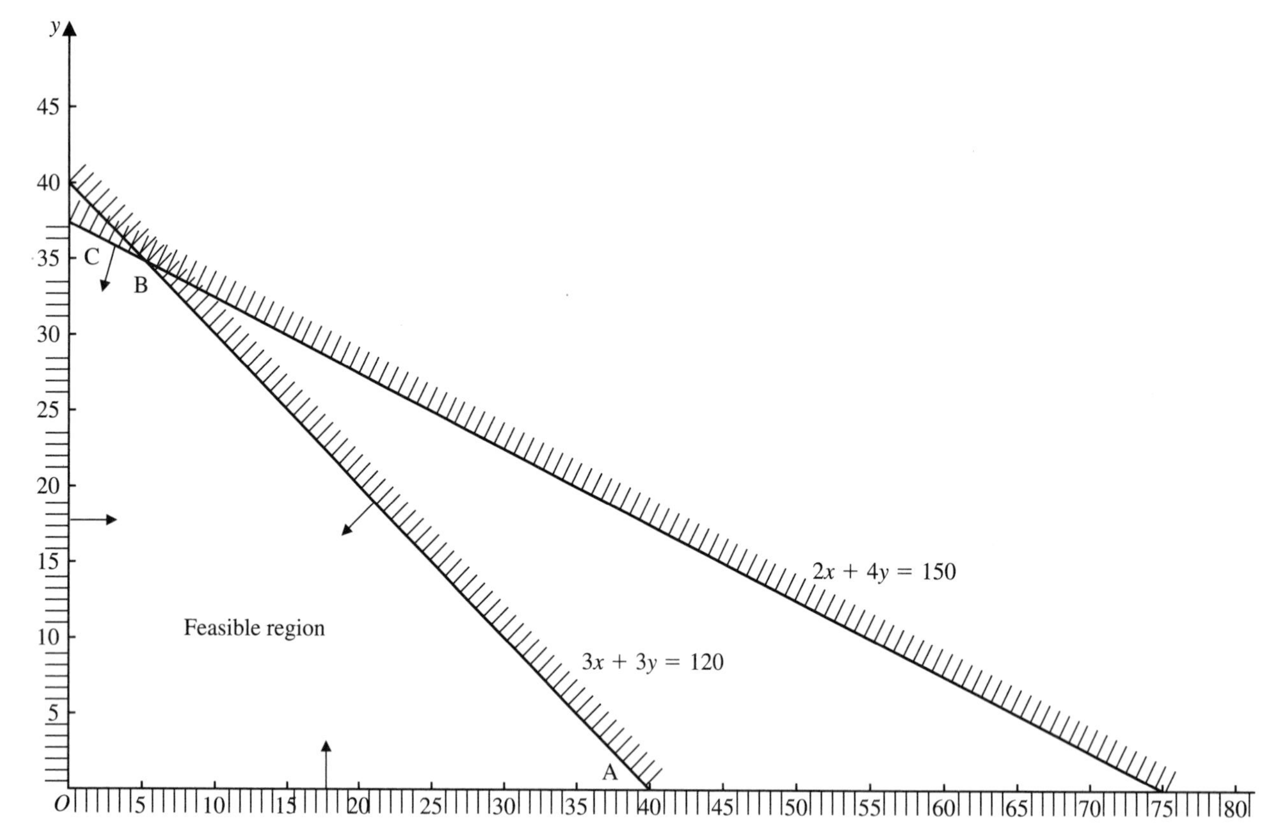

It is convenient for present purposes, and also for future work, to convert inequalities (i) and (ii) into equations. There are standard methods for dealing with systems of equations. If you replace inequality (i) by the equation

$$3x + 3y + s = 120$$

then

$$s = 120 - (3x + 3y)$$

and from inequality (i) this means $s \geqslant 0$.

Similarly, the inequality (ii) is replaced by the equation

$$2x + 4y + t = 150$$

Then

$$t = 150 - (2x + 4y)$$

and from inequality (ii) this means $t \geqslant 0$.

In terms of x, y, s and t you now have:

Equation of OA is $y = 0$

Equation of AB is $s = 0$

Equation of BC is $t = 0$

Equation of OC is $x = 0$

and therefore at O: $x = 0, \quad y = 0$

at A: $y = 0, \quad s = 0$

at B: $s = 0, \quad t = 0$

at C: $x = 0, \quad t = 0$

- **So at a vertex of the feasible region two of the variables x, y, s and t are zero.**

The variable s is called the **slack variable** for inequality (i). It gives the difference between the amount of resource available (120) and the amount used $(3x + 3y)$.

Similarly the variable t is called the **slack variable** for inequality (ii). It gives the difference between the amount of resource available (150) and the amount used $(2x + 4y)$.

The coordinates of the vertices of the feasible region are now easy to handle. It makes the process of considering extreme points very straightforward. These ideas are easily generalised to problems involving more than two decision variables.

6.8 Basic solutions

When we converted the inequalities (i) and (ii) in the previous problem, we obtained the two equations

$$3x + 3y + s = 120$$

$$2x + 4y + t = 150$$

These two equations contain four unknowns: x, y, s and t. In this case,

$$(\text{number of variables}) - (\text{number of equations}) = 4 - 2 = 2$$

If you take this number (that is, 2, in this case) of the variables and set them to zero, then solve the equations, the solution you obtain is called a **basic solution**.

The two variables set to zero are called **non-basic variables**.

The variables solved for are called **basic variables**.

A basic solution which is also a feasible solution is called a **basic feasible solution**. The table overleaf summarises the position for this problem. The equations have been solved after setting two variables at a time to zero.

Vertex	x	y	s	t	Type of solution	Value of $z = 10x + 12y$	(x, y)
O	0	0	120	150	basic feasible	0	(0, 0)
C	0	$37\frac{1}{2}$	$7\frac{1}{2}$	0	basic feasible	450	$(0, 37\frac{1}{2})$
B	5	35	0	0	basic feasible	470	(5, 35)
	0	40	0	−10	basic non-feasible	—	(0, 40)
A	40	0	0	70	basic feasible	400	(40, 0)
	75	0	−75	0	basic non-feasible	—	(75, 0)

6.9 Simplex method

The algebraic method for solving linear programming problems is called the **simplex method** and was developed by George Dantzig in 1947. The **simplex algorithm** which is used consists of two steps:

1. A way of finding out whether a given solution, corresponding to an extreme point of the feasible region, is an optimal solution.
2. A way of obtaining an adjacent extreme point with a larger value for the objective function.

It is useful at this point to define a **standard form** for a linear programming problem. We shall say that

- **a linear programming problem is in standard form if**
  (a) the objective function $\alpha x + \beta y + \gamma z$ is to be maximised
  (b) all the constraints, other than the non-negativity conditions, are of the form

$$ax + by + cz \leqslant d$$

A problem which is not in standard form may be rewritten in standard form using the following observation:

$$\text{Minimise}\,(\alpha x + \beta y + \gamma z) = -\text{Maximise}\,(-\alpha x - \beta y - \gamma z)$$

That is, to minimise the objective function you could maximise its negative instead.

The calculations involved in using the simplex algorithm are recorded in a sequence of tables which are known as **simplex tableaux.** The method will be illustrated by solving example 1 using the simplex algorithm.

To use the simplex method as we shall describe it, the problem must be written in standard form.

Example 13

Use the simplex method to solve example 1.

The **first step** is to **add slack variables to the inequalities**, other than the non-negativity conditions, as described earlier.

For example 1:

$$3x + 3y \leqslant 120 \Rightarrow 3x + 3y + s = 120$$
$$2x + 4y \leqslant 150 \Rightarrow 2x + 4y + t = 150$$

The **second step** is to **rewrite the linear programming problem so that each equation contains all variables x, y, s, t.**

The problem now becomes:
Maximise $z = 10x + 12y + 0s + 0t$
subject to the constraints:

$$3x + 3y + 1s + 0t = 120$$
$$2x + 4y + 0s + 1t = 150$$
$$x \geqslant 0, y \geqslant 0, s \geqslant 0, t \geqslant 0$$

The **third step** is to put all this information in the form of a table which is known as the **initial tableau**. In order to do this, write the objective function in the form

$$z - 10x - 12y - 0s - 0t = 0$$

Now draw up this table:

Tableau 1

Basic variable	x	y	s	t	Value	
s	3	3	1	0	120	
t	2	4	0	1	150	
z	−10	−12	0	0	0	← objective row
		↑				

The first row corresponds to the first constraint.
The second row corresponds to the second constraint.
The third row corresponds to the objective function.

In drawing up this table we have taken as the initial basic feasible solution $x = 0$, $y = 0$, $s = 120$ and $t = 150$.

So s and t are basic variables and x and y are non-basic variables. Since s appears only in the first inequality and its value is obtained from this, it appears as the basic variable for that row. For a similar reason t appears at the beginning of row 2.

Notice that basic variables, s and t, each appear in *only one row* and they appear there with a coefficient of 1.

The column which the basic variable labels has all zeros except for the 1 in the row which it labels.

To interpret the solution that corresponds to this tableau, look at the first and last columns.

$$s = 120,\ t = 150 \Rightarrow x = y = 0 \quad \text{and} \quad z = 0$$

Note: An equivalent form of the tableau above, which will be used in London D1 examinations, is

z	x	y	s	t	Value	
1	−10	−12	0	0	0	← objective function
0	3	3	1	0	120	← first constraint
0	2	4	0	1	150	← second constraint

The numbers in the tableau now simply correspond to the coefficients of each variable.

The **optimality condition** states:

- **If the objective row of a tableau has zero entries in the columns labelled by basic variables and no negative entries in the columns labelled by non-basic variables then the solution represented by the tableau is optimal.**

Clearly this condition is not satisfied by Tableau 1 and so we do not have an optimal solution. We must therefore construct a new tableau corresponding to an adjacent extreme point. We will describe the steps for doing this without going into the theoretical details.

The first step in the process is to choose the non-basic variable which is to become a basic variable. (This is often called **entering the basis.**) The most common rule for selecting this variable is to select the variable with the *most negative entry* in the objective function row. In our case this is (-12) and the corresponding variable is y. We usually indicate this with a vertical arrow as we have done in Tableau 1.

The new basic variable is called the **entering variable** and the column it is in is called the **pivotal column**.

The second step in the process is to choose *which variable is to leave the basis* (the **leaving variable**). In order to do this we calculate **θ-values.** These are calculated for each row, other than the objective row. You find the θ-value for a row by dividing the entry in the value column, which must always be positive, by the entry in the pivotal column. Only θ-values for rows in which the entries in the pivotal column are *positive* are used.

From Tableau 1 you get:

Row (i) $\theta = \dfrac{120}{3} = 40$ Row (ii) $\theta = \dfrac{150}{4} = 37\frac{1}{2}$

The row with the smallest θ-value is called the **pivotal row**. Here, the pivotal row is row (ii).

The entry at the intersection of the pivotal row and the pivotal column is called the **pivot**. It is usually ringed.

In practice all the above information is included in Tableau 1:

Basic variable	x	y	s	t	Value	
s	3	3	1	0	120	$\theta = \frac{120}{3} = 40$
t	2	④	0	1	150	$\theta = \frac{150}{4} = 37\frac{1}{2}$ ← pivotal row
z	−10	−12	0	0	0	

pivot (④) ↑ pivotal column (y)

Forming a new tableau

The first step in forming a new tableau is to divide the pivotal row by the pivot so that the pivot now becomes 1:

Basic variable	x	y	s	t	Value
s	3	3	1	0	120
t	$\frac{1}{2}$	①	0	$\frac{1}{4}$	$37\frac{1}{2}$
z	−10	−12	0	0	0

The second step is to add suitable multiples of the new pivotal row to all other rows, including the objective row, so that all other elements in the pivotal column become zero.

1st row To get the required zero in the y column, take row (i) − 3 × row (ii):

x	y	s	t	Value
$3 - 1\frac{1}{2}$ $= 1\frac{1}{2}$	$3 - 3$ $= 0$	$1 - 0$ $= 1$	$0 - \frac{3}{4}$ $= -\frac{3}{4}$	$120 - 3 \times 37\frac{1}{2}$ $= 7\frac{1}{2}$

3rd row To get the required zero in the y column, take row (iii) + 12 × row (ii):

x	y	s	t	Value
$-10 + 6$ $= -4$	$-12 + 12$ $= 0$	$0 + 0$ $= 0$	$0 + 3$ $= 3$	$0 + 450$ $= 450$

To complete the new tableau, replace the label on the pivotal row by the entering variable (in this case, y).

The second tableau is then

Tableau 2

Basic variable	x	y	s	t	Value	
s	$(1\frac{1}{2})$	0	1	$-\frac{3}{4}$	$7\frac{1}{2}$	←
(y)	$\frac{1}{2}$	1	0	$\frac{1}{4}$	$37\frac{1}{2}$	
z	-4	0	0	3	450	
	↑					

This tableau corresponds to the solution $s = 7\frac{1}{2}$, $y = 37\frac{1}{2}$, $x = 0$, $t = 0$ with $z = 450$.

If you apply the optimality condition to this tableau you will see that it is not optimal – the objective row (z-row) contains a negative entry. You must therefore form a new tableau by pivoting.

Step 1 Choose the pivotal column. As there is only one negative entry (-4) in the z-row, choose the corresponding variable x as the entering variable. The pivotal column is indicated by the arrow ↑.

Step 2 Calculate θ-values:

Row (i) $\frac{(7\frac{1}{2})}{(1\frac{1}{2})} = 5$

Row (ii) $\frac{(37\frac{1}{2})}{(\frac{1}{2})} = 75$

The θ-value for row (i) is the smaller and the pivotal row is therefore row (i), shown thus ←. The **leaving variable** is therefore s. The pivot is therefore $1\frac{1}{2}$ at the intersection of the pivotal row and pivotal column. It is shown ringed.

Step 3 Divide the pivotal row by the pivot:

Basic variable	x	y	s	t	Value
s	(1)	0	$\frac{2}{3}$	$-\frac{1}{2}$	5
y	$\frac{1}{2}$	1	0	$\frac{1}{4}$	$37\frac{1}{2}$
z	-4	0	0	3	450

Step 4 Add multiples of the new pivotal row to all other rows so that all other elements in the pivotal column become zero.

Tableau 3

Basic variable	x	y	s	t	Value	
ⓧ	1	0	$\frac{2}{3}$	$-\frac{1}{2}$	5	
y	0	1	$-\frac{1}{3}$	$\frac{1}{2}$	35	$\leftarrow$ row (ii) $-\frac{1}{2}$ row (i)
z	0	0	$\frac{8}{3}$	1	470	$\leftarrow$ row (iii) $+ 4$ row (i)

This tableau corresponds to the solution

$$x = 5, y = 35, s = 0, t = 0 \quad \text{and} \quad z = 470$$

Since there are no negative entries in the z-row this solution is optimal. This final tableau is called the **optimal tableau**.

The simplex method can be readily applied to problems when there are more than two decision variables, as is illustrated by the following example.

Example 14

Use the simplex method to solve the linear programming problem:

Maximise $P = 8x + 9y + 5z$
subject to the constraints:

$$2x + 3y + 4z \leqslant 3$$
$$6x + 6y + 2z \leqslant 8$$
$$x \geqslant 0, y \geqslant 0, z \geqslant 0$$

The problem is in standard form.
First introduce slack variables s and t into the constraints to produce the required form. The problem then becomes

Maximise $P = 8x + 9y + 5z + 0s + 0t$
subject to the constraints:

$$2x + 3y + 4z + s = 3$$
$$6x + 6y + 2z + t = 8$$
$$x \geqslant 0, y \geqslant 0, z \geqslant 0, s \geqslant 0, t \geqslant 0$$

A basic solution in this case has $(5 - 2) = 3$ variables zero.

The initial tableau is then

Tableau 1

Basic variable	x	y	z	s	t	Value	
s	2	③	4	1	0	3	$\theta = \frac{3}{3} = 1 \leftarrow$
t	6	6	2	0	1	8	$\theta = \frac{8}{6} = 1\frac{1}{3}$
P	−8	−9	−5	0	0	0	
		$\uparrow$					

This corresponds to $s = 3,\ t = 8,\ x = 0,\ y = 0,\ z = 0,\ P = 0$. The pivotal column, the θ-values giving the pivotal row and the pivot are indicated on this initial tableau. The entering variable is y and the leaving variable is s.

Dividing the pivotal row by the pivot gives

Basic variable	x	y	z	s	t	Value
s	$\frac{2}{3}$	①	$\frac{4}{3}$	$\frac{1}{3}$	0	1
t	6	6	2	0	1	8
P	−8	−9	−5	0	0	0

The second tableau is then

Tableau 2

Basic variable	x	y	z	s	t	Value	
ⓨ	$\frac{2}{3}$	1	$\frac{4}{3}$	$\frac{1}{3}$	0	1	
t	②	0	−6	−2	1	2	row (ii) − 6 row (i) ←
P	−2	0	7	3	0	9	row (iii) + 9 row (i)
	↑						

This corresponds to the solution

$$y = 1,\ t = 2,\ x = 0,\ z = 0,\ s = 0 \quad \text{and} \quad P = 9$$

It is not optimal as there is a negative entry in the objective row. We therefore form a new tableau.

The variable to enter the basis is now x. The θ values are

row (i) $\dfrac{1}{\left(\frac{2}{3}\right)} = \frac{3}{2}$

row (ii) $\frac{2}{2} = 1$

and so row (ii) is the pivotal row and the ringed ② is the pivot. The leaving variable is t.

Dividing the pivotal row by the pivot gives

Basic variable	x	y	z	s	t	Value
y	$\frac{2}{3}$	1	$\frac{4}{3}$	$\frac{1}{3}$	0	1
t	①	0	−3	−1	$\frac{1}{2}$	1
P	−2	0	7	3	0	9

Adding multiples of the new pivotal row to all other rows so that all other elements in the pivotal column become zero gives the tableau:

Tableau 3

Basic variable	x	y	z	s	t	Value	
y	0	1	$\frac{10}{3}$	1	$-\frac{1}{3}$	$\frac{1}{3}$	row (i) $-\frac{2}{3}$ row (ii)
ⓧ	1	0	-3	-1	$\frac{1}{2}$	1	
P	0	0	1	1	1	11	row (iii) + 2 row (ii)

This corresponds to the solution

$$y = \tfrac{1}{3},\ x = 1,\ z = 0,\ s = 0,\ t = 0 \quad \text{and} \quad P = 11$$

It is the optimal solution as there are no negative entries in the objective row.

Example 15

As a final example, and for comparison, apply the simplex method to example 9:

Maximise $z = 16x + 24y$
subject to the constraints:

$$2x + 3y \leqslant 24$$
$$2x + y \leqslant 16$$
$$y \leqslant 6$$
$$x \geqslant 0,\ y \geqslant 0$$

As there are three constraints, other than the non-negativity condition, we need three slack variables s, t, u. Introducing these gives:

Maximise $z = 16x + 24y + 0s + 0t + 0u$
subject to the constraints:

$$2x + 3y + s = 24$$
$$2x + y + t = 16$$
$$y + u = 6$$
$$x \geqslant 0,\ y \geqslant 0,\ s \geqslant 0,\ t \geqslant 0,\ u \geqslant 0$$

The initial tableau is:

Basic variable	x	y	s	t	u	Value	
s	2	3	1	0	0	24	$\theta = \frac{24}{3} = 8$
t	2	1	0	1	0	16	$\theta = \frac{16}{1} = 16$
u	0	①	0	0	1	6	$\theta = \frac{6}{1} = 6 \leftarrow$
z	-16	-24	0	0	0	0	
		↑					

The entering variable is y and the leaving variable is u. This corresponds to the solution $s = 24$, $t = 16$, $u = 6$, $x = 0$, $y = 0$ and $z = 0$.

Since the pivot is 1 you can proceed immediately to produce the next tableau by adding suitable multiples of row (iii) to the others:

Basic variable	x	y	s	t	u	Value	
s	②	0	1	0	−3	6	row (i) − 3 row (iii) ←
t	2	0	0	1	−1	10	row (ii) − row (iii)
ⓨ	0	1	0	0	1	6	
z	−16	0	0	0	24	144	row (iv) + 24 row (iii)
	↑						

The θ values are

$$\text{row (i) } \tfrac{6}{2} = 3$$

$$\text{row (ii) } \tfrac{10}{2} = 5$$

and so the entering variable is x and the leaving variable is s, the pivot being the ringed ②.

Dividing the pivotal row by the pivot gives

Basic variable	x	y	s	t	u	Value
s	①	0	$\frac{1}{2}$	0	$-\frac{3}{2}$	3
t	2	0	0	1	−1	10
y	0	1	0	0	1	6
z	−16	0	0	0	24	144

You can now produce the next tableau by proceeding in the usual way:

Basic variable	x	y	s	t	u	Value	
ⓧ	1	0	$\frac{1}{2}$	0	$-\frac{3}{2}$	3	
t	0	0	−1	1	2	4	row (ii) − 2 row (i)
y	0	1	0	0	1	6	row (iii) + 0 row (i)
z	0	0	8	0	0	192	row (iv) + 16 row (i)

(Notice that row (iii) remains unchanged as there is already a zero in the pivot column in this row.)

This tableau corresponds to the solution

$$x = 3,\ t = 4,\ y = 6 \quad \text{and} \quad z = 192$$

Compare this with the previous results (on page 179). Notice that in addition to telling us that the maximum value of z occurs when $x = 3$ and $y = 6$ it also tells us that there is a non-zero slack of 4 in the second inequality. This corresponds to unused time or material, depending on the context.

Exercise 6F

Use the simplex algorithm to solve the following linear programming problems.

1 Maximise $z = 2x + 3y$
subject to the constraints:

$$3x + y \leqslant 30$$
$$x + 2y \leqslant 30$$
$$x \geqslant 0,\ y \geqslant 0$$

2 Maximise $z = 30x + 15y$
subject to the constraints:

$$3x + 4y \leqslant 84$$
$$4x + 3y \leqslant 84$$
$$x \leqslant 15$$
$$x \geqslant 0,\ y \geqslant 0$$

3 Maximise $P = 3x + 6y + 2z$
subject to the constraints:

$$3x + 4y + z \leqslant 20$$
$$x + 3y + 2z \leqslant 10$$
$$x \geqslant 0,\ y \geqslant 0,\ z \geqslant 0$$

4 A company manufactures two kinds of cloth A and B and uses three different colours of wool. The material required to make a unit length of each type of cloth and the total amount of wool of each colour that is available are shown in the table.

Requirements for unit length of cloth of type		Colour of wool	Wool available (kg)
A (kg)	B (kg)		
4	1	Red	56
5	3	Green	105
1	2	Blue	56

The manufacturer makes a profit of £12 on a unit length of cloth A and a profit of £15 on a unit length of cloth B. How should he use the available material so as to make the largest possible profit?

(a) Formulate this as a linear programming problem.

(b) Solve the resulting problem by using the simplex algorithm.

(c) Confirm your answer to (b) by solving the problem graphically.

SUMMARY OF KEY POINTS

1 Any pair of values of x and y which satisfy all the constraints in a linear programming problem is called a feasible solution.

2 The region which contains all feasible solutions is called the feasible region.

3 The optimal solution of a linear programming problem, if it exists, will occur at one or more of the extreme points (vertices) of the feasible region.

4 The simplex method is an algebraic method for solving linear programming problems.

(i) The column which contains the entering variable is called the pivotal column.

(ii) The row with the smallest θ-value is called the pivotal row.

(iii) The entry at the intersection of the pivotal row and the pivotal column is called the pivot.

5 Optimality condition:

If the objective row of a tableau has zero entries in the columns labelled by basic variables and no negative entries in the columns labelled by non-basic variables then the solution represented by the tableau is optimal.

Matchings

7

The matching problem occurs in many situations, such as matching staff to classes in a school timetable, and matching classes to rooms. The problem is how to match resources (such as people with different skills) to tasks (which have differing skill requirements).

This chapter shows you how to find a match, and how to use graph theory to improve it if it is not the best match. But first we need to look at graphs in which the vertices are split into two sets, and in which the edges run from one set to the other.

7.1 Modelling using a bipartite graph

Chapter 5 (page 137) introduces the concept of a **cut**. A cut is a way of partitioning a set. It divides the members of the set into two subsets. Every member of the set is in one or other of the subsets. No member is in both subsets: the subsets are mutually exclusive and exhaustive.

A partition divides the members of the set of graph vertices into two subsets:

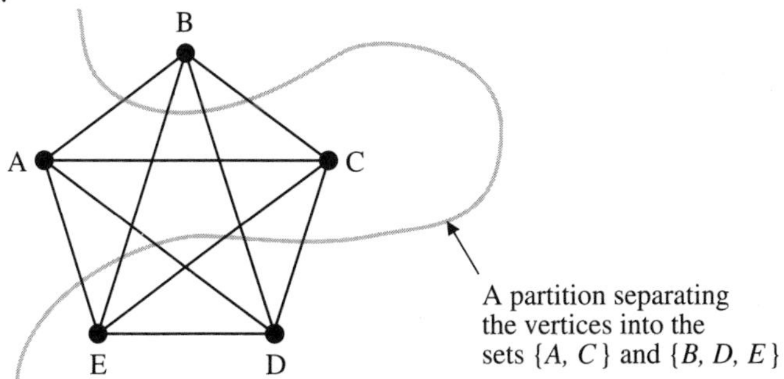

Graphs in which the vertices are partitioned, and in which the edges always run from one of the vertex sets to the other, are often useful in modelling.

- **A bipartite graph is one in which the vertices are partitioned in some way, and in which all edges connect a vertex in one of the two subsets to a vertex in the other subset.**

Here is an example of a bipartite graph.

The two vertex subsets are {A, B, C} and {W, X, Y, Z}. Notice that there are no edges connecting any of A, B or C, and likewise none connecting any of W, X, Y or Z.

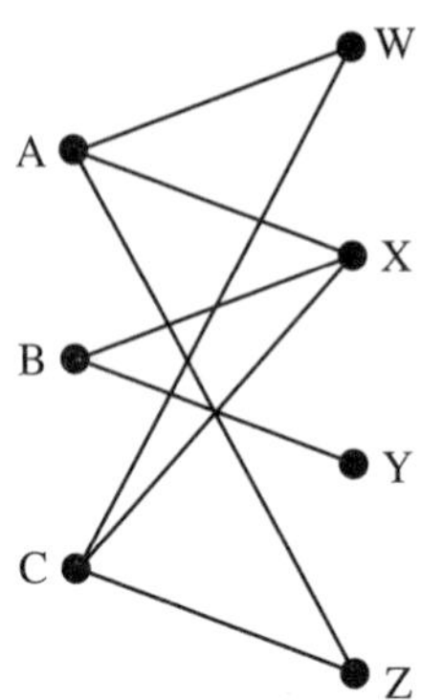

Example 1

Five teachers are available to teach mathematics, statistics, science and economics to a class. The teachers and the subjects which they are qualified to teach are as follows:

Mrs A:	mathematics, statistics and economics
Mr B:	statistics and economics
Miss C:	statistics and science
Mr D:	mathematics and science
Mr E:	science

How can we match the teachers to the class so that each subject is taught by just one teacher?

The information about teachers and subjects may be represented by a bipartite graph with the teachers as one set of vertices and the subjects as the other. The edges show which subject can be taught by which teacher:

The problem is to find a suitable subgraph which is a matching. Furthermore you want your matching to have four edges – all of the subjects have to be taught!

Notice that you will not be able to use one of the teachers.

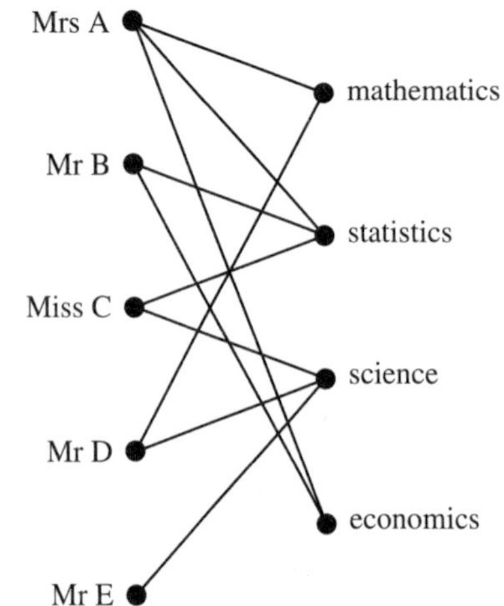

Example 2

A party has been arranged and the guests have been asked to state their dietary restrictions. Here are some of their replies:

Joan won't eat meat.
John eats only white meat and potato products.
Jane eats anything.
Jack doesn't eat pasta.

The dishes will include:

Beef sandwiches; a green salad; a pasta salad; potato crisps; chicken drumsticks

Draw bipartite graphs to represent:

(a) what the guests will eat
(b) what they won't eat.

(a) This bipartite graph represents what the guests *will* eat:

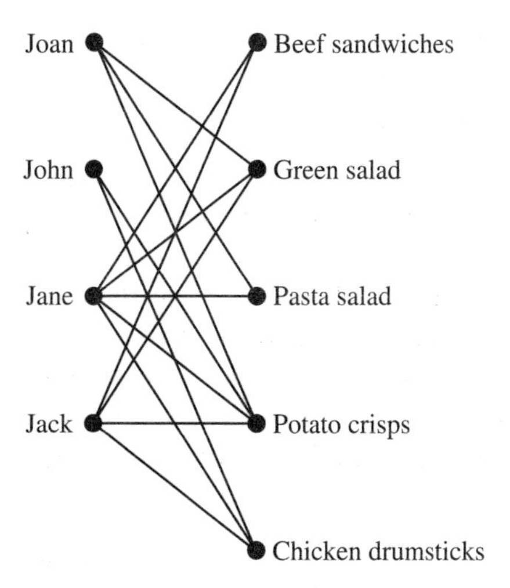

(b) This bipartite graph represents what the guests *won't* eat:

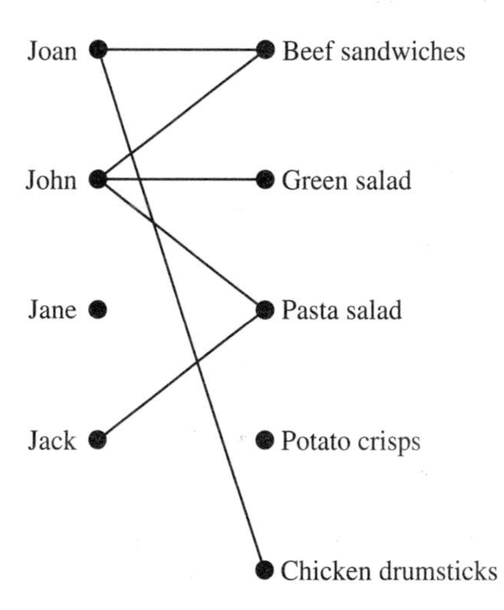

Example 3

At a party some handshaking takes place. Some people shake hands with many other people, some shake hands with just a few others, and there may be some people who shake hands with no-one. (But no-one shakes hands with another person more than once, nor with him/her self!)

Use a bipartite graph to model this situation, and show that there must be two people at the party who shake hands with the same number of people.

Here is a description of the graph that can help solve the problem:

First vertex set:	the people at the party
Second vertex set:	positive whole numbers
Edges:	one for each person connnecting him/her to the number of people whose hands he/she has shaken

For example, if person 2 shakes hands with 5 people, then there would be an edge from person 2 (in the first vertex set) to the number 5 (in the second vertex set).

Suppose that the party was a quiet, private affair with (say) six people attending. Then there would be six vertices in the first vertex set. The second vertex set could contain any of the integers, 0, 1, 2, 3, 4 or 5 but **not both** 0 and 5. This is because the largest number of hands that can be shaken by one person is 5, and if one person shakes the hand of 5 others then nobody has shaken 0 hands. On the other hand (!) if there is someone who shakes 0 hands, then the largest number of hands that can be shaken by anyone else is 4.

Every vertex in the first vertex set must have an edge leaving it, but we have seen that there are at most five destinations for those six edges. So there must be at least two edges converging on the same vertex in the second set:

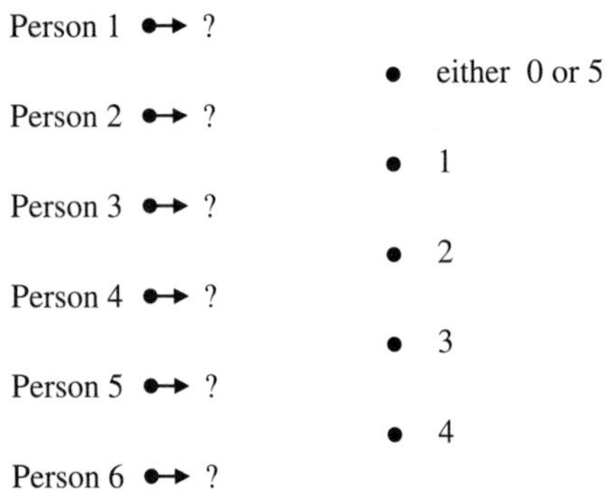

Person 1 •→ ?
• either 0 or 5
Person 2 •→ ?
• 1
Person 3 •→ ?
• 2
Person 4 •→ ?
• 3
Person 5 •→ ?
• 4
Person 6 •→ ?

This shows what happens in the case of six people. But the same applies whatever the (finite) number of people.

Exercise 7A

1 This matrix represents a bipartite graph. Draw the graph.

	A	B	C	D	E	F	G	H	I	J	K
A	–	1	–	–	1	–	–	–	–	–	1
B	1	–	1	–	–	–	1	–	–	–	–
C	–	1	–	1	–	–	–	–	–	–	–
D	–	–	1	–	–	–	–	–	–	–	–
E	1	–	–	–	–	–	1	–	–	–	–
F	–	–	–	–	–	–	–	1	1	–	1
G	–	1	–	–	1	–	–	1	–	–	1
H	–	–	–	–	–	1	1	–	–	–	–
I	–	–	–	–	–	1	–	–	–	1	–
J	–	–	–	–	–	–	–	–	1	–	–
K	1	–	–	–	–	1	1	–	–	–	–

2 Which of the following represent bipartite graphs?

(a)

	A	B	C	D	E
A	–	1	–	–	–
B	1	–	1	1	–
C	–	1	–	–	1
D	–	1	–	–	1
E	–	–	1	1	–

(b)

	A	B	C	D
A	–	–	1	1
B	–	–	–	1
C	1	–	–	1
D	1	1	1	–

(c) vertex set = {A, B, C, D, E}
edge set = {(A, B), (B, D), (A, C), (C, E)}

3 (a) Draw a bipartite graph to represent the following information:

Airport	Direct flights to:
Heathrow	Bristol, Birmingham, E. Midlands, Manchester, Newcastle
Gatwick	E. Midlands, Manchester, Newcastle
Stanstead	Bristol, Manchester, Newcastle
Luton	Bristol, Birmingham, Manchester, Newcastle

(b) If in addition to the above flights a helicopter connection is introduced linking Heathrow and Gatwick, will it then be possible to represent direct flights by a bipartite graph?

4 Small washers, sizes 4 and smaller, can be used with bolts of the same size or of one size smaller.
Larger washers, sizes 5 and above, can be used with bolts of the same size, of one size smaller, or of two sizes smaller.
Use a bipartite graph to show with which sizes of bolts from 2 to 8, washers of sizes 2 to 8 can be used.

7.2 Matchings

The mathematical concept of a matching is a formal way of expressing the everyday idea of a matching. In a matching some of the members of one set (perhaps all of them) are linked, one-to-one, with some (or all) of the members of a second set. For instance, there is a matching between the members of an audience attending a formal concert and the coats left in the cloakroom. The linkage here is provided by cloakroom tickets, the person holding ticket number 42 being the owner of the coat on peg number 42. In this example every coat is linked to a member of the audience, but there may be members of the audience who have not come in coats.

- **A matching is a bipartite graph in which every vertex (in either vertex set) has at most one edge connected to it.**

Here are three possible matchings from the teaching problem in example 1. The first seems odd as it has no edges, but it is still a matching according to the definition:

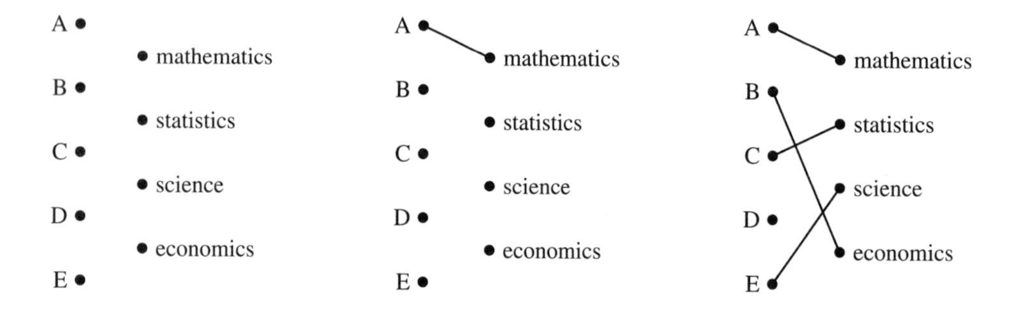

- **A maximal matching is a matching in which the number of edges is as large as possible.**

In a bipartite graph with n_1 vertices in the first vertex set and n_2 vertices in the second vertex set, the number of edges in a maximal matching cannot exceed the smaller of n_1 and n_2.

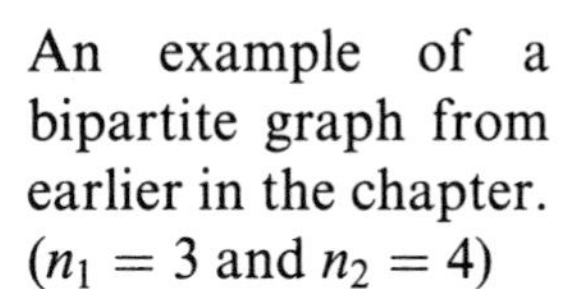

An example of a bipartite graph from earlier in the chapter. ($n_1 = 3$ and $n_2 = 4$)

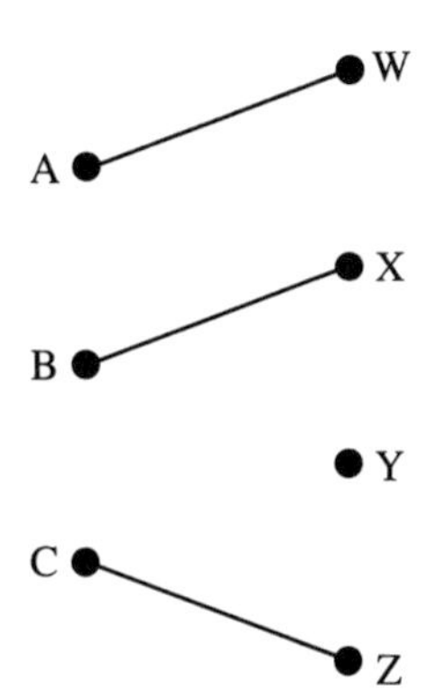

A maximal matching. (There are many others.) In this example there are 3 edges in every maximal matching.

- **A complete matching is a matching in which there are *n* vertices in each vertex set, and in which the number of edges is also *n*.**

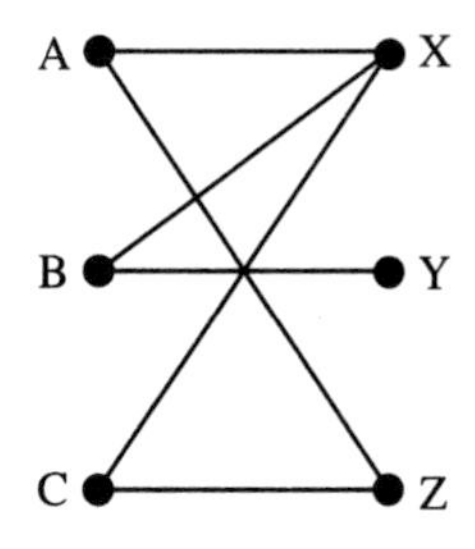

An example of a bipartite graph in which $n_1 = 3$ and $n_2 = 3$.

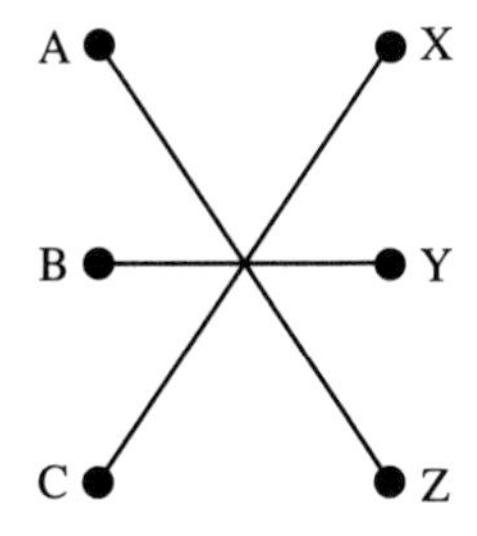

A complete matching. (There is one other.) There are 3 edges in the matching.

So since we have more teachers than subjects in example 1 a complete matching will not be possible. At least one teacher will end up without a subject.

Conversely if there are more subjects than teachers then at least one subject will end up without a teacher.

If there are the same numbers of subjects and teachers then it might be possible to achieve a complete matching, but this will depend on the mix of requirements and skills.

Exercise 7B

1 Four teachers, Mr Adams, Mrs Brown, Mr Charles and Ms Davis, are to be assigned to teach languages in the sixth form. The languages to be taught are French (1), German (2), Spanish (3) and Russian (4). The teachers are not all qualified to teach all the languages. Mr Adams is qualified to teach (1) and (2). Mrs Brown is qualified to teach all the languages (1), (2), (3) and (4). Mr Charles is qualified to teach (1) and (2) but Ms Davis is only qualified to teach (1). Each teacher teaches only one language.

(a) Represent this situation as a bipartite graph.

(b) Find the best possible matching and explain why it cannot be improved.

2 Five people in an office were asked which of five jobs they would like to do. Their replies are summarised in the table:

Person	Jobs
1	A, C
2	A, C
3	A, C, E
4	B, D, E
5	C, D

(a) Represent their replies by a bipartite graph.

(b) Hence decide if they can all be satisfied. If they can be satisfied, show how. If they cannot be satisfied, give a reason.

3 Mr Gardiner has four plots of land available to grow potatoes (P), carrots (C), sweetcorn (S) and broccoli (B). The soil in these plots varies and so the plots can only be used for certain vegetables. The information is summarised in this table:

Plot	Vegetables which may be grown
1	P, B
2	P, C
3	P, C, S
4	C

By drawing a bipartite graph and finding a matching, decide if all the crops can be grown.

4 A maths department has four committees: Teaching (T), Careers (C), Administration (A) and University Entrance (U). The members of the committees are shown in the table.

Committee	Members
T	1, 6
C	1, 2, 6
A	2, 6
U	3, 4, 5

where Mr Brown is 1, Miss Evans is 2, Mr McNichol is 3, Ms Dolan is 4, Mr Schack is 5 and Mrs Wilson is 6.

The principal wants four people for a new committee – all different and one from each of the above committees. Decide if this is possible, giving your reasons.

5 Find a maximal matching in the following bipartite graph:

How many different maximal matchings are there?

6 Find a complete matching in the following bipartite graph:

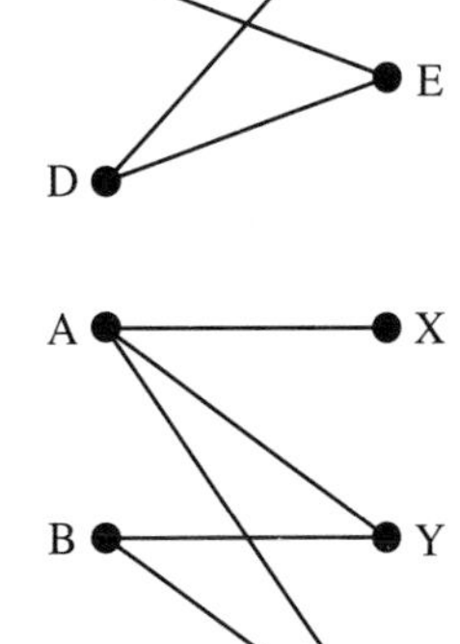

7 Show that for the bolts and washers in Exercise 7A, question 4, there is only one complete matching.

8 Here is the bipartite graph relating teachers to subjects from example 1:

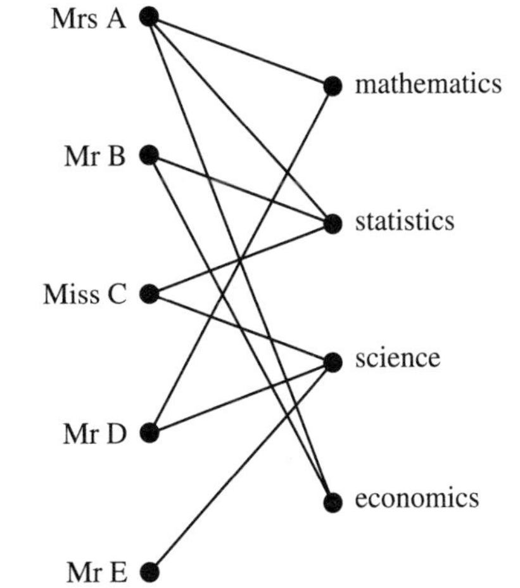

(a) Draw a tree diagram to show that there are 82 possible matchings.

(b) Count how many of these are maximal matchings.

7.3 Improving a matching using an alternating path: the matching improvement algorithm

Counting the number of matchings in the teaching problem in example 1 (as in question 8 of Exercise 7B) is very time consuming. This shows that the complete enumeration approach is not a practicable method for finding a maximal matching, except in the simplest of situations. Fortunately an algorithm exists which enables a larger matching to be constructed from a smaller one, provided that a larger matching exists. So a maximal matching can be achieved by repeatedly applying this algorithm.

Let us begin with a non-maximal matching from example 1:

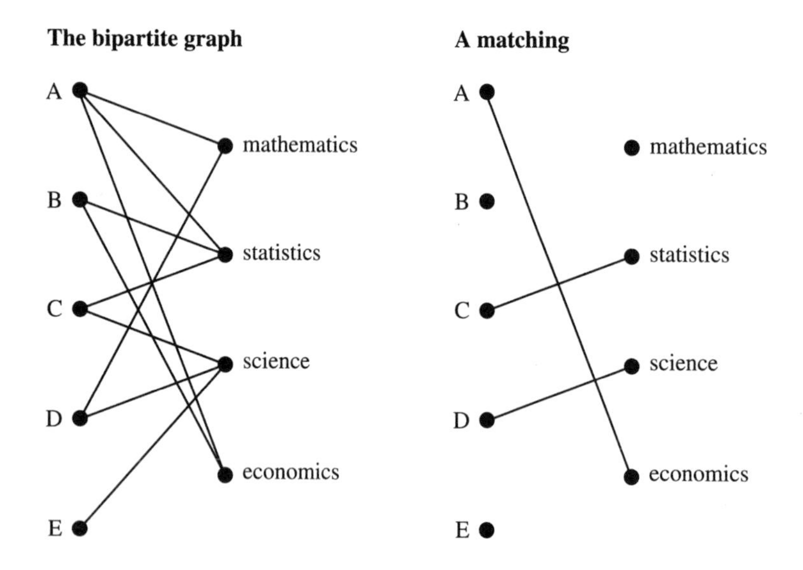

Notice that there is no teacher allocated to mathematics in this matching. Let us start from mathematics and construct a tree, alternately using edges which *are not* in the matching and edges which *are* in the matching:

Stage 1
The edges which have mathematics as a vertex are drawn. These are not in the current matching.

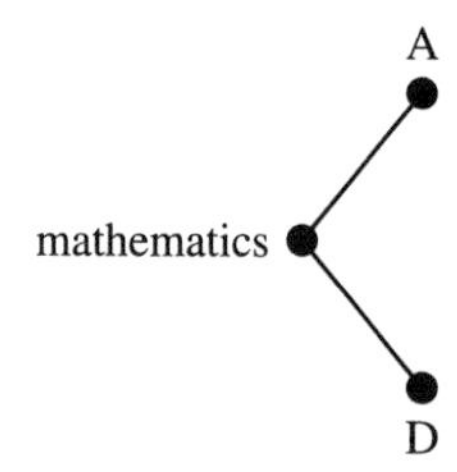

Stage 2
For each vertex reached so far there is one edge in the current matching, taking us to a subject. These edges from the current matching are drawn in, using a different colour or shading or a thick line.

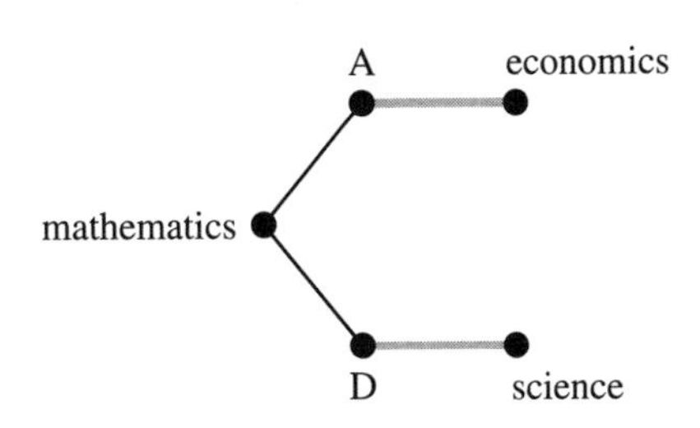

Stage 3
Now you can step back to different staff using edges that are in the graph but are not in the current matching. Draw in these edges using the original colour or shading.

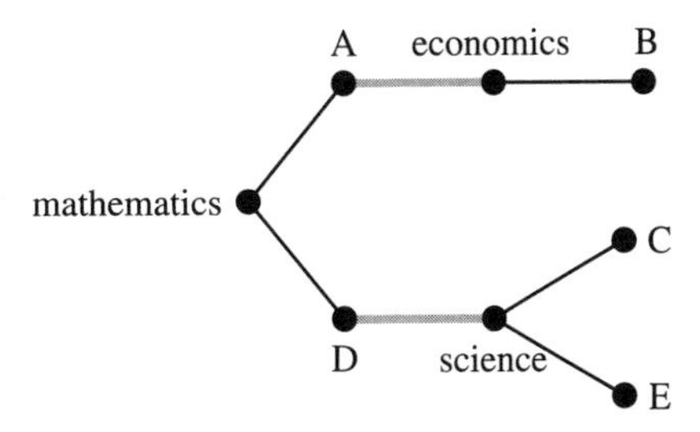

Continue in this way either until you reach Mr B or Mr E who are not matched (which is called 'breakthrough'), or until you can add no further edges. If the latter happens then you can conclude that no improvement is possible and that the current matching is maximal.

Notice that at each stage in the process you **alternate** between adding edges that *are not* in the current matching, and then adding edges that *are* in the current matching.

We build up what is called an **alternating path** until we reach breakthrough.

■ **An alternating path is made up of edges of the bipartite graph which alternately *are* in and *are not* in the current matching and whose start and finish vertices are not in the matching.**

Having achieved breakthrough to B or E you can now create a new improved matching by:

- discarding those edges which were already in the current matching
- adding in those edges which were not already in the current matching.

You can do this by changing the shading of the edges in the paths you have drawn. First let's look at the breakthrough to B:

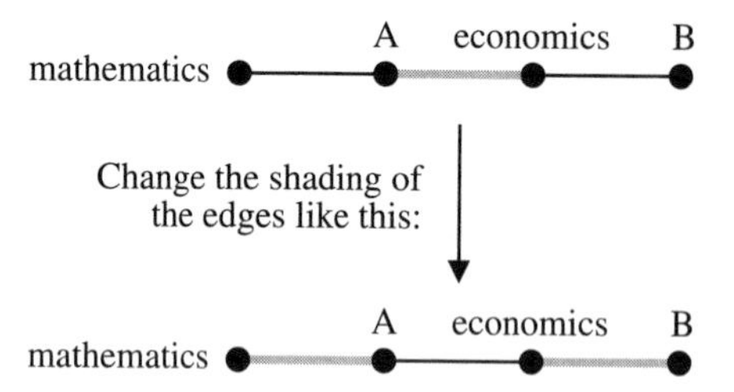

This removes the edge between A and economics, and adds edges between mathematics and A, and between economics and B. This process is called **changing the status** of the edges.

Now add the shaded edges only to the original matching. It produces the improved (and maximal) matching:

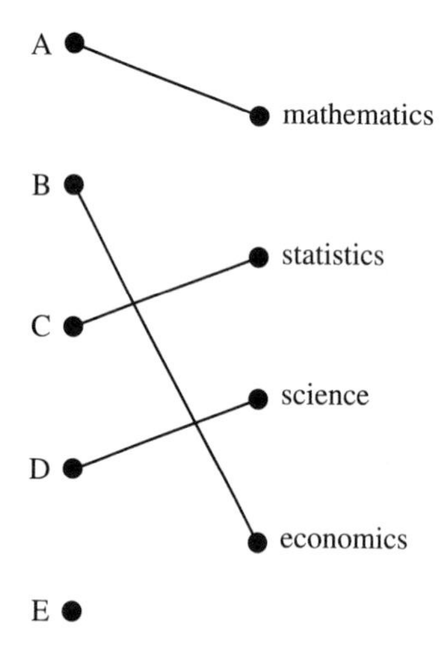

You can do the same with the breakthrough to E:

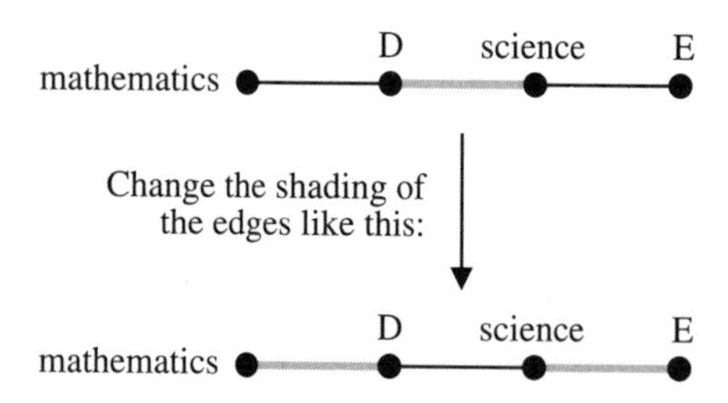

This removes the edge between D and science, and adds edges between mathematics and D, and between science and E. Now add the shaded edges only to the original matching. This produces a different maximal matching:

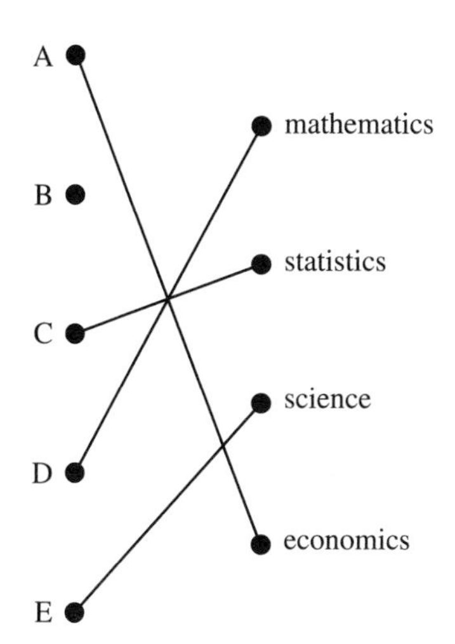

Notice that our alternating path(s) began in the second vertex set and ended in the first. This is not a requirement. You could just as easily have started in the first vertex set and ended in the second.

Our final matching is maximal, but not complete, since Mr B is not timetabled to teach any class.

- **The matching improvement algorithm improves an existing matching, if improvement is possible, by establishing an alternating path between vertices not in the current matching.**

- **The status of each edge in the alternating path is changed to improve the matching. If the current matching is maximal then no alternating path will be found.**

Changing the status

The process represented by changing the shading of a path is called **changing the status** of the edges. For example changing status here:

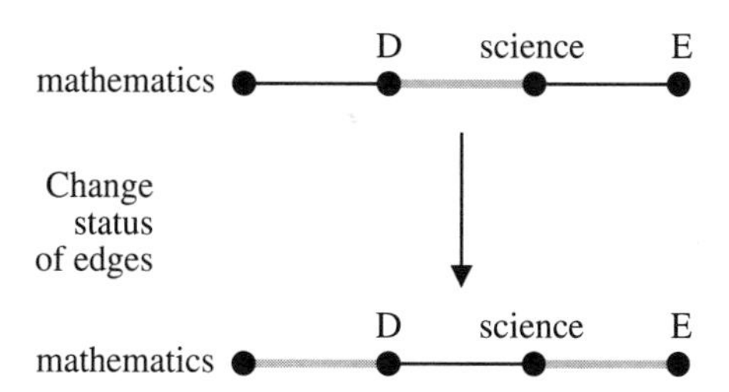

removes the link between D and science and creates links between maths and D and between science and E.

You will see in example 4 that an alternating path might contain only one edge.

Example 4

Construct a maximal matching in the bipartite graph shown, from an initial matching in which there is only one edge connecting B to X.

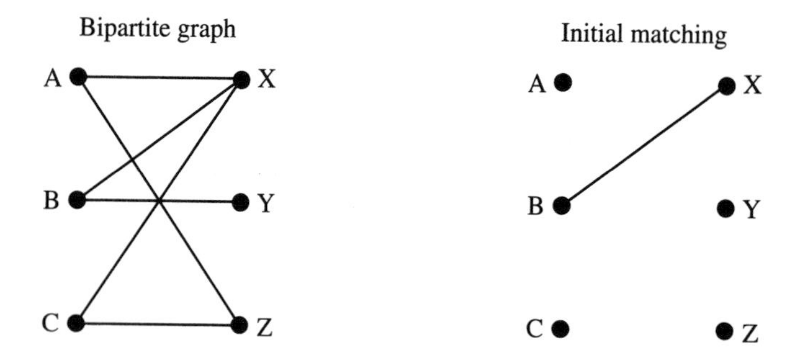

Starting with A, which is not in the current matching, you have breakthrough at Z, which is also not in the matching. This gives you an alternating path containing only one edge.

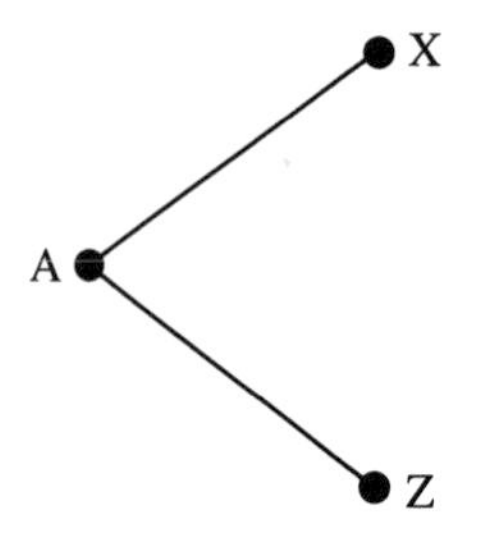

Changing status brings edge AZ into the current matching:

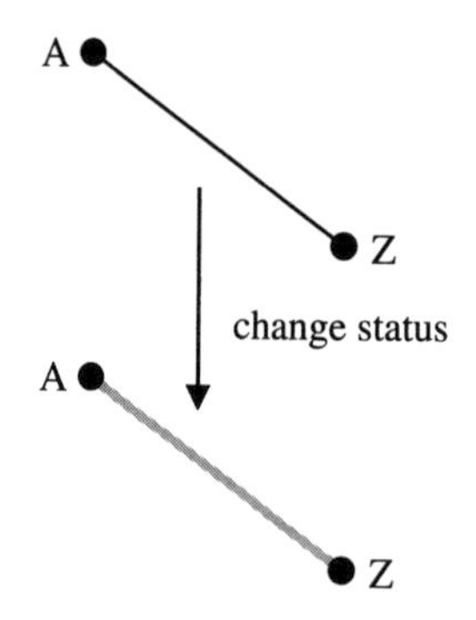

There are now two edges, BX and AZ in the matching:

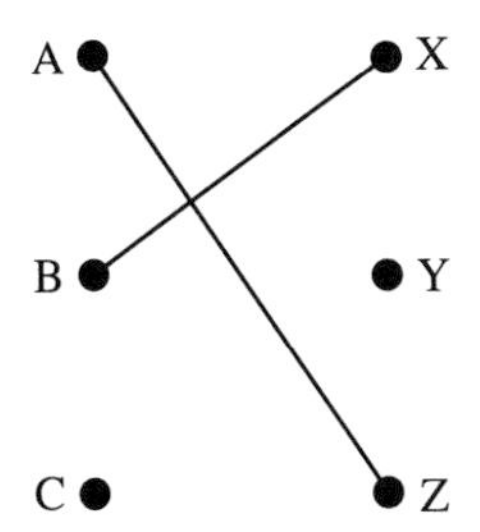

Starting at C, which is not in the matching, produce a tree. One branch ends at A. (This is because A can only be connected to X and Z, and both have appeared in the tree by the time we have included A in the tree.) The other branch achieves breakthrough at Y:

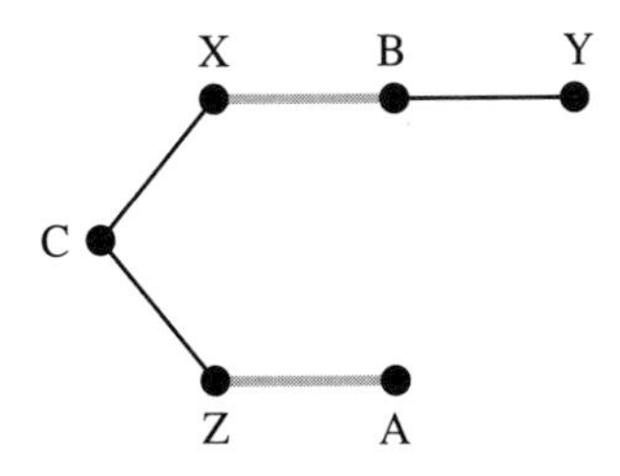

Changing status removes BX from the matching, but brings in CX and BY:

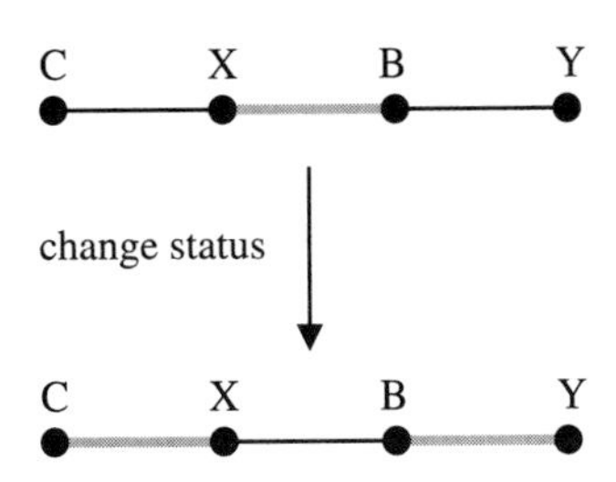

The matching is now maximal and also complete:

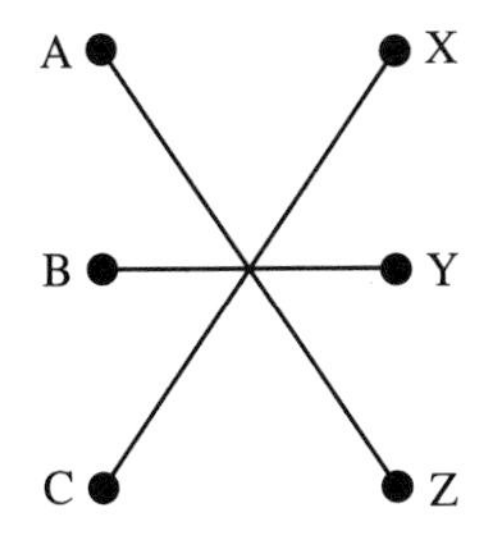

Exercise 7C

1 Consider the following bipartite graph:

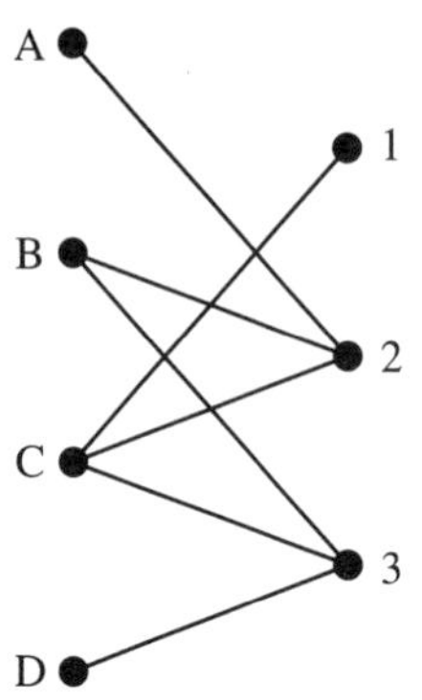

Starting with C connected to 2 and B connected to 3, use the matching improvement algorithm to establish a maximal matching.

2 Consider the following bipartite graph:

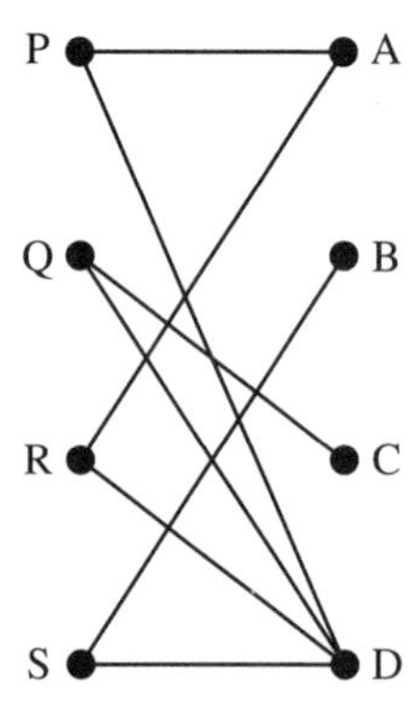

Starting with A connected to P and D connected to Q, use the matching improvement algorithm to establish a complete matching.

3 Starting from 4 connected to X, use the matching improvement algorithm to establish a complete matching in this bipartite graph:

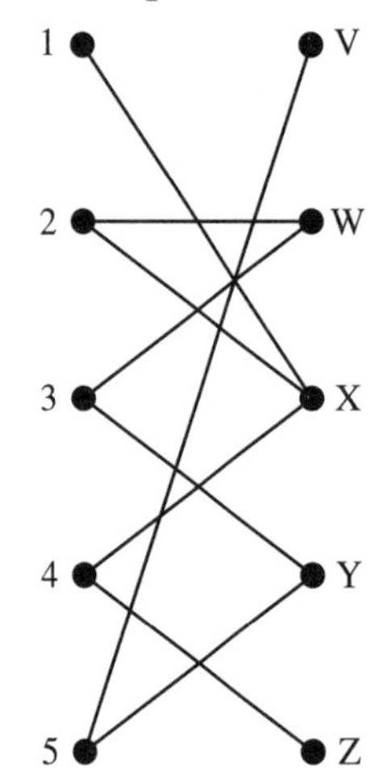

4 Use the matching improvement algorithm to show that there is no complete matching within this bipartite graph:

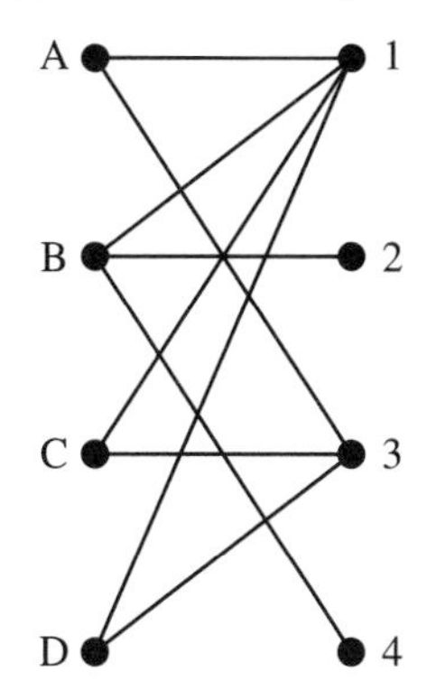

5 Five managers are to be appointed to company offices in London, Manchester, Nottingham, Oldham and Plymouth. Anne would prefer London or Manchester. Brian would be happy with London or Nottingham. Colin supports Oldham Athletic Football Club, and would like to live in Oldham. Denise can cover Oldham or Plymouth. Elinor would like to live either in Manchester, since she supports Manchester United or in Plymouth, where she has relatives.
Starting from Anne in London, Denise in Oldham and Elinor in Plymouth, use the matching improvement algorithm to find a complete matching.

6 Five classes, C1, C2, C3, C4 and C5 are to be timetabled in five rooms, R1, R2, R3, R4 and R5. Class requirements and room facilities in terms of size, computing availability and laboratory space, dictate that the following allocations are possible:

C1 can be timetabled in R4 or R5 only.
C2 can be timetabled in R3 or R4 only.
C3 can be timetabled in R2 or R4 only.
C4 can be timetabled in any room.
C5 can be timetabled in R2 only.

By repeatedly applying the matching improvement algorithm, find a complete matching.

7 Marriages are to be arranged between members of the set $\{A, B, C, D\}$ and members of $\{W, X, Y, Z\}$. W is very popular and would be an acceptable partner for A, B, C or D. Equally W would be happy with any of A, B, C or D as partners. Other compatible pairs are A and Y, C and Z, D and X, D and Z.

(a) As matchmaker you receive a fee for each marriage that you negotiate. It has been suggested that A and Y should marry, that B and W should marry, and that D and Z should marry. Use the algorithm to improve on this solution, both to maximise your fees and for the sakes of C and X.

(b) During the negotiation period C and X become mutually compatible, and C and Z fall out and vow never to speak to each other again. Go back to the original suggestion (A–Y, B–W and D–Z) and from it use the algorithm to construct a new improved solution.

(c) Fickle as ever, C rejects X and discovers shared common ground with Y. Go back to the original suggestion and from it use the algorithm to prove that there is now no complete matching.

8 The network is the bipartite graph from page 204 together with two other vertices, S and T. Extra edges connect S to each member of the first vertex set, and connect each member of the second vertex set to T.

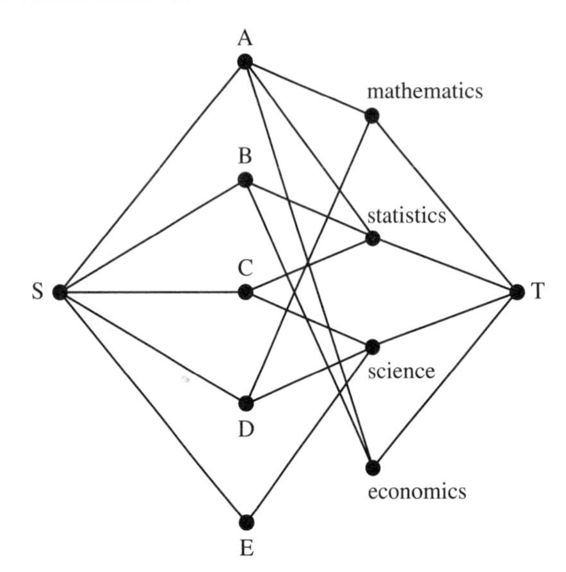

(a) Regard this as a transmission network in which each edge has capacity 1. Find a maximum flow from S to T.

(b) Relate your solution to part (a) to the matching problem of example 3.

SUMMARY OF KEY POINTS

1. A bipartite graph is one in which the vertices are partitioned in some way, and in which all edges connect a vertex of one of the two subsets to a vertex of the other subset.
2. A matching is a bipartite graph in which every vertex has at most one edge.
3. A maximal matching is a matching in which the number of edges is as large as possible.
4. A complete matching is a matching in which there are n vertices in each vertex set, and in which the number of edges is also n.
5. An alternating path is made up of edges of the bipartite graph which alternately are in and are not in the current matching and whose start and finish vertices are not in the matching.

Appendix: Further work

Linear programming

This section shows how the linear programming methods introduced in chapter 6 can be applied to the network concepts covered in chapters 2, 5 and 7.

This work is included for interest and completeness **but will not be examined in your D1 examination.**

Linear programming in networks

Most (and possibly all) network problems can be formulated as linear programming problems, although the appropriate specialised network algorithm may be more efficient in solving the problem. To achieve this we'll need to use the matrix representation of our network. Let the matrix be $\mathbf{C}$. This has elements $c_{ij} (i = 1, 2, \ldots, n; j = 1, 2, \ldots, n)$, c_{ij} being the cost on the edge connnecting vertex i to vertex j.

We'll also need a second matrix $\mathbf{X}$. This contains elements x_{ij}, each being 0 or 1 to indicate whether or not we are using the edge from vertex i to vertex j.

To illustrate, we will show how to use linear programming to find the shortest path between two vertices. (We start with the shortest path problem because the minimum connector problem is, surprisingly, one of the more difficult network problems to model using linear programming.)

Shortest path

Suppose that we need to find the cheapest or shortest path from vertex 1 to vertex n in an n-vertex network. The linear programming formulation for this is:

Minimise: $\sum_{i \neq j} c_{ij} x_{ij}$

subject to: $\sum_{j} x_{1j} = 1$ — You must leave vertex 1.

$\sum_{i \neq 2} x_{i2} = \sum_{j \neq 2} x_{2j}$ — If you arrive at 2 then you must leave it.

$\sum_{i \neq 3} x_{i3} = \sum_{j \neq 3} x_{3j}$ — If you arrive at 3 then you must leave it.

$\vdots$

$\sum_{i \neq n-1} x_{i(n-1)} = \sum_{j \neq n-1} x_{(n-1)j}$ — If you arrive at $(n-1)$ then you must leave it.

$\sum_{i} x_{in} = 1$ — You must arrive at vertex n.

$x_{ij} \geqslant 0$ for all i and j

You might think that the above formulation is inadequate to model our requirements fully as we need each of the $x_{i,j}$ to be either 0 or 1, and this requirement makes the problem look rather like an integer programming problem. In fact it can be proved (and is perfectly believable if you think carefully about it) that the formulation is completely adequate, and that the optimal answer will indeed have only 0s and 1s for the x_{ij}s.

Example

The cost matrix for a network is:

$$\begin{bmatrix} 100 & 5 & 3 & 1 & 6 \\ 5 & 100 & 7 & 2 & 1 \\ 3 & 7 & 100 & 100 & 8 \\ 1 & 2 & 100 & 100 & 3 \\ 6 & 1 & 8 & 3 & 100 \end{bmatrix}$$

Notice that we can use a large cost where there is no edge.

A linear programming problem to find the shortest path from vertex 3 to vertex 2 is:

Minimise:

$$\begin{aligned} & 5x_{12} + 3x_{13} + x_{14} + 6x_{15} \\ & + 5x_{21} + 7x_{23} + 2x_{24} + x_{25} \\ & + 3x_{31} + 7x_{32} + 8x_{35} \\ & + x_{41} + 2x_{42} + 3x_{45} \\ & + 6x_{51} + x_{52} + 8x_{53} + 3x_{54} \end{aligned}$$

subject to: $x_{31} + x_{32} + x_{34} + x_{35} = 1$
$x_{21} + x_{31} + x_{41} + x_{51} - x_{12} - x_{13} - x_{14} - x_{15} = 0$
$x_{14} + x_{24} + x_{34} + x_{54} - x_{41} - x_{42} - x_{43} - x_{45} = 0$
$x_{15} + x_{25} + x_{35} + x_{45} - x_{51} - x_{52} - x_{53} - x_{54} = 0$
$x_{12} + x_{32} + x_{42} + x_{52} = 1$

all $x_{ij} \geqslant 0$.

When this was submitted to a linear programming computer package the following answer was produced:

$$x_{14} = 1; \quad x_{31} = 1; \quad x_{42} = 1; \quad \text{all other } x_{ij} = 0.$$

The minimum cost was 6 (i.e. $1 + 3 + 2$).

(You might like to consider what would happen if some of the costs were negative, i.e. if there were profits available.)

Linear programming in transmission networks

We can use linear programming to solve maximum flow problems. As on page 223 we will need a **C** matrix and an **X** matrix. In this context the **C** matrix has elements $c_{ij} (i = 1, 2, \ldots, n; j = 1, 2, \ldots, n)$, c_{ij} being the capacity of the edge connecting vertex i to vertex j. The second matrix **X** contains elements x_{ij}, indicating the flow along the edge from vertex i to vertex j.

c_{ij} is put equal to 0 where there is no edge, or where the edge is directed from j to i (in which case c_{ji} is greater than 0).

Suppose that you need to find the maximum flow from source to sink in a transmission network with n vertices in which vertex 1 is the source and vertex n the sink. The linear programming formulation for this is:

Maximise: $\sum_j x_{1j}$ (the total flow out of the source)

subject to: $x_{ij} \leqslant c_{ij}$ for all i and j (capacity constraints)

$\sum_i x_{i2} = \sum_j x_{2j}$ What comes into vertex 2 must leave it.

$\sum_i x_{i3} = \sum_j x_{3j}$ What comes into vertex 3 must leave it.

...

$\sum_i x_{i(n-1)} = \sum_j x_{(n-1)j}$ What comes into vertex $(n - 1)$ must leave it.

$x_{ij} \geqslant 0$ for all i and j

Remember that the x_{ij} are no longer indicator variables telling you whether or not you are using an edge, as on page 223. Instead they record the amount flowing in each edge.

Applying this to the network from page 147 gives:

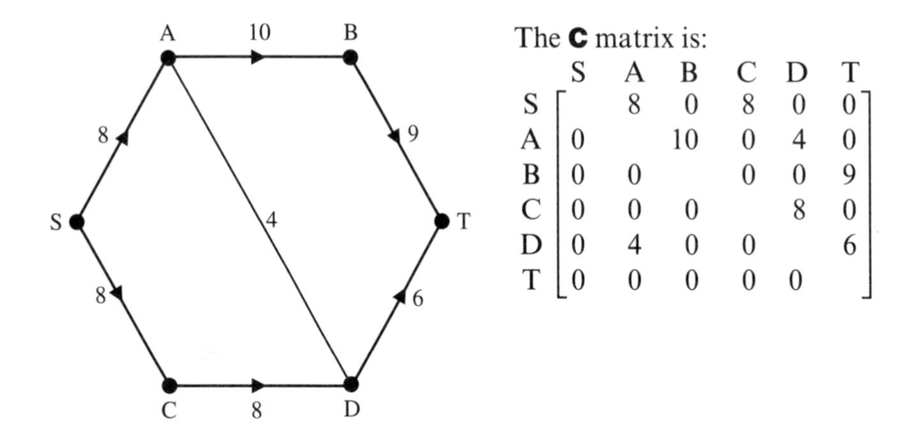

The **C** matrix is:

	S	A	B	C	D	T
S		8	0	8	0	0
A	0		10	0	4	0
B	0	0		0	0	9
C	0	0	0		8	0
D	0	4	0	0		6
T	0	0	0	0	0	

The linear programming formulation is:

Maximise: $x_{SA} + x_{SC}$

subject to:

$$x_{SA} + x_{DA} - x_{AB} - x_{AD} = 0$$
$$x_{AB} - x_{BT} = 0$$
$$x_{SC} - x_{CD} = 0$$
$$x_{CD} + x_{AD} - x_{DA} - x_{DT} = 0$$
$$x_{SA} \leqslant 8 \quad x_{SC} \leqslant 8 \quad x_{AB} \leqslant 10 \quad x_{AD} \leqslant 4$$
$$x_{BT} \leqslant 9 \quad x_{CD} \leqslant 8 \quad x_{DA} \leqslant 4 \quad x_{DT} \leqslant 6$$
$$\text{all } x \text{ values} \geqslant 0$$

We have not written down terms with zero coefficients.

When submitted to a linear programming computer package the following answer was produced:

$$x_{SA} = 7 \quad x_{SC} = 8 \quad x_{AB} = 9 \quad x_{AD} = 0$$
$$x_{BT} = 9 \quad x_{CD} = 8 \quad x_{DA} = 2 \quad x_{DT} = 6$$

giving a maximum flow of 15.

Allocation and transportation

In matching problems it is often the case that there is more than one, maximal matching. Sometimes there are many. As far as establishing matchings is concerned these are all equivalent. However, in many practical problems there is a cost to be paid or a benefit to be gained in linking a vertex from the first vertex set to a vertex from the second, and these costs and benefits will vary from vertex pair to vertex pair.

As an example consider the problem of matching 6 people to 6 tasks where each person can do each task, but where their differing skills mean that the times they take vary from person to person, and from task to task. There are $6! = 720$ different matchings. But suppose we are paying for these tasks by the time taken, and that we want to minimise our expenditure. Or suppose that the tasks have to be performed one after the other and that we want the total time taken to be minimised.

We can model these situations by adding edge weights to our bipartite graph. We then need to be able to choose the best out of all of the possible matchings, i.e. the matching with the minimum or maximum total edge weight.

These problems are called **allocation problems**. There is a well-known algorithm called the Hungarian algorithm which produces a solution. Unfortunately there is a step in the middle of the algorithm which is hard to execute on large problems – it is as difficult computationally as the original allocation problem! So the Hungarian algorithm is not of much practical use.

Unsurprisingly it is possible, in fact rather easy, to formulate an allocation problem as an integer programming problem in which the variables can take the values 0 or 1. Furthermore the resulting problem is one of that family of integer programming problems for which the integer formulation is unnecessary – the linear programming formulation automatically produces 0 or 1 solutions.

Generalising further, consider the problem faced by a large retail company in supplying quantities of an item to its shops from a number of warehouses. This looks as if it might be an allocation problem – they will need to allocate shops to warehouses. But there is an extra complication. Associated with each shop there is a particular demand. Associated with each warehouse there is a particular supply. So there may be a need to supply from more than one warehouse, and there will almost certainly be the need to supply more than one shop from a given warehouse.

The problem is how much to send from which warehouse to which shop to satisfy demands, within the constraints of supplies, at minimum cost. This is known as a **transportation problem**.

We can model this type of problem by adding vertex weights to our bipartite graph. The weights on the vertices of the first vertex set will represent supplies. The weights on the vertices of the second vertex set will represent demands. An efficient algorithm exists to solve such problems, and again they can be modelled as integer programming problems. The integer program will not be a 0 or 1 problem, but is still a member of the family for which a linear programming formulation will produce an integer solution.

Linear programming in matching

The problem of matching teachers to subjects from page 204 can also be solved by linear programming.

The linear program to find a maximal matching is as follows:

$$\mathbf{C} = \begin{array}{c} \\ A \\ B \\ C \\ D \\ E \end{array} \begin{array}{cccc} \text{mathematics} & \text{statistics} & \text{science} & \text{economics} \\ \end{array}$$

	mathematics	statistics	science	economics
A	1	1	0	1
B	0	1	0	1
C	0	1	1	0
D	1	0	1	0
E	0	0	1	0

The **C** matrix captures the information held within the bipartite graph. A '1' indicates the presence of an edge; a '0' indicates the absence.

The **X** matrix is similar in structure, with '1' indicating the presence of an edge in the matching, and '0' the absence of an edge in the matching.

The formulation is:

Maximise:	$\sum_{ij} x_{ij}$	the number of matchings
subject to:	$x_{ij} \leqslant c_{ij}$ for all i and j	this makes **X** a subgraph of **C**
	$\sum_{j} x_{ij} \leqslant 1$ for all i	At most one edge leaves each vertex of the first vertex set.
	$\sum_{i} x_{ij} \leqslant 1$ for all j	At most one edge arrives at each vertex of the second vertex set.
	$x_{ij} = 0$ or 1 for all i and j	We can ignore this integer 0/1 requirement – the structure of the rest of the problem ensures that it will be met automatically.

Applying this to the teacher–subject example we obtain:

Maximise:

$$\begin{aligned} x_{Am} + x_{Ast} \quad & + x_{Ae} \\ + x_{Bst} \quad & + x_{Be} \\ + x_{Cst} + x_{Csc} & \\ + x_{Dm} \quad + x_{Dsc} & \\ + x_{Esc} & \end{aligned}$$

(We have not written down terms which are forced to be zero at the outset.)

subject to:
$$x_{\mathrm{Am}} \leqslant 1 \quad x_{\mathrm{Ast}} \leqslant 1 \quad x_{\mathrm{Ae}} \leqslant 1 \quad x_{\mathrm{Bst}} \leqslant 1 \quad x_{\mathrm{Be}} \leqslant 1$$
$$x_{\mathrm{Cst}} \leqslant 1 \quad x_{\mathrm{Csc}} \leqslant 1 \quad x_{\mathrm{Dm}} \leqslant 1 \quad x_{\mathrm{Dsc}} \leqslant 1 \quad x_{\mathrm{Esc}} \leqslant 1$$
$$x_{\mathrm{Am}} + x_{\mathrm{Ast}} + x_{\mathrm{Ae}} \leqslant 1$$
$$x_{\mathrm{Bst}} + x_{\mathrm{Be}} \leqslant 1$$
$$x_{\mathrm{Cst}} + x_{\mathrm{Csc}} \leqslant 1$$
$$x_{\mathrm{Dm}} + x_{\mathrm{Dsc}} \leqslant 1$$
$$x_{\mathrm{Esc}} \leqslant 1$$
$$x_{\mathrm{Am}} + x_{\mathrm{Dm}} \leqslant 1$$
$$x_{\mathrm{Ast}} + x_{\mathrm{Bst}} + x_{\mathrm{Cst}} \leqslant 1$$
$$x_{\mathrm{Csc}} + x_{\mathrm{Dsc}} + x_{\mathrm{Esc}} \leqslant 1$$
$$x_{\mathrm{Ae}} + x_{\mathrm{Be}} \leqslant 1$$
$$\text{all } x \geqslant 0$$

When submitted to a linear programming computer package the following answer was produced: $x_{\mathrm{Be}} = 1$, $x_{\mathrm{Cst}} = 1$, $x_{\mathrm{Dm}} = 1$, $x_{\mathrm{Esc}} = 1$, remaining x values $= 0$.

This corresponds to the matching:

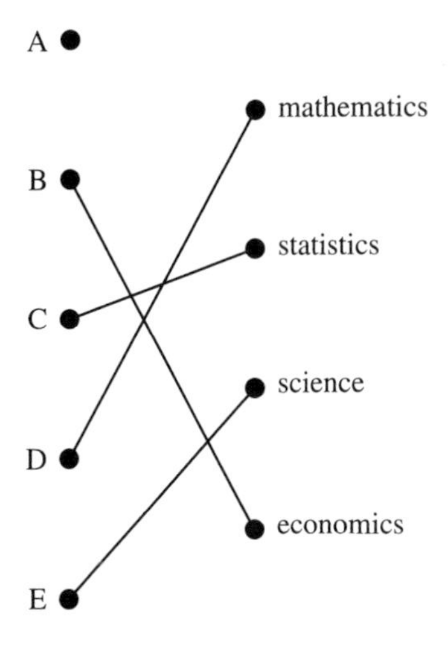

Decision making in graphs

This section extends the work on the travelling salesman problem and the route inspection algorithm (Chinese postman problem) from chapter 3. It is not examinable but it is included for interest and completeness.

Converting a practical problem into its classsical equivalent

Here is a simple, practical problem and its classical equivalent:

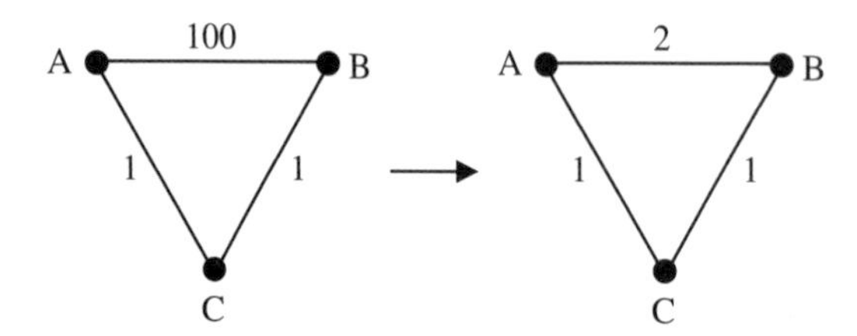

In the original network there is only one Hamilton cycle starting and ending at A (or two if you count ACBA as being different to ABCA). Its length is 102. But the tour ACBCA is only of length 4. It visits every vertex, but visits C twice.

By replacing the 100 weight on AB by the shortest distance between A and B we transform the network into a new one in which the minimum Hamilton cycle is of length 4. (We would have to remember that the arcs now represent links which might be via other vertices.)

Note that the original network does *not* satisfy the triangle inequality (page 38).

Determining whether or not a Hamilton cycle exists is not always easy. Here are two examples of networks in which there is no Hamilton cycle:

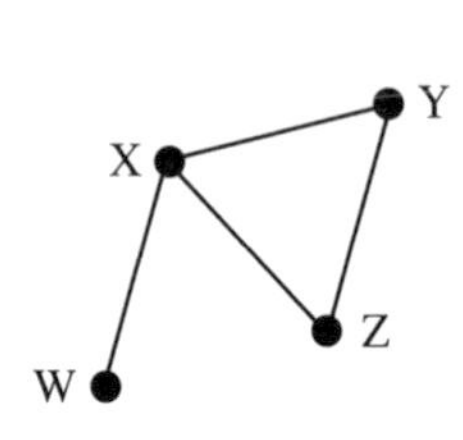

In this graph W is a 'dangle', i.e. a vertex of order 1. There can be no Hamilton cycle in a network containing a dangle.

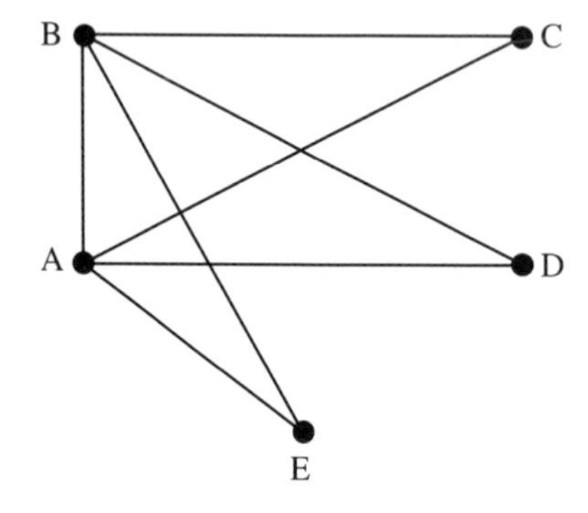

This graph is the graph from page 74 with two edges, CD and DE, missing. It is not straightforward to see that there is no Hamilton cycle.

Notice that in both cases, even though there is no Hamilton cycle, our practical sales representative would have no difficulty in planning a tour.

Example 1

Convert this network to a complete network showing the shortest distance between each pair of vertices.

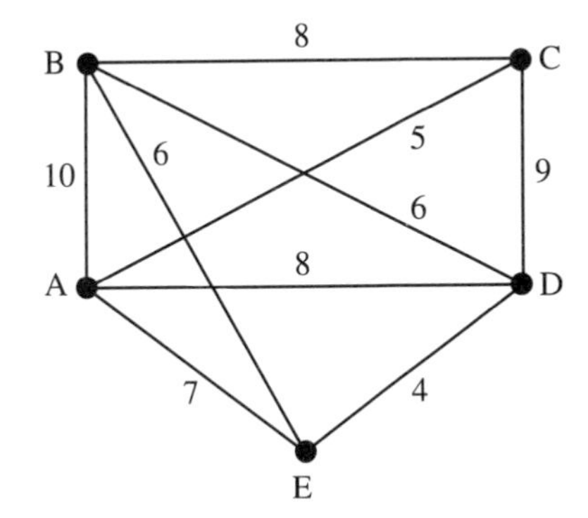

The only change you need to make is to include a new edge CE of length 12.

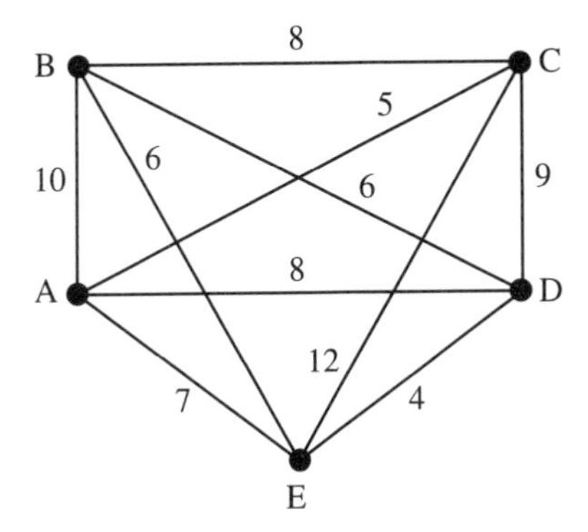

The new edge represents the shortest distance from C to E in the original network, which is via A. In all other cases the original edges give the shortest distances, so no conversion is needed.

Solving the classical problem in the converted network is equivalent to solving the practical problem in the original network.

Exercise A1

1 (a) Explain why a connected network with a dangle (a vertex of order 1) cannot contain a Hamilton cycle.

(b) Find a tour visiting every vertex in the following network, and give the length of your tour.

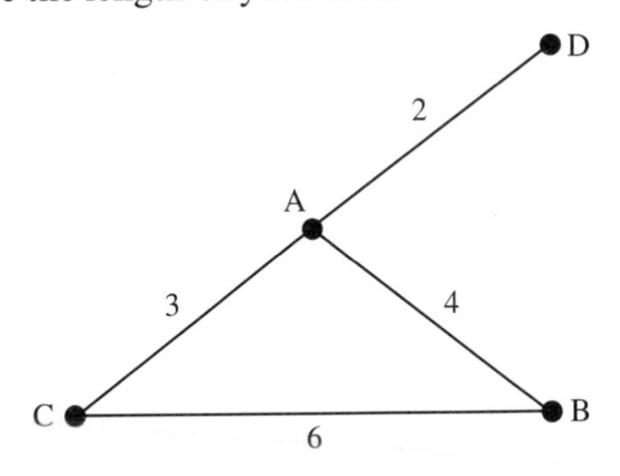

(c) The following graph is the complete graph on the vertices ABCD. By inspection find the shortest distances between each pair of vertices in the original network, and mark these as edge weights in the complete graph.

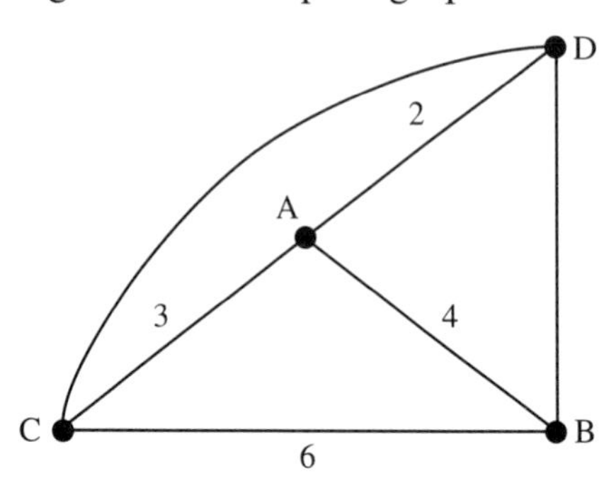

(d) Find a Hamilton cycle in the complete network from part (c). Give its length and find the tour to which it corresponds in the original network.

2 The network gives the costs of air flights (in £) between five cities. Not all pairs of cities are connected by a direct air service.

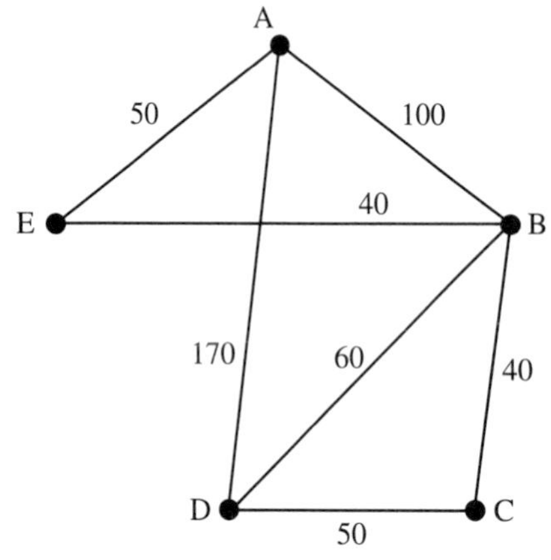

(a) There are only two Hamilton cycles starting and ending at A (really the same cycle in opposite directions).
Give the cost of visiting each city using the edges of this cycle.

(b) Allowing cities to be visited more than once, find a tour visiting each city which is cheaper than that given in part (a).

(c) By inspection create a complete network on the five vertices A, B, C, D, E in which the edge weights are the lowest costs of flying between the cities. Where a cost is not the cost of a direct flight, make a note of the intermediate city on the cheapest route.

(d) By inspection find the shortest Hamilton cycle in the network from part (c).
Check that the shortest Hamilton cycle gives the same tour as that in part (b).

Complexity of the TSP

Suppose that it takes 10 minutes to execute the complete enumeration algorithm for the TSP on a complete network on 5 vertices:

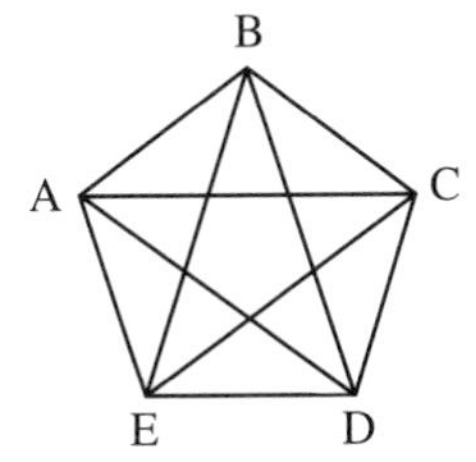

We saw in chapter 3 that there are $4! = 4 \times 3 \times 2 \times 1 = 24$ cycles to consider.

If you want to tackle a complete vertex on 6 vertices, then there will be $5! = 120$ cycles to list – hence approximately 5 times the work and 5 times the time (not allowing for the extra task of choosing the smallest out of 120, which will not take 5 times as long as choosing the smallest out of 24). So it will take about 50 minutes to do this.

For a complete network on 10 vertices the task will take $\frac{9!}{4!} \times 10 = 151\,200$ minutes = 105 days (without sleep!).

Chapter 2 covered Dijkstra's algorithm for finding a shortest path through a network. Doubling the number of vertices in a complete network causes the time taken to complete a worst-case application of Dijkstra's algorithm to increase roughly by a factor of 4. Trebling will cause, roughly, an increase by a factor of 9. This gives a measure of the complexity of the algorithm, and Dijkstra's algorithm is said to have quadratic (i.e. squaring) complexity because the time taken to solve a problem using the algorithm increases in proportion to the square of the size of the problem.

Repeatedly using Dijkstra to find all of the shortest routes in a network has cubic complexity.

Algorithms with quadratic or cubic complexities are examples of algorithms with polynomial complexities. These can be solved in reasonable amounts of time. We say they are **viable**. Our complete enumeration algorithm for the travelling salesman problem has factorial complexity and is not viable.

The travelling salesman problem is well known to mathematicians since it represents an apparently simple, yet unsolved problem. It is one of a class of problems known as the **NP complete problems**. (NP stands for Non-deterministic Polynomial time.) Remarkably, it has been proved that if an algorithm with polynomial complexity can be constructed to solve any one of them, then it could be used to construct an optimal polynomial algorithm for *any* NP problem. Most mathematicians believe that no such algorithm can be constructed, but no-one has succeeded in proving that it cannot be done.

More about upper and lower bounds

Upper bounds

As we pointed out in chapter 3 the 'twice the minimum connector plus shortcuts' approach can produce good (ie. low) upper bounds, but it is difficult to make it algorithmic – in a large network it would be difficult to find a good set of shortcuts. Furthermore, the approach is only guaranteed to produce an upper bound to the practical problem.

There are alternative heuristic approaches which are algorithmic. One such is the **nearest neighbour algorithm**. The algorithm is a heuristic algorithm. Such an algorithm represents a sensible, ordered approach to solving the problem, producing good results in most applications, but not guaranteeing optimality.

Heuristic means 'to find out or investigate'.

The first step is to convert the original network into the complete network of shortest distances. The method then produces a Hamilton cycle in this transformed network, which in turn, implies a practical tour. In large and complex networks the algorithm generally produces Hamilton cycles with good (low) total lengths.

The algorithm is:

Step 1 Select any starting vertex.

Step 2 Go to the nearest vertex which has not yet been visited. (There may be a choice here.)

Step 3 Repeat step 2 until all vertices have been visited and then return to the start vertex.

Note that if the network is not complete the algorithm could terminate at step 2 without finding a Hamilton cycle. If there is no Hamilton cycle then the algorithm certainly will terminate at step 2.

Warning — a common error

The nearest neighbour algorithm for the travelling salesman problem and Prim's algorithm for the minimum connector (see chapter 2) are often confused. In the nearest neighbour the next vertex is the nearest unvisited vertex to the current vertex. In Prim the next vertex is the nearest unconnected vertex to the current connected **set** of vertices.

Applying the algorithm to the problem in example 1 on page 231 produces the following results:

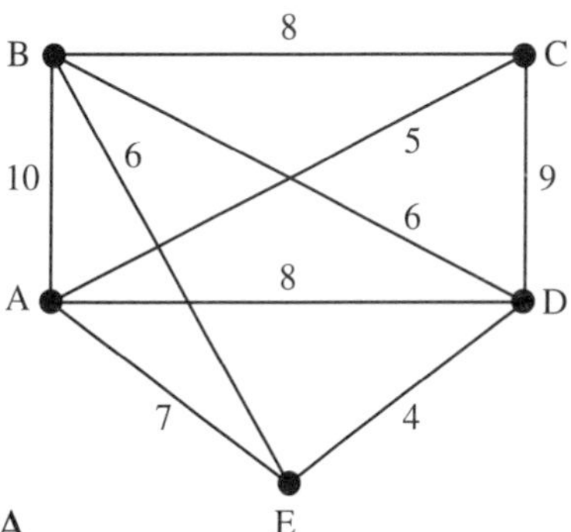

Step 1 Start at A.

Step 2 Go to C (nearest unvisited vertex to A).

Step 2 Go to B (nearest unvisited vertex to C).

Step 2 Go to D (or to E – there is a choice here).

Step 2 Go to E (or D depending on the choice above).

Step 3 All vertices have now been visited, so return to A.

So there are two nearest neighbour tours starting and ending at A. They are ACBDEA (length = 30) and ACBEDA (length = 31).

The tour produced is dependent on the starting vertex, so it is possible, though not usually necessary, to repeat the algorithm for each starting vertex in turn. The shortest of these tours can then be chosen to give an upper bound. Applying this to the network of example 1 gives:

Starting vertex	Nearest neighbour tour(s)	Length
A	ACBDEA	30
	ACBEDA	31
B	BDEACB	30
	BEDACB	31
C	CAEDBC	30
D	DEBCAD	31
E	EDBCAE	30

The nearest neighbour algorithm and the minimum spanning tree algorithms are all examples of 'greedy' algorithms. They split into stages, and at each stage there is a choice to be made. Greedy algorithms take the best available choice at each stage. This does not always guarantee to produce the best result. In the case of Prim and Kruskal (see chapter 2) it does. In the nearest neighbour algorithm it might not.

Lower bounds

Notice that this method for producing a lower bound is *not* guaranteed to work for the practical problem. In the classical problem, when we delete a vertex and all of its edges from the network, we can be certain that what will be left of a Hamilton cycle will be a tree connecting the remaining vertices. In the practical problem, when the minimum tour can re-visit vertices, we cannot guarantee this. It is then possible that we could be left with an undeleted portion of the tour *which is disconnected.* This means that we cannot be confident that the minimum connector of the undeleted vertices is not greater than the undeleted part of the tour.

For example:

Suppose that in this network...

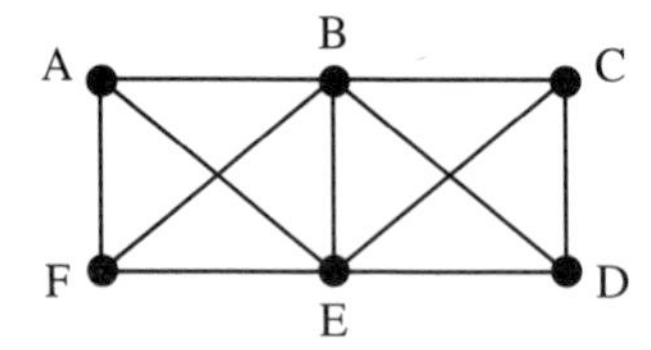

...the best solution to the practical problem is the tour AFABEBCDCBA:

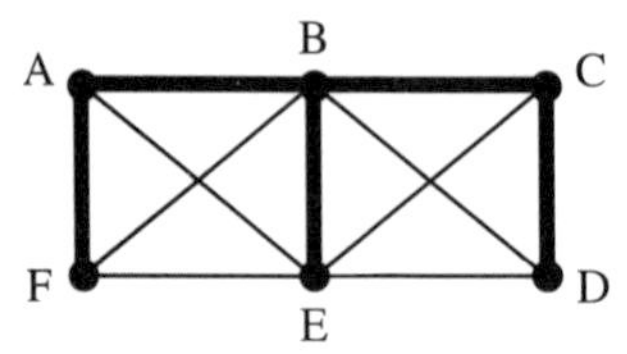

Then deleting B and all connected edges leaves:

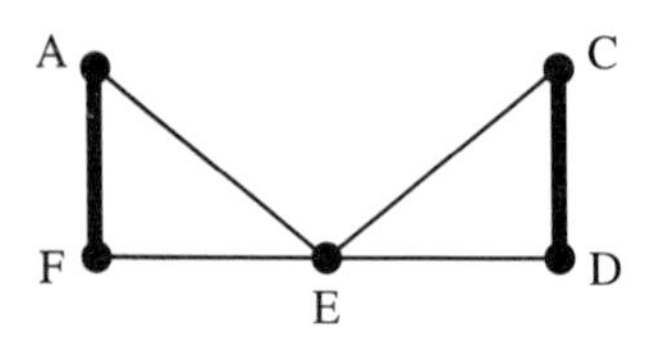

Exercise A2

1

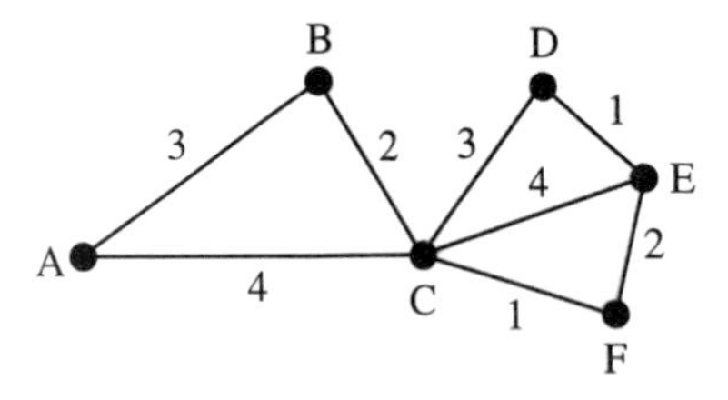

(a) Apply the nearest neighbour algorithm to the above network, starting at A.

(b) Does the network contain a Hamilton cycle?

(c) Add an edge BD, of length 5, to the network. Does the new network contain a Hamilton cycle?

(d) Apply the nearest neighbour algorithm to the new network, starting at A.

(e) Apply the nearest neighbour algorithm to the new network, starting at B.

2 In a complete network the nearest neighbour algorithm will always find a Hamilton cycle. Consider the following complete network on 4 vertices:

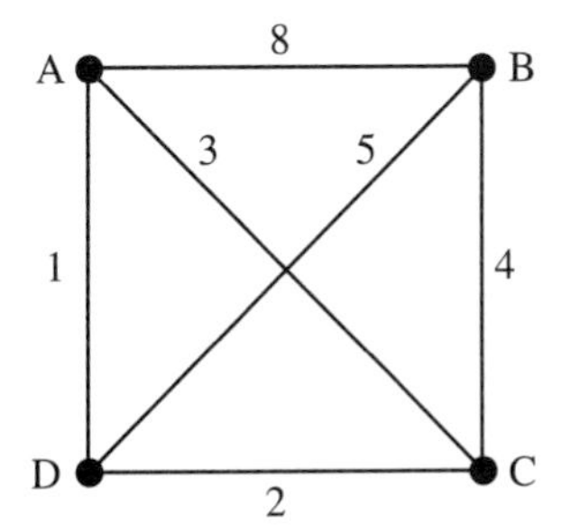

Apply the nearest neighbour algorithm to the network, starting from vertex A. Repeat, starting from vertices B, C and D in turn. Give a Hamilton cycle of least total weight.

3 The matrix represents distances in a geographical network.

$$\begin{bmatrix} 0 & 12 & 23 & 15 & 18 & 10 \\ 12 & 0 & 20 & 21 & 24 & 13 \\ 23 & 20 & 0 & 29 & 31 & 14 \\ 15 & 21 & 29 & 0 & 9 & 25 \\ 18 & 24 & 31 & 9 & 0 & 17 \\ 10 & 13 & 14 & 25 & 17 & 0 \end{bmatrix}$$

Use the nearest neighbour algorithm to find a short Hamilton cycle.

4 The diagram shows a technician's desk, T, together with the positions of six dials, A, B, C, D, E and F, that have to be inspected every ten minutes. The technician can take the direct line route between any pair of points.

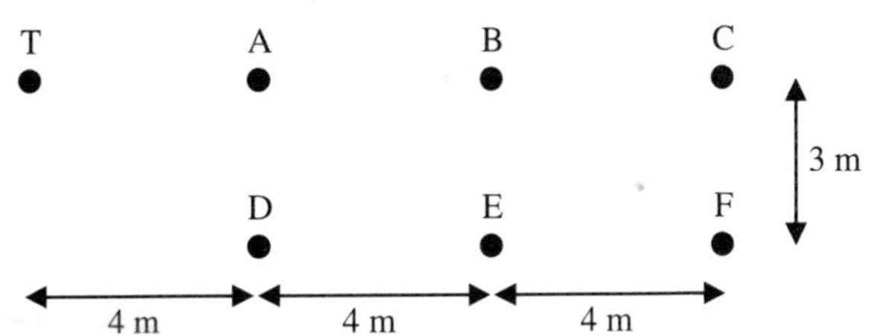

(a) Use the nearest neighbour algorithm to find a route for the technician, and give the length of your route.

(b) Use the minimum connector/shortcuts method from page 78 to find a shorter route than that given by the nearest neighbour algorithm, and give the length of your route.

(c) Suggest a new position for the technician's desk which leads to a reduction in the distance given in part (b).

5 The following matrix gives the costs of airline flight tickets for direct flights between six connected cities:

To

From		A	B	C	D	E	F
	A	—	40	55	53	85	140
	B	40	—	62	20	78	95
	C	55	65	—	45	65	315
	D	45	20	45	—	30	205
	E	95	75	65	30	—	67
	F	140	105	295	170	75	—

Use a greedy algorithm to find a Hamilton cycle, starting and finishing at A, with a low associated cost. Show that the algorithm has not produced the minimum cost tour.

6 The diagram shows road connections and distances (in km) between some major archaeological sites in Crete.

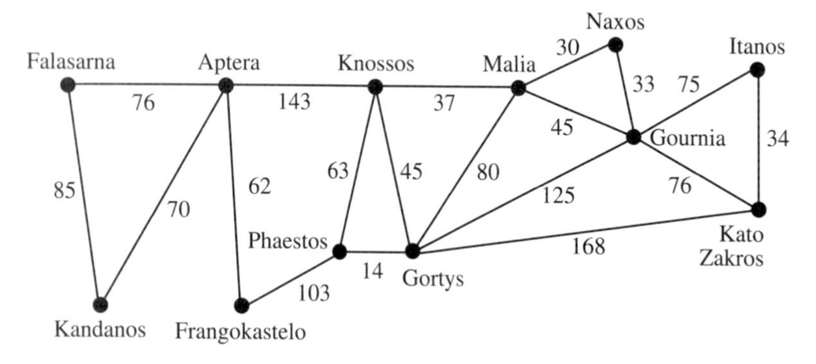

(a) Show that there is no Hamilton cycle in this network.

(b) Use the nearest neighbour algorithm to find a short tour, starting and ending at Knossos, of sites other than Falasarna and Kandanos. Use this to construct a complete tour.

(c) Find the minimum length tour.

Examination style paper D1a

Answer all questions **Time 90 minutes**

1. Use the quick-sort algorithm to sort the following list of names into alphabetical order:

Takashima, Jones, Evans, Chung, Basra, Smith, Adams.

Show the state of the list after each rearrangement.

(6 marks)

2. The dependence table for a particular project is given in Table 1.

Activity	Depends on
A	—
B	—
C	—
D	A
E	B, C
F	B, C
G	C
H	D, F, G
I	D, E
J	D

Table 1

Draw an activity network for this project. Your network should include a minimum number of dummy activities.

(8 marks)

3. Figure 1 shows a capacitated network to be used for transporting goods from A to G. The numbers on the arcs indicate the capacity of the arcs.

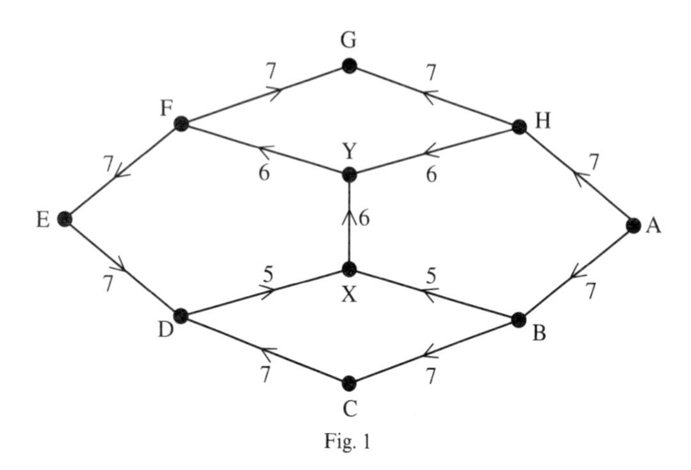

Fig. 1

(*a*) Use a suitable algorithm to find the maximum flow from A to G, showing all your labelling on a copy of the network.

(*b*) Draw a network showing the maximum flow.

(*c*) Verify your answer using the maximum flow–minimum cut theorem, listing the arcs that your minimum cut passes through. **(10 marks)**

4. In the network shown in Fig. 2 the edge weights represent distances in miles.

There are 4 tours which start at A, visit each vertex once and once only, and return to A.

(*a*) List the vertices in these 4 tours and give the total length of each.

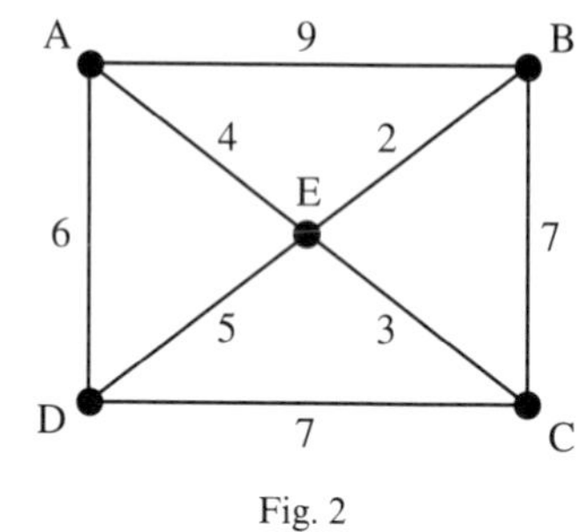

Fig. 2

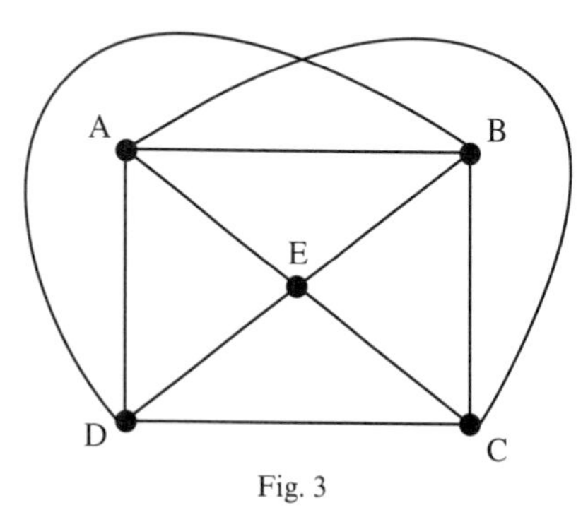

Fig. 3

(*b*) Copy the graph shown in Fig. 3, in which each vertex A, B, C, D, E is joined to every other vertex. Give each edge a weight equal to the *minimum* distance between its end vertices in Fig. 2 by any possible route. (You may do this by inspection. An algorithm is not required.)

Algorithm

Step 1 Let A be the current vertex.

Step 2 Find the nearest unvisited vertex to the current vertex, move to that vertex and call it the current vertex.

Step 3 Repeat step 2 until all vertices have been visited, and then return to A.

(*c*) Use this algorithm to find a tour starting from A, visiting each vertex once and only once, and returning to A, in the network shown in Fig. 3. State the total length of this tour.

(*d*) Interpret the tour you found in (*c*) in terms of a tour found in part (*a*), but in which vertices may be revisited. **(11 marks)**

5. A coach has to select her netball team, in which there are 7 positions, from among 8 players. The positions and the players are represented by the columns and the rows in the table below. A '√' in a cell indicates that the corresponding player can play in the corresponding position.

	Goal Shooter (GS)	Goal Attack (GA)	Wing Attack (WA)	Centre (CE)	Wing Defence (WD)	Goal Defence (GD)	Goal Keeper (GK)
Aretha	√	√	√				
Bhavni			√	√	√	√	
Cathy				√	√	√	√
Deepna	√						
Eve			√	√	√		
Freda	√	√	√	√	√		
Gaynor		√	√		√		
Helen			√	√	√	√	√

The coach would like to play the following in the indicated positions:

Aretha→GS; Bhavni→WA; Cathy→CE;
Eve→WD; Gaynor→GA; Helen→GD.

However, this would leave her without a Goal Keeper.

(*a*) Find an alternating path from Goal Keeper to an unused player. Use your alternating path to produce an improved matching, corresponding to a full team.

The coach starts the game with the following team:

Aretha→GS; Bhavni→WA; Cathy→GK;
Eve→WD; Freda→CE; Gaynor→GA; Helen→GD

Part way through the game Helen sustains an injury, so the coach has to bring Deepna on to play.

(*b*) Find an alternating path, starting from Deepna, which will give the coach a rearrangement of positions allowing Deepna to play as Goal Shooter.
List the full set of team positions implied by your path.

(13 marks)

6. A new pen has just been introduced. It comes in two versions, 'standard' and 'luxury'. The minimum order for standard pens is 24 and the minimum order for luxury pens is 6. In the light of previous experience Mrs Black orders at least twice as many standard pens as luxury pens. She pays £2.50 for a standard pen and £7.50 for a luxury pen. The maximum she is allowed to spend on ordering this new pen is £200. If she purchases x standard pens and y luxury pens,

(*a*) write down the inequalities satisfied by x and y.

The standard pen sells for £4 and the luxury pen sells for £15.

(*b*) Determine graphically how many of each pen Mrs Black should purchase in order to maximise her profit, given that all pens purchased are sold in a given week. State this profit.

(14 marks)

7. A gardener wishes to connect sprinklers A, B, C, D, E, F and G to a water hydrant at H. The possible routes for hoses are shown in Fig. 4 and the associated distances, in tens of metres, are given in Table 2.

	A	B	C	D	E	F	G	H
A	—	20	—	—	—	—	15	50
B	20	—	15	—	—	—	12	14
C	—	15	—	—	—	—	—	25
D	—	—	—	—	18	—	—	12
E	—	—	—	18	—	11	—	10
F	—	—	—	—	11	—	20	23
G	15	12	—	—	—	20	—	—
H	50	14	25	12	10	23	—	—

Table 2

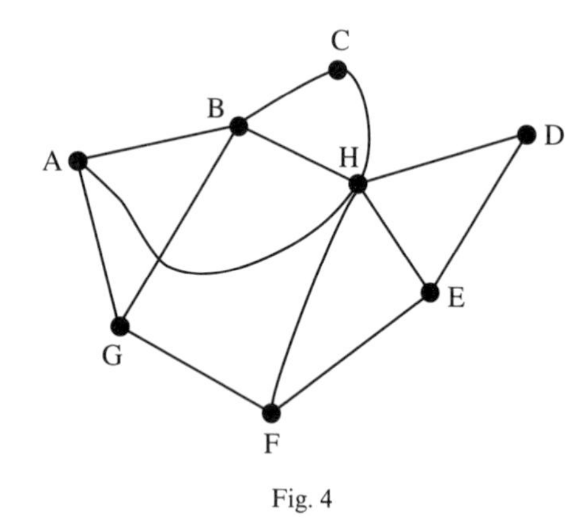

Fig. 4

(*a*) Use the matrix form of Prim's algorithm to find a minimum spanning tree (minimum connector), MST, for the network. Show your minimum spanning tree on a diagram and state the minimum length of hose needed by the gardener.

The gardener decides that the minimum connector solution will lead to a capacity problem in the section of hose connecting H to B.

(*b*) Find an alternative plan for the hose as follows:
  (i) Use Dijkstra's algorithm to find the shortest routes from H to all other vertices.
  (ii) Lay hoses along these routes serving intermediate sprinklers en-route.

(*c*) Show this hose plan on a diagram.

(*d*) What extra length of hose will be needed to implement this solution? **(18 marks)**

8.

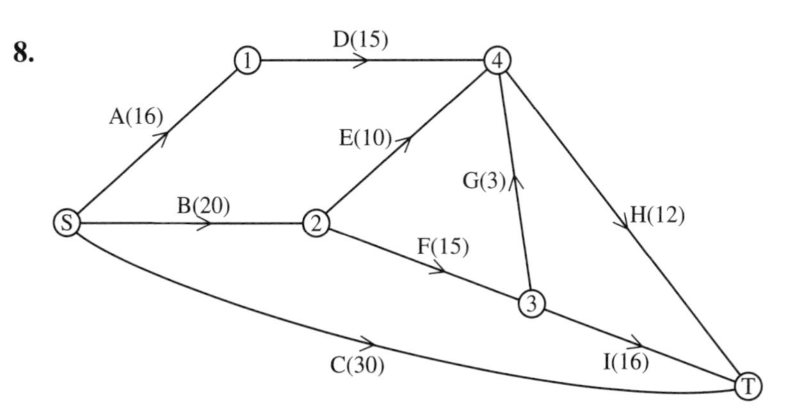

Fig. 5

The activity network for an industrial project is shown in Fig. 5. The arcs are labelled with the time, in days, required to complete the corresponding activity. The start vertex is S and the terminal vertex is T.

(*a*) Find the earliest and latest time for each event and hence determine the length of the critical path.

(*b*) Determine the critical events, the critical activities and the critical path.

(*c*) Draw up time schedules for the project
  (i) with non-critical activities as early as possible,
  (ii) with non-critical activities as late as possible.

Given that each activity requires one worker,

(*d*) determine the minimum number of workers required to complete the project in the critical time. **(20 marks)**

Examination style paper

D1b

Answer all questions **Time 90 minutes**

1.

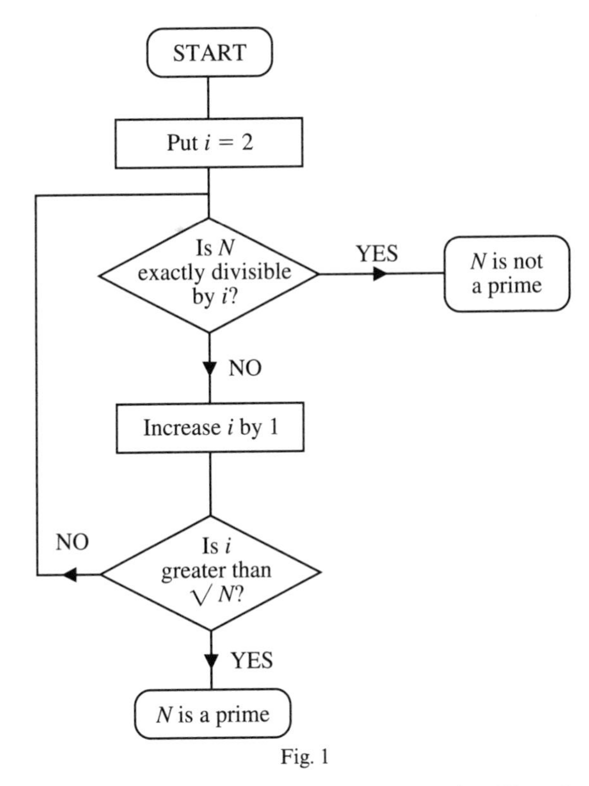

Fig. 1

The algorithm given by the flow chart in Fig. 1 determines whether a number N is a prime number. Use this algorithm to determine whether 119 is prime or not. Show the result of each step of the algorithm. **(7 marks)**

2. (*a*) Describe briefly the first fit decreasing algorithm for bin packing.

The lanes on the car deck of a ferry are 20 m long. Vehicles with the following lengths, in metres, have to be fitted into these lanes:

9, 6, 15, 7, 3, 4, 12, 3

(*b*) Show that at least 3 lanes will be required.

(*c*) Carry out the algorithm you have described in part (*a*) and show that in fact 4 lanes are required. **(9 marks)**

3. The secretary of a football league is in the process of arranging next year's fixture list. There are 6 teams in the league, and they play each other twice, once at home and once away. The secretary has arranged fixtures for weeks 1 to 6 of the season as indicated in the matrix:

		away team					
		A	**B**	**C**	**D**	**E**	**F**
home team	**A**	—		4	2	6	
	B	1	—	6		5	
	C	3	2	—	1		
	D	5			—	3	6
	E		4			—	2
	F		3	5	4	1	—

He is now trying to organise week 7. He has pencilled in C to play *away* at E and A to be at *home* to B. However, this would leave D to play F, and that match has already been staged twice – home and away.

(*a*) Draw a bipartite graph to represent the current position for weeks 1 to 6. Let the first vertex set be {A, B, C, D, E, F}, representing home teams, and let the second vertex set be {A, B, C, D, E, F}, representing away teams. The edges of the graph should represent blanks in the current fixture matrix. Mark those edges representing the pencilled-in part matching for week 7.

(*b*) The edge from home team D to away team C to home team E to away team D is an alternating path. Why does it not lead to an improved matching?

(*c*) Find an alternating path from home team D which does lead to an improved matching. Hence produce an improved matching and interpret that matching in terms of fixtures for week 7. **(10 marks)**

4. Every item produced in a factory has to go through two stages, grinding first and then polishing. The factory has two production lines. The first and older line has two grinders, each of which can operate at 20 items per hour. These feed one polisher, which operates at 30 items per hour.

The newer production line has a fast grinder, which can handle 80 items per hour, and which feeds two modern polishers, each with capacity of 40 items per hour.
There is also a small transporter which can take part-finished items from the new grinder to the old polisher, or from the old grinders to the new polishers. This can handle up to 20 items per hour.

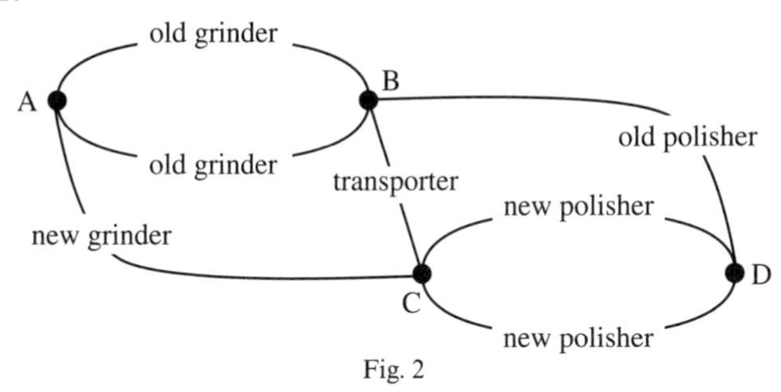

Fig. 2

The new grinder and the old polisher have been undergoing maintenance, and one of the old grinders has been feeding 20 items per hour to one of the new polishers via the conveyor. The machines are now all available and the current flow pattern needs to be changed.

(*a*) Explain why the following capacitated graph can be used as a model of the factory:

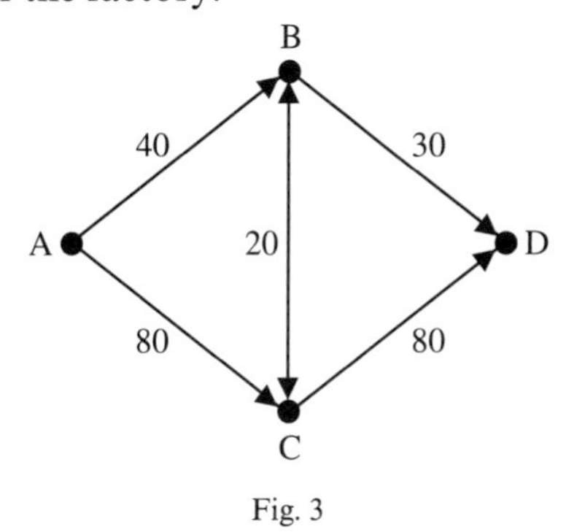

Fig. 3

(*b*) Mark the current flows on a copy of the graph, and show the unused capacity and back capacity of each edge.

(*c*) Use a suitable algorithm to establish a maximum flow through the network.

At each stage show your new flows and updated unused capacities and back capacities on a fresh copy of the graph. (Marks will not be awarded for a solution which does not show the steps of the algorithm.)

(*d*) State the maximum total flow through the factory, and give a cut which proves that it is the maximum flow.

(12 marks)

5. A minibus operator is contracted to transport 50 workers to their place of employment. He has available 3 type A minibuses and 4 type B minibuses. A type A minibus carries 15 workers and a type B minibus carries 10 workers. At this time of the day only 5 drivers are available. It costs £50 to operate a type A minibus and £40 to operate a type B minibus. Let x be the number of type A minibuses used and y the number of type B minibuses used.

(*a*) Write down four inequalities satisfied by x and y, other than $x \geqslant 0$, $y \geqslant 0$.

(*b*) Display these inequalities on a graph and label clearly the feasible region.

Given that x and y must be integers,

(*c*) determine the possible combinations of (x,y) which satisfy the constraints,

(*d*) obtain the minimum cost of the operation and the number of type A and type B minibuses used in this case.

(13 marks)

6.

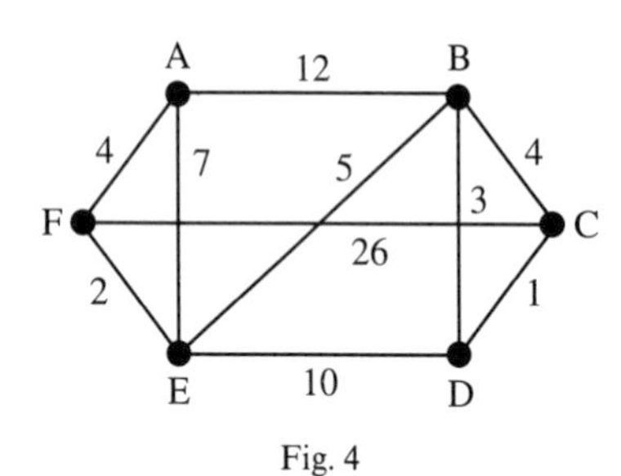

Fig. 4

(*a*) Use Dijkstra's algorithm to find the shortest route from A to D in the network shown in Fig. 4. The weights on the edges represent distances.

(*b*) Use Kruskal's algorithm to find a minimum spanning tree for the network. Give the order in which you include edges and the total length of your tree.

(14 marks)

7. A firm estimates that its profit £P is dependent on 3 variables x, y and z and is given by the equation

$$P = x + 4y + 10z.$$

The variables x, y and z are all non-negative and also satisfy the constraints

$$x + 4y + 2z \leqslant 40,$$
$$x \qquad + 4z \leqslant 8.$$

Use the simplex algorithm to determine the maximum value of P and the values of x, y and z for which it occurs.

(17 marks)

8. The logo shown below is to be sewn on to a large number of garments. The sewing therefore needs to be done as efficiently as possible. The sewing machine needle can only move along the edges that are shown, although it need not be sewing all of the time, so it does not have to stitch any edge twice. It must start and finish at the same vertex.

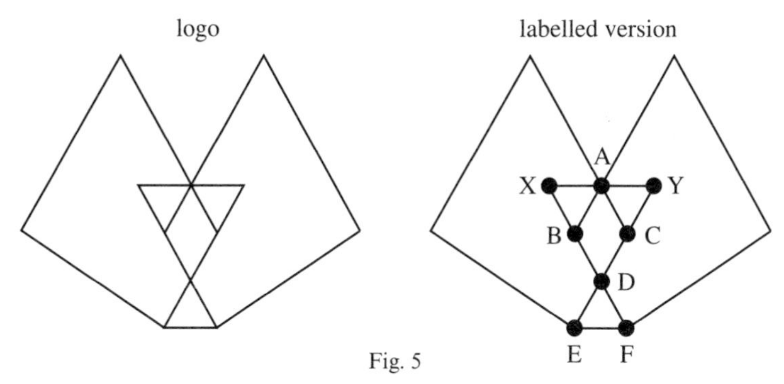

Fig. 5

Vertex-to-vertex distances (in mm) are as follows:

	A	B	C	D	E	F	X	Y
A	—	10	10	—	80	80	10	10
B	10	—	—	10	—	—	10	—
C	10	—	—	10	—	—	—	10
D	—	10	10	—	10	10	—	—
E	80	—	—	10	—	10	—	—
F	80	—	—	10	10	—	—	—
X	10	10	—	—	—	—	—	—
Y	10	—	10	—	—	—	—	—

(*a*) Use the route inspection algorithm to find an optimal route for the sewing machine needle. Show your working and give your route, indicating when the needle is stitching and when it is not stitching. Give the length of your route.

An improved machine control program is installed which means that the needle does not have to follow the shown edges when it is not stitching, and that it does not have to finish at the vertex where it started.

(*b*) How will your solution to part (*a*) be affected by the fact that the needle does not have to follow the shown edges when it is not stitching?

(*c*) Describe how you would modify the route inspection algorithm to allow for a route visiting every edge not to have to end at the starting vertex.

(*d*) Give an optimal route for the needle under the new control program. Give the length of your route. **(18 marks)**

Answers

Exercise 1A

1 (a) $x = 2, x = -1$ (b) $x = -\frac{1}{3}, x = 5$
(c) No real solutions

2 (a) 13 (b)18

3 1, 4, 9, 16, 25, 36, 49, 64, 81, 100
All square numbers less than or equal to 100

4 1, 2, 4, 5, 10, 20, 25, 50, 100
Works out all factors of 100

5

1
1 1
1 2 1
1 3 3 1
1 4 6 4 1
1 5 10 10 5 1
1 6 15 20 15 6 1
1 7 21 35 35 21 7 1

6 1, 1, 2, 3, 5, 8, 13, 21, 34, 55, 89, 144

Exercise 1B

1 (a) 42, 45, 50, 55, 68, 70
(b) 70, 68, 55, 50, 45, 42

2 2, 4, 5, 6, 8, 9, 10

3 Farmer, Jones, Monro, Patel, Wilson

4 –5, –3, –1, 0, 2, 4, 5

5 (a) & (b) 8.9, 9.0, 9.1, 9.2, 9.6, 9.7, 9.8
(c) quick sort

Exercise 1C

1 (a) 5 workers (b) 5 workers (c) No

2 Yes

3 2 workers

4 3 lengths

5 (a) 4 (b) 3 (c) 3

Exercise 2A

1 (a) vertex B has degree 3, vertex G has degree 4
(b) ABHGI or ABCHGI or ABHCGI or ABHCDEFGI or ABHCDEFI or ABHGCDEFI or ABHGFI or ABCHGFI
(c) ABCDEFI or ABCDEFGI or ABHCDEFI or ABHGCDEFGI or ABHGCDEFI or ABHCDEFGI
(d) No. Because EF and DA are directed a cycle must include . . . CDE . . . But that means that neither CA nor DA can be in the cycle. That only leaves AB for getting into and out of A, which would mean visiting B twice.

2

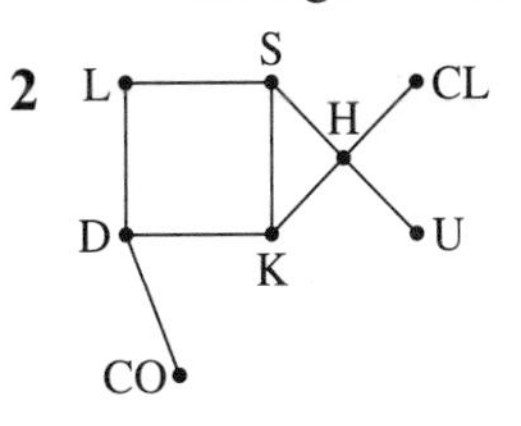

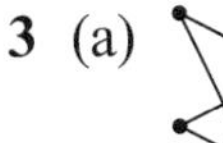

3 (a)

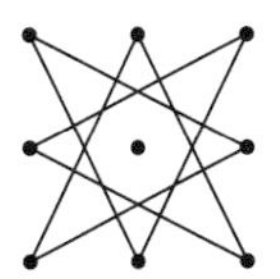

(b)

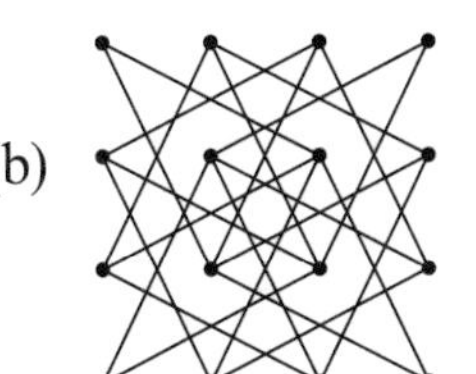

4 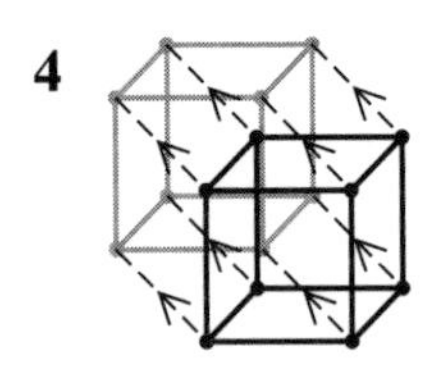

5 (a) e.g. ABEBCDCFGFDEFEA or GFEDFCDCBABEA

(b) e.g. GFEDFCBEAB

(c) e.g. GFDCBEA

6 (a) 3 (b) 4

Exercise 2B

1 (a)

	A	B	C	D	E
A	0	1	1	1	1
B	1	0	1	1	0
C	1	1	0	0	0
D	1	1	0	0	0
E	1	0	0	0	0

(b)

	A	B	C	D	E
A	0	1	1	1	1
B	1	2	1	1	0
C	1	1	2	1	0
D	1	1	1	0	1
E	1	0	0	1	0

(c)

	P	Q	R	S
P	0	2	0	1
Q	0	0	2	0
R	1	2	0	1
S	1	0	0	0

2

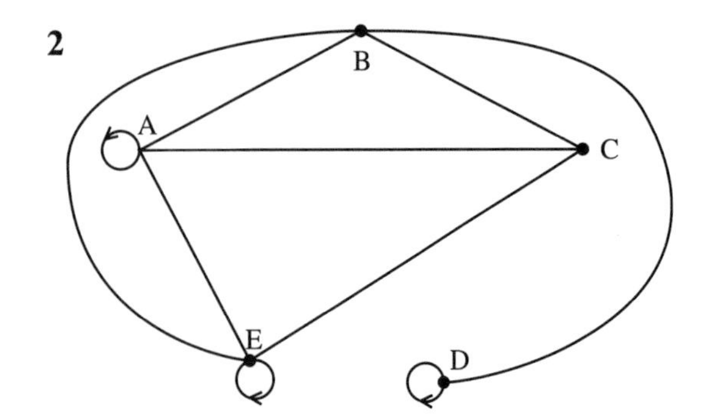

3 (a)

	L	D	S	H	CL	CO	K	U
L	0	1	1	0	0	0	0	0
D	1	0	0	0	0	1	1	0
S	1	0	0	1	0	0	1	0
H	0	0	1	0	1	0	1	1
CL	0	0	0	1	0	0	0	0
CO	0	1	0	0	0	0	0	0
K	0	1	1	1	0	0	0	0
U	0	0	0	1	0	0	0	0

(b) If the vertices are labelled:

1 2 3 4

5 6 7 8

9 10 11 12

13 14 15 16

the matrix is:

	1	2	3	4	5	6	7	8	9	10	11	12	13	14	15	16
1	0	0	0	0	0	0	1	0	0	1	0	0	0	0	0	0
2	0	0	0	0	0	0	0	1	1	0	0	0	0	0	0	0
3	0	0	0	0	1	0	0	0	0	0	0	1	0	0	0	0
4	0	0	0	0	0	1	0	0	0	0	1	0	0	0	0	0
5	0	0	1	0	0	0	0	0	0	0	0	0	0	1	0	0
6	0	0	0	1	0	0	0	0	0	0	0	0	1	0	0	0
7	1	0	0	0	0	0	0	0	0	0	0	0	0	0	0	1
8	0	1	0	0	0	0	0	0	0	①	0	0	0	0	1	0
9	0	1	0	0	0	0	0	0	0	0	0	0	0	0	1	0
10	1	0	0	0	0	0	0	①	0	0	0	0	0	0	0	1
11	0	0	0	1	0	0	0	0	0	0	0	0	1	0	0	0
12	0	0	1	0	0	0	0	0	0	0	0	0	0	1	0	0
13	0	0	0	0	0	1	0	0	0	0	1	0	0	0	0	0
14	0	0	0	0	1	0	0	0	0	0	0	1	0	0	0	0
15	0	0	0	0	0	0	0	1	1	0	0	0	0	0	0	0
16	0	0	0	0	0	0	1	0	0	1	0	0	0	0	0	0

4

		to				
		V	W	X	Y	Z
	V	0	1	1	1	1
	W	1	0	0	1	0
from	X	1	1	1	0	1
	Y	1	1	0	0	1
	Z	0	1	0	1	0

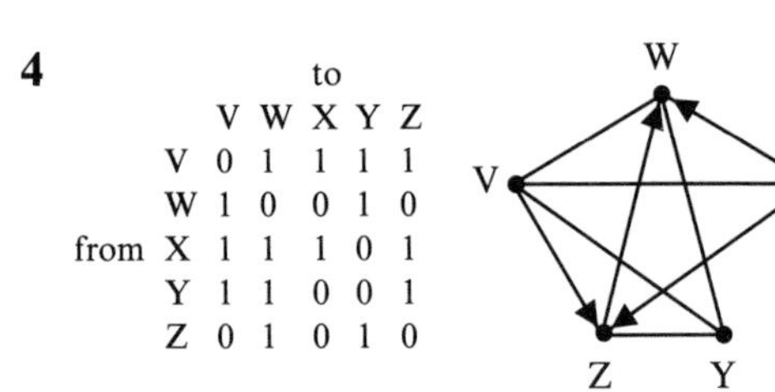

5 (a) Because every edge starts and ends at a vertex, and so contributes 2 to the sum of the degrees of vertices.

(b) The number of odd vertices must be even.

(c) Draw a graph with nine vertices (people). Join by an edge pairs who are friends. Sum of vertex degrees is even. If each was friends with 5 others, all vertex degrees would be 5. But 9×5 would then be the sum of vertex degrees – and 45 is not even.

6

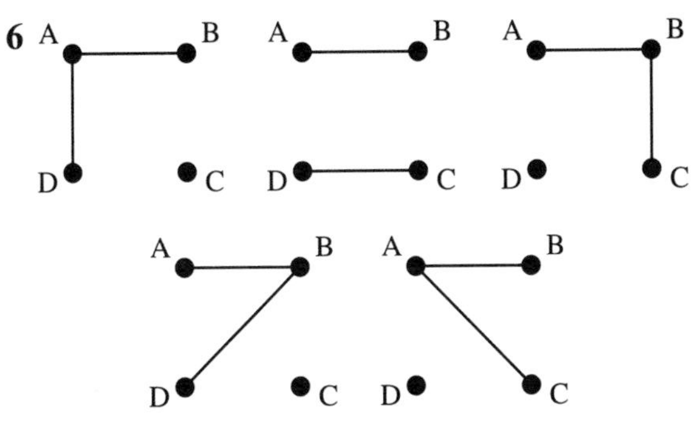

7 (a) not possible

(b)

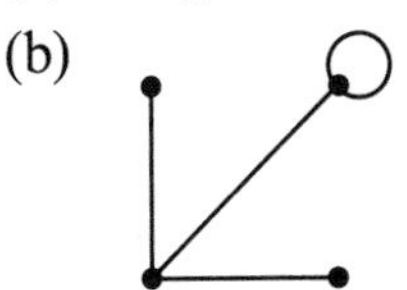

(c) not possible

8 (a)

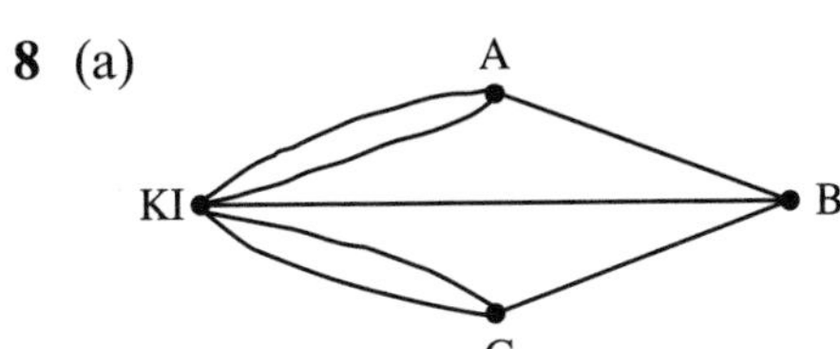

(b) Yes – would have to start and end with the two odd vertices.

(c) Cannot be done since there are 4 odd vertices:

Vertex	A	B	C	KI
Degree	3	3	3	5

9

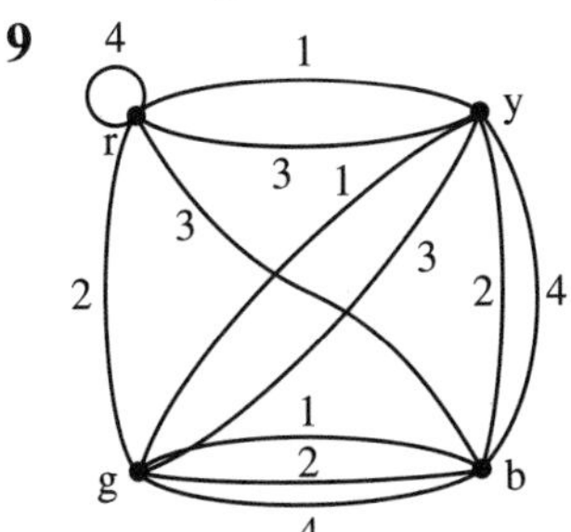

This has, for example, the following two subgraphs:

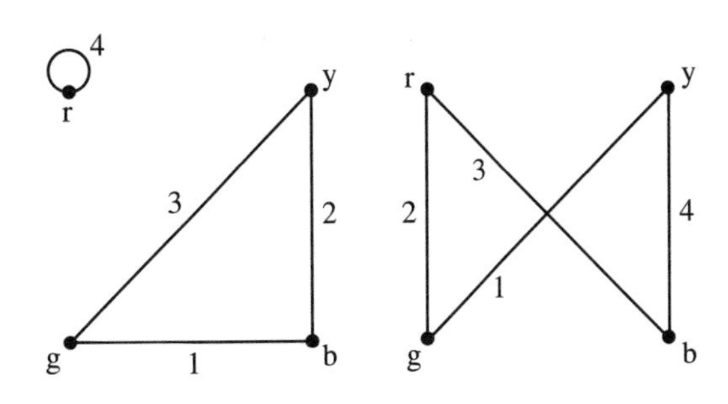

Each has all 4 lines appearing once and only once.

Each has all vertices of degree 2.

The first subgraph can be thought of as front and back colours.

The second subgraph represents top and bottom colours.

e.g.

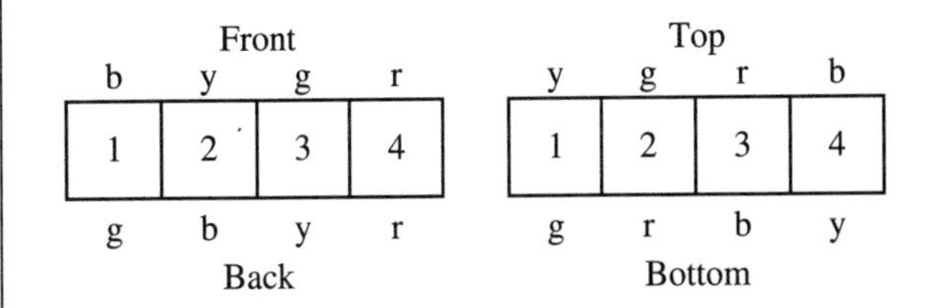

Exercise 2C

1 ABCA, 10; ABDA, 17; ABCDA, 17; ACDA, 11; BCDB, 8

2 ACDEB, 15 ADCB, 11
ACEDB, 26 ADEB, 6
ADCEB, 18 AECB, 16
ADECB, 17 AEDB, 14
AECDB, 26 ACB, 10
AEDCB, 14 ADB, 11
ACDB, 20 AEB, 5
ACEB, 17 AB, 6

3 (a)

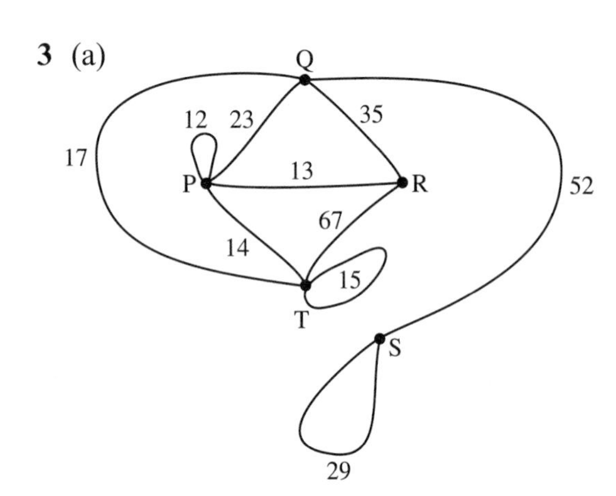

(b) Vertex P Q R S T
Order 5 4 3 3 5

(c) It is not possible to get into and out of S without revisiting Q.

(d) S Q R P T Q S
$52 + 35 + 13 + 14 + 17 + 52 = 183$

4 (a) (i) no (ii) yes (iii) yes (iv) no

(b) no. of edges + 1 = no. of vertices

5 (a)

1 1 0	0 0	0 1	0 1	1 1 1 1	1 1 1 0	1 0	0 0	1 0
W	E	L	L	C	O	D	E	D

6 (a)

- ~A: [0 | 0]
  - ~B: [0 | 0]
    - ~C: [0 | 0]
      - ~D: [0 | 0]
        - ~E: [0 | 0]
        - E: [6 | 5]
      - D: [3 | 4]
        - ~E: [3 | 4]
        - E: [9 | ×]
    - C: [2 | 2]
      - ~D: [2 | 2]
        - ~E: [2 | 2]
        - E: [8 | 7]
      - D: [5 | 6]
        - ~E: [5 | 6]
        - E: [11 | ×]
  - B: [7 | 10]
    - ~C: [7 | 10]
      - ~D: [7 | 10]
        - ~E: [7 | 10]
        - E: [13 | ×]
      - D: [10 | ×]
    - C: [9 | ×]
- A: [2 | 5]
  - ~B: [2 | 5]
    - ~C: [2 | 5]
      - ~D: [2 | 5]
        - ~E: [2 | 5]
        - E: [8 | 10]
      - D: [5 | 9]
        - ~E: [5 | 9]
        - E: [11 | ×]
    - C: [4 | 7]
      - ~D: [4 | 7]
        - ~E: [4 | 7]
        - E: [10 | ×]
      - D: [7 | 11]
        - ~E: [7 | 11]
        - E: [13 | ×]
  - B: [9 | ×]

(b) Best solution is to carry A, C and D, with weight 7 and value 11.

7 (a) 2 units of A and 1 unit of C, with a weight of 950 kg and a fee of £1020

(b) 2 units of A and 1 unit of D, with a weight of 900 kg and a fee of £940

Exercise 2D

1

	1	5	3	2	4
	A	B	C	D	E
A		3	4	2	
B	3		3		(1)
C	4	3		(1)	3
D	(2)		1		1
E		1	3	(1)	

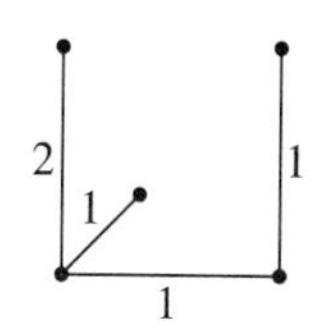

Total weight = 5

2

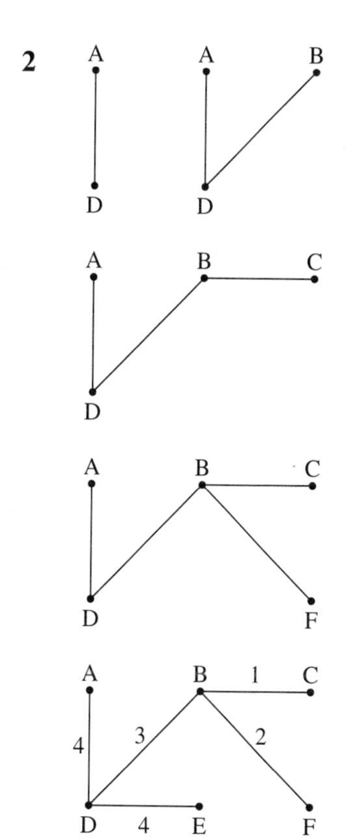

Total weight = 14
(There is an alternative answer, also of total weight 14.)

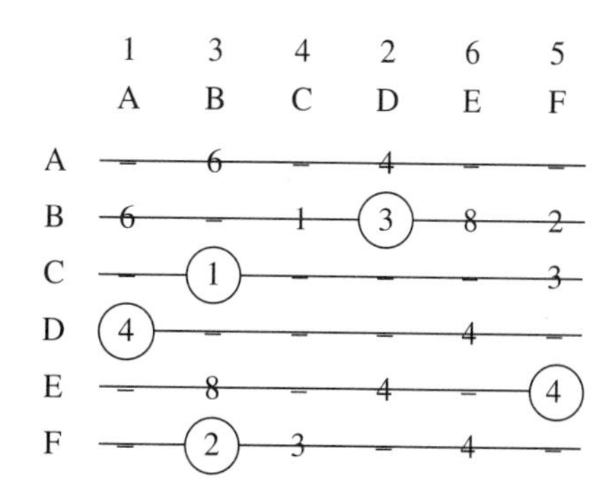

	1	3	4	2	6	5
	A	B	C	D	E	F
A	–	6	–	4	–	–
B	6	–	1	(3)	8	2
C	–	(1)	–	–	–	3
D	(4)	–	–	–	4	–
E	–	8	–	4	–	(4)
F	–	(2)	3	–	4	–

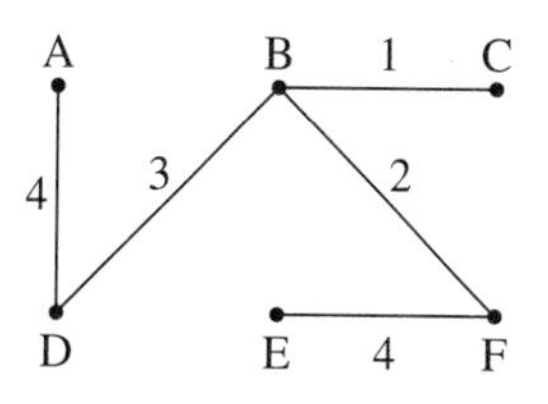

3

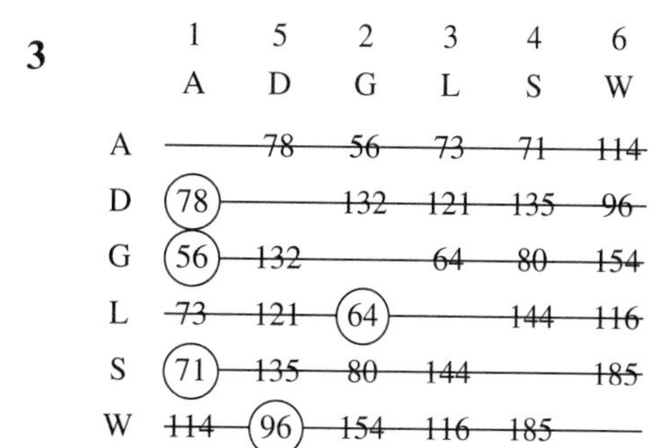

	1	5	2	3	4	6
	A	D	G	L	S	W
A		78	56	73	71	114
D	(78)		132	121	135	96
G	(56)	132		64	80	154
L	73	121	(64)		144	116
S	(71)	135	80	144		185
W	114	(96)	154	116	185	

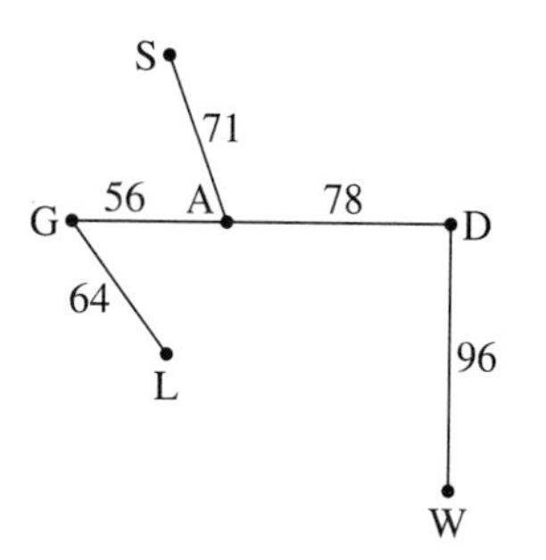

Total length = 365 miles

4

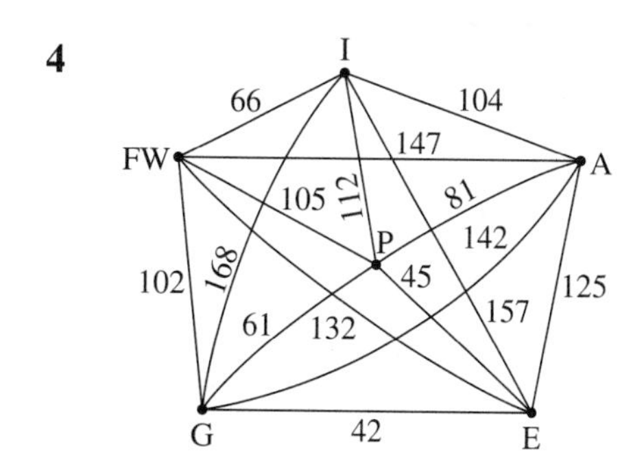

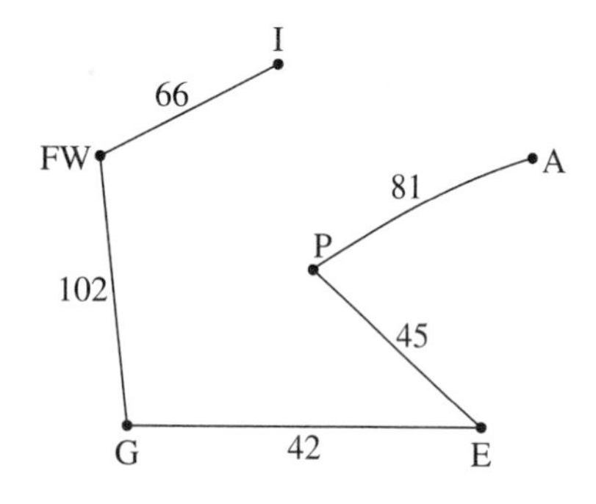

Total length = 336 miles

2

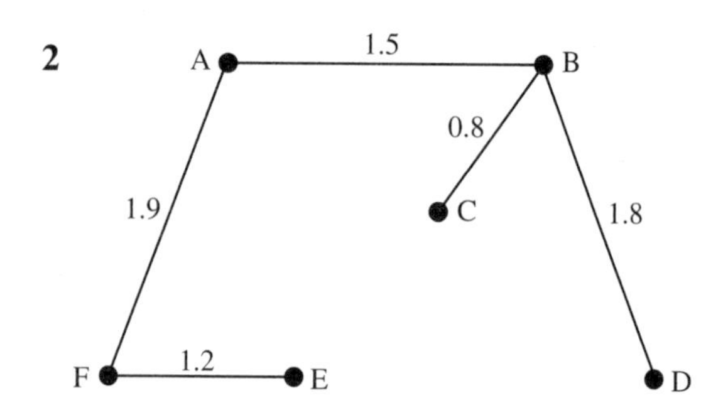

order of inclusion: A, B, C, D, F, E
total cost = £7.2 million

3

	1	2	3	4	6	5
	A	B	C	D	E	F
A	–	~~1.5~~	~~1.6~~	–	~~2.5~~	~~1.9~~
B	(1.5)	–	~~0.8~~	~~1.8~~	–	–
C	~~1.6~~	(0.8)	–	~~2.7~~	~~2.2~~	~~2.9~~
D	–	(1.8)	~~2.7~~	–	~~2.1~~	–
E	2.5	–	2.2	2.1	–	(1.2)
F	(1.9)	–	~~2.9~~	–	~~1.2~~	–

Exercise 2E

1

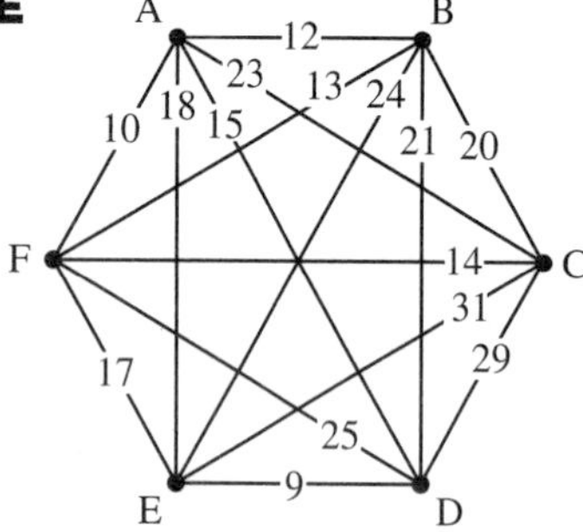

(a)

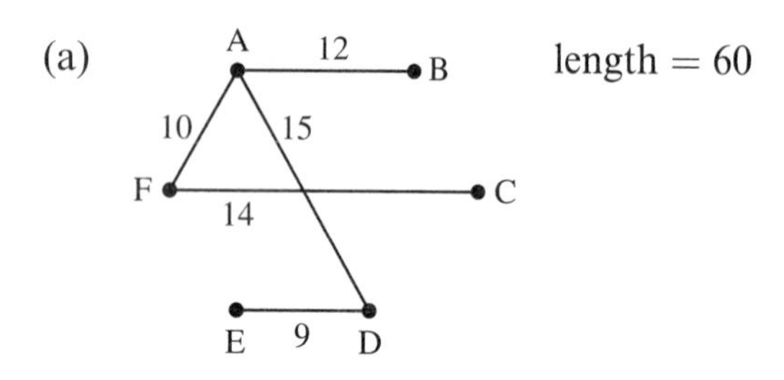

length = 60

(b)

	1	3	4	5	6	2
	A	B	C	D	E	F
A	~~0~~	~~12~~	~~23~~	~~15~~	~~18~~	~~10~~
B	(12)	~~0~~	~~20~~	~~21~~	~~24~~	~~13~~
C	~~23~~	~~20~~	~~0~~	~~29~~	~~31~~	(14)
D	(15)	~~21~~	~~29~~	~~0~~	~~9~~	~~25~~
E	~~18~~	~~24~~	~~31~~	(9)	~~0~~	~~17~~
F	(10)	~~13~~	~~14~~	~~25~~	~~17~~	~~0~~

4 (a) e.g.

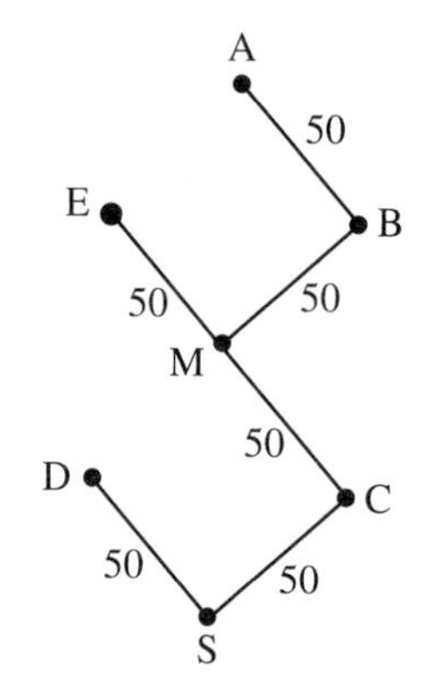

Length = $6 \times 50 = 300$ m
This drains correctly (others would not, e.g. if EA was used instead of EM).

(b)

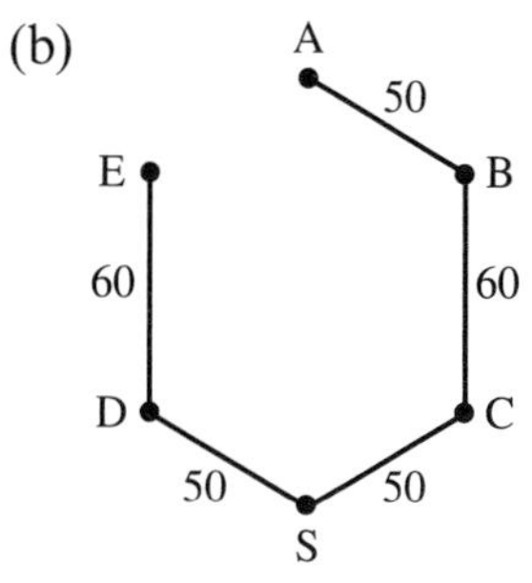

ED needed rather than EA for correct drainage

Total length = 270 m

Manhole not needed

5

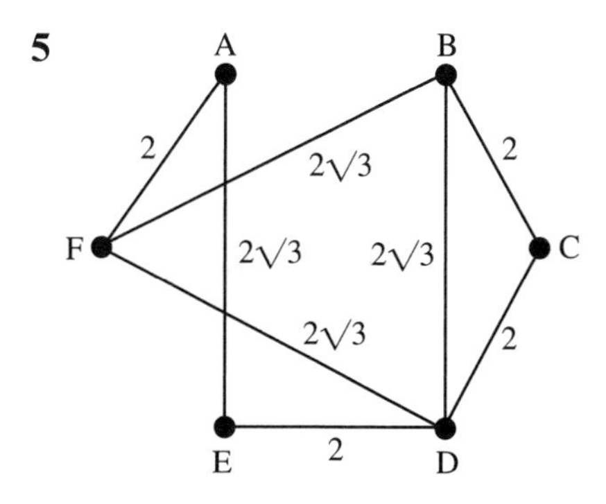

AF FB BC CD DE
AF FD DE DC CB
AF AE ED DC CB

length $= 8 + 2\sqrt{3}$

6

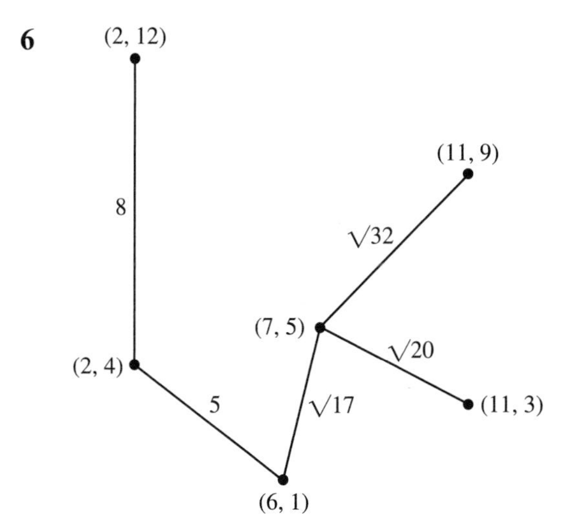

length $= 8 + 5 + \sqrt{17} + \sqrt{20} + \sqrt{32} \approx 27.25$

(b) (i) e.g.

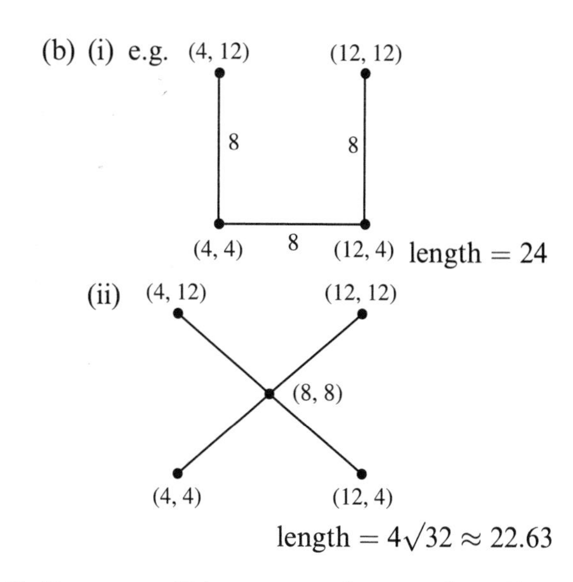

length = 24

length $= 4\sqrt{32} \approx 22.63$

7 To execute Prim on complete graph on 5 vertices need to first choose
smallest of 4 edges
then smallest of $2 \times 3 = 6$ edges
then smallest of $3 \times 2 = 6$ edges
then smallest of $4 \times 1 = 4$ edges
Total comparisons = 20

On the complete graph on 6 vertices the number of comparisons is

$5 + (2 \times 4) + (3 \times 3) + (4 \times 2) + (5 \times 1) = 35$

So time $\approx \frac{35}{20} \times$ time for 5
$= 1.75 \times$ time for 5

For 50 vertices no. of comparisons is 20 825

So time $\approx \frac{20\,285}{20} \times$ time for 5
$\approx 1000 \times$ time for 5

8 (a)

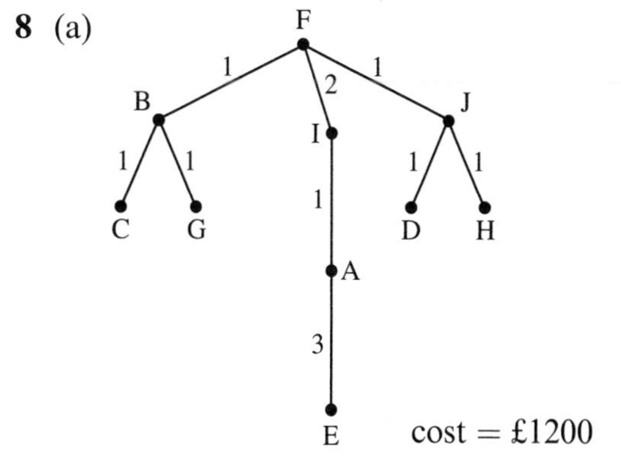

cost = £1200

(b) No – cost of translating out of C generally greater than cost of translating into C (though the minimum spanning tree must be found before we can be sure that this makes an overall difference to the cost).

(c) Because D $\leftrightarrow x \leftrightarrow$ A costs less than 10 for all $x =$ B, C, E, H, I or J. Sack the A $\leftrightarrow$ D translator.

Exercise 2F

1

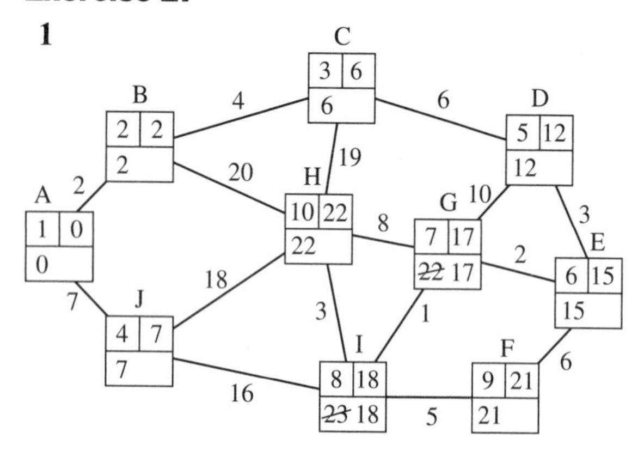

(a) ABCDE – 15 (b) EGIH – 8

(c) ABH – 22

2 (a)

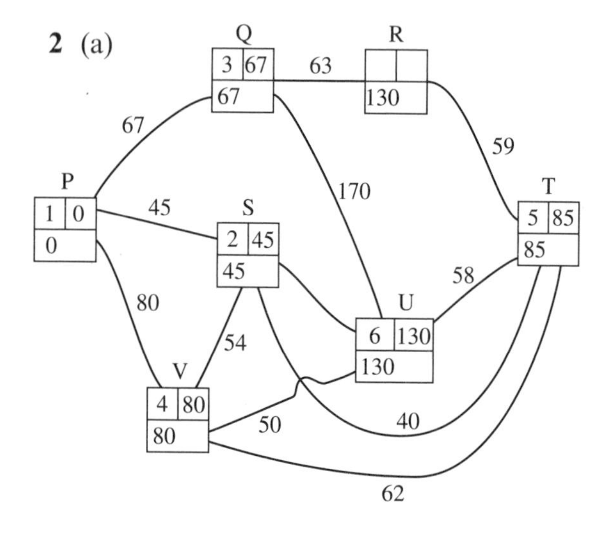

(b) PVU – 130p

3

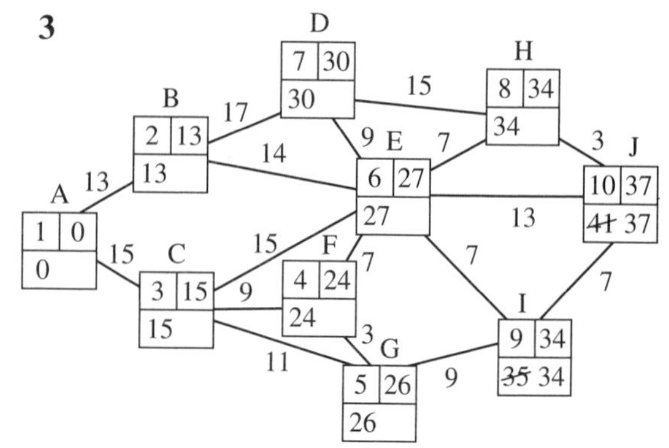

(a) ABEHJ – 37 km

(b) distances $\rightarrow$ times in minutes
Either add 5 (minutes) to all edges incident on E (ensuring that entering and leaving E collects a delay of 10 min) *or* delete E and compare fastest time from A to J in reduced network to $37 + 10 = 47$ min (fastest time through E).
ACGIJ – 42 min

4 (a) shortest route: Frangokastelo; Phaestos; Gortys; (Malia;) Gournia; Itanos distance = 317 km

(b) fastest route: Frangokastelo; Aptera; Knossos; Malia; Gournia; Itanos time = 4.99 hours

(c) shortest route Kandanos to Itanos is: Kandanos; Aptera; Knossos; Malia; Gournia; Itanos – 370 km distance saved by ferry $= 370 - 317 = 53$ km

5

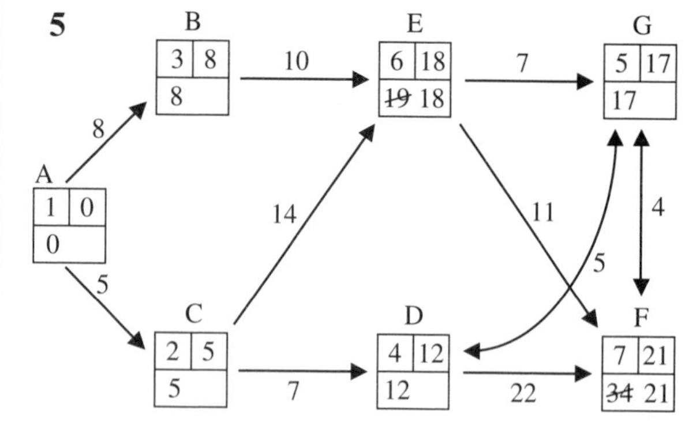

(a) ACDGF – 21

(b)

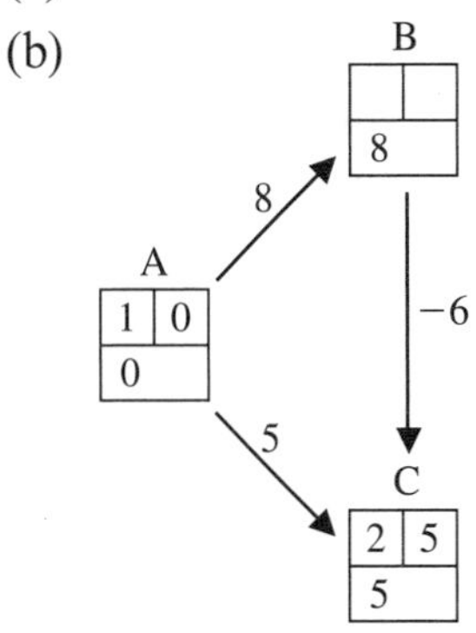

C is labelled second, with a label of 5. But the best route is ABC at a cost of $8 - 6 = 2$.

If we could return to B at a cost of less than 6 we could cycle back and forth with ever increasing profit.

6 (a) Working values are computed by the following:

Compute min(label, link) and replace working value if this is bigger than the current working value. The vertex to be labelled is that with the largest working value. Fastest chain is ABCFG, with a speed of 9.

(b) Slowest chain must include DF, e.g. ADFG with a speed of 2.

7 (a) In his first step Colin has to compute 4 working values. Having labelled a vertex he then has to consider 3 working value computations. In all he has $4 + 3 + 2 + 1 = \dfrac{5 \times 4}{2} = 10$ working value computations.

In a complete graph on 10 vertices he will have $\dfrac{10 \times 9}{2} = 45$ such computations.

So it will take him approximately 45 minutes.

For 100 vertices it will take him $\dfrac{100 \times 99}{2} = 4950$ minutes ($82\frac{1}{2}$ hours).

(b) $22\frac{1}{2}$ minutes

(c) One application of Dijkstra will find all shortest routes from (say) A. So to find all shortest routes in a complete network on 5 vertices will require (in the worst case) 4 applications of Dijkstra. (We can be certain that the 5th application will not be needed since, after the 4th, all shortest routes involving the 5th vertex will be known.) Thus:

No of vertices	5	10	100
Colin's time	$4 \times 10 = 40$ min	$9 \times 45 = 405$ min	99×4950 min ≈ 340 days
Joan's time	$4 \times 5 = 20$ min	$9 \times 22\frac{1}{2} = 202\frac{1}{2}$ min	99×2475 min ≈ 170 days

8

Month	1	2	3	4	5	6
Product	D	D or B	B	B or D	B	A

Profits less costs of switching = £245 000

Exercise 3A

1 (a)

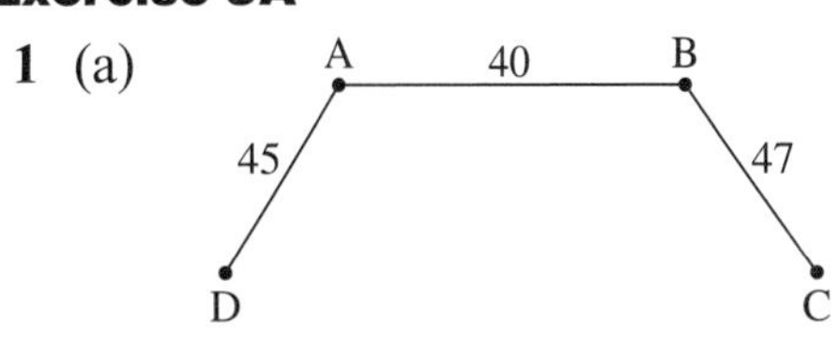

length 132

Hence an upper bound is $2 \times 132 = 264$ for DABCBAD

(b) If on reaching C we go directly to D we have a reduction of

$$(45 + 40 + 47) - 90(\text{CD}) = 42$$

giving an upper bound of 222.

2 (i)

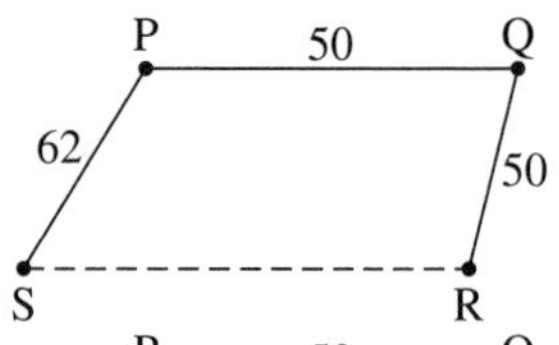

(ii)

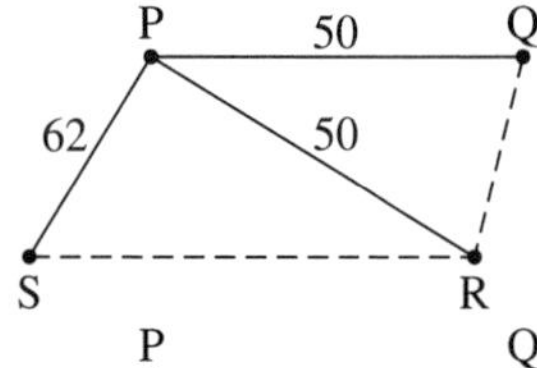

(iii)

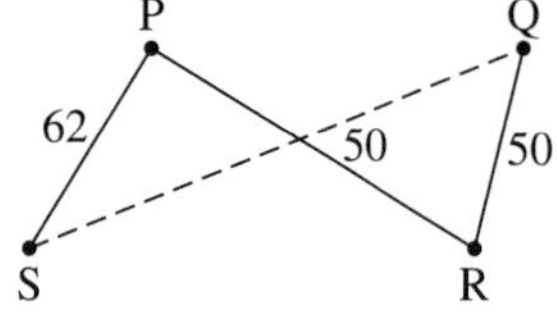

Upper bound is $2 \times 162 = 324$

(i) Using RS gives

$62 + 50 + 50 + \text{RS}(70) = 232$

(ii) Using QR and SR also gives 232

(iii) Using only QS gives

$62 + 50 + 50 + 104 = 266$

3 (a) (i)

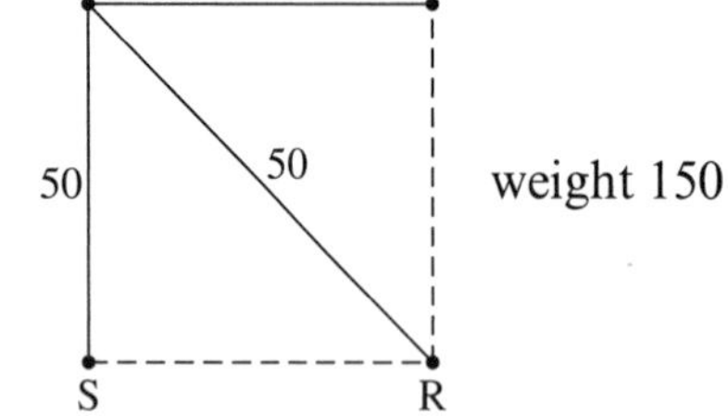

weight 150

(ii)

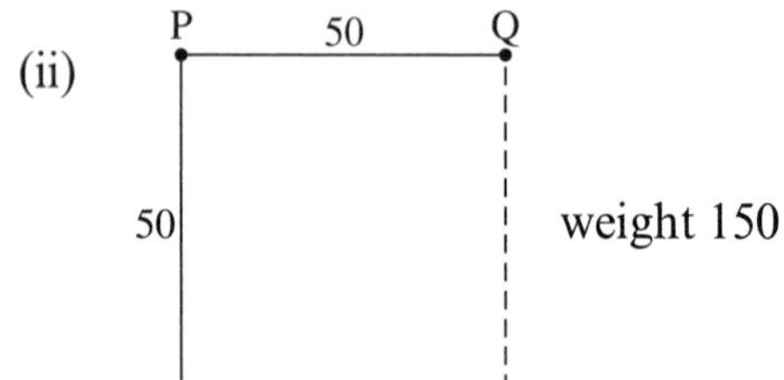

weight 150

(iii)

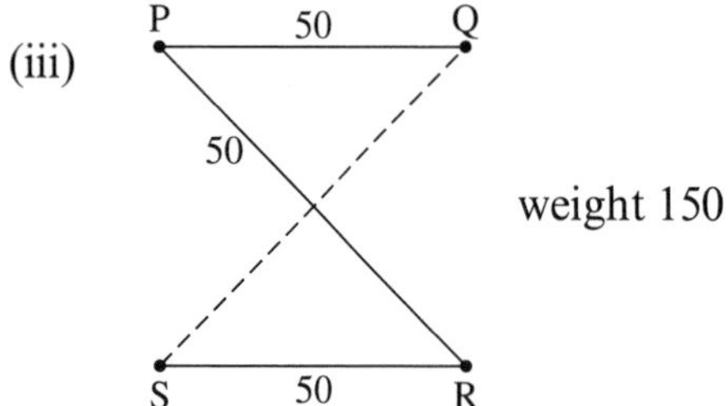

weight 150

Upper bound $2 \times 150 = 300$

(b) (i) Using cuts shown 217

(ii) Using cut shown 217

(iii) Using cut shown 235

4 (a) Two possible MSTs:

(i)

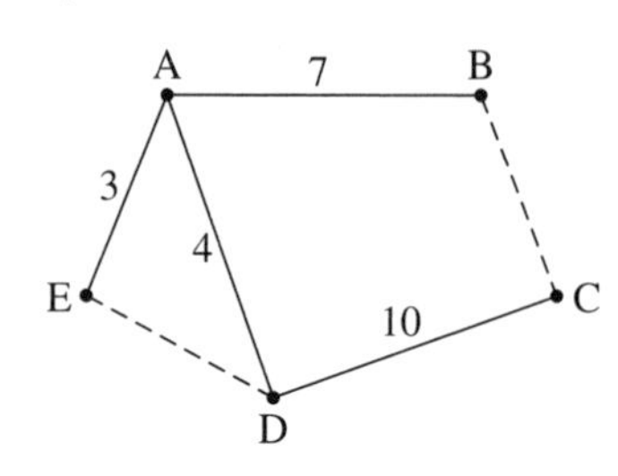

(ii)

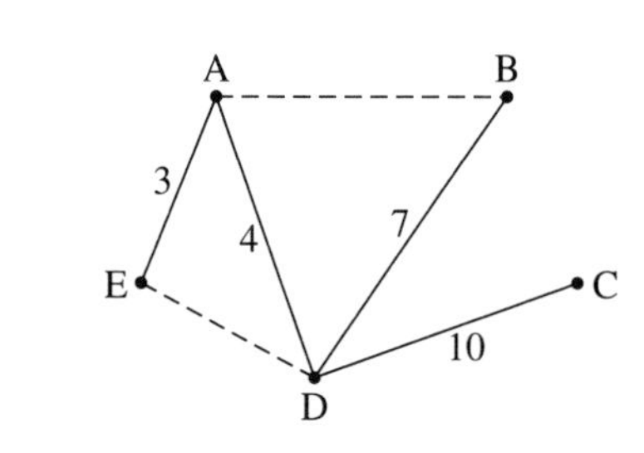

weight 24

Hence an upper bound is $2 \times 24 = 48$

(b) (i) Using shortcuts BC and ED in (i) reduces this by $(21 - 17) = 4$ and $(7 - 5) = 2$ and so gives 42.

(ii) Using BC does not improve upper bound but using AB does by $(11 - 7) = 4$ and ED as before reduces by 2, again giving 42.

5 (a)

	1	4	3	2	5
	A	B	C	D	E
A	—	17	10	9	12
B	17	–	(8)	14	5
C	10	8	–	(7)	11
D	(9)	14	7	–	11
E	12	(5)	11	11	–

AD(9), DC(7), CB(8), BE(5) length 29

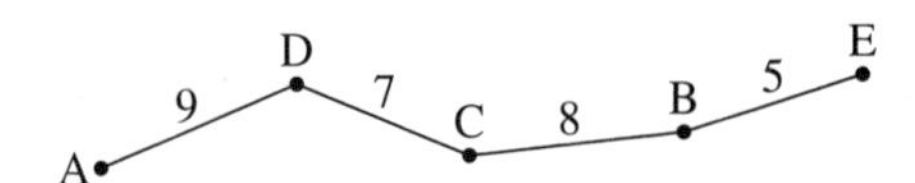

(b) Hence an upper bound is $2 \times 29 = 58$ (ABCBEBCDA).

(c) If when we reach E we shortcut back to A, the length is
$9 + 7 + 8 + 5 + AE(12) = 41$, which is a better upper bound.

6 (a) A minimum spanning tree is

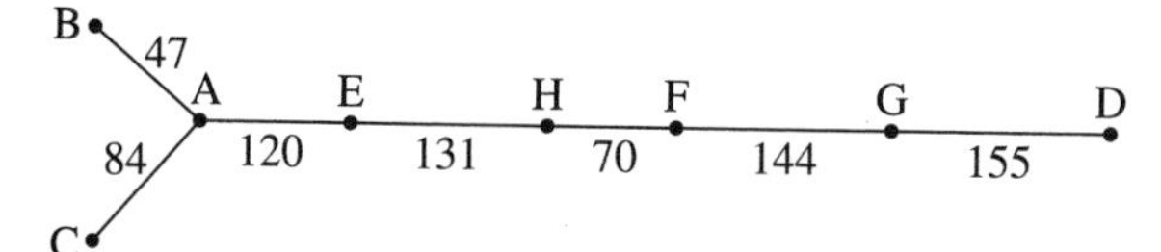

weight 751

(b) An upper bound is
$2 \times$ weight of MST $= 1502$
BACAEHFGDGFHEAB

(c) If when we reach D we cut directly to B we save
$(47 + 120 + 131 + 70 + 144 + 155)$
$-BD(402) = 265$
This reduces upper bound to 1237 miles.

Exercise 3B

1 (a) Weight of MST(124) $+ 40 + 45 = 209$
(b) Weight of MST(120) $+ 40 + 47 = 207$
(c) Weight of MST(85) $+ 47 + 75 = 207$
(d) Weight of MST(87) $+ 45 + 77 = 209$

2 (a) Weight of MST(112) $+ 50 + 50 = 212$
(b) Weight of MST(100) $+ 62 + 70 = 232$
(c) From Exercise 3A, question 2: length of tour $\leqslant 232$
From (a) and (b): length of tour $\geqslant 232$
$\therefore$ length of minimum tour $= 232$
From (b) above this is PQRSP.

3 (a) Weight of MST(22) $+ 3 + 4 = 29$
(b) Weight of MST(17) $+ 7 + 7 = 31$
(c) Weight of MST(14) $+ 10 + 13 = 37$
(d) Weight of MST(23) $+ 5 + 4 = 32$
(e) Weight of MST(21) $+ 5 + 3 = 29$
(f) $37 \leqslant L \leqslant 42$
(g) ABCDEA; $7 + 17 + 10 + 5 + 3 = 42$

4 (a) Weight of MST(20) $+ 9 + 10 = 39$
(b) Weight of MST(27) $+ 8 + 5 = 40$
(c) Weight of MST(25) $+ 8 + 7 = 40$
(d) Weight of MST(23) $+ 9 + 7 = 39$
(e) Weight of MST(24) $+ 5 + 11 = 40$
(f) From (e) and answer to question 5 of Exercise 3A: $40 \leqslant L \leqslant 41$

In (e) we have

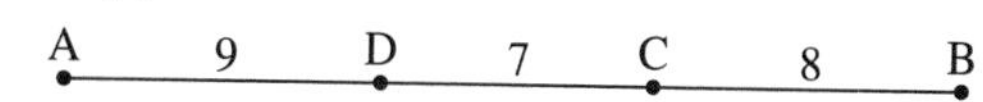

+EB(5) and either EC or ED(11).
EB joins vertex E to one end of chain.
Neither EC nor ED does, but EA does. We then have
EADCBE of length
$12 + 9 + 7 + 8 + 5 = 41$
This is the minimum tour.

5 The MST found previously was:

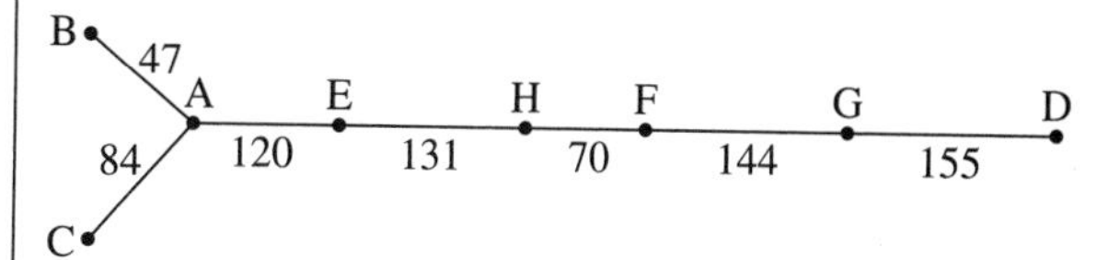

of weight 751.

(a) If we delete B then MST is obtained by removing edge AB from above, giving an MST of weight 704.
The two edges of smallest weight at B are BA(47) and BC(121).
Lower bound is $704 + 47 + 121 = 872$

(b) As above,
(weight of MST) $= 751 - 84 = 667$
Weight of edges to be added $= CA(84) + CB(121)$
Lower bound is $667 + 84 + 121 = 872$

(c) As above,
(weight of MST) = 751 − 155 = 596
Weight of edges to be added = DF(220) + DG(155)
Lower bound is 596 + 220 + 155 = 971

(d) L satisfies $971 \leqslant L \leqslant 1237$

6 (a) best upper bound 53
(b) best lower bound 52 (removing A or D)
(c) $52 \leqslant L \leqslant 53$
(d) shortest possible route: ABEFCDA, length 53

Exercise 3C

1 Edge AC has to be repeated.
Minimum weight route ABCACDA of length 9 + 8 + 10 + 10 + 12 + 7 = 56

2 (a)

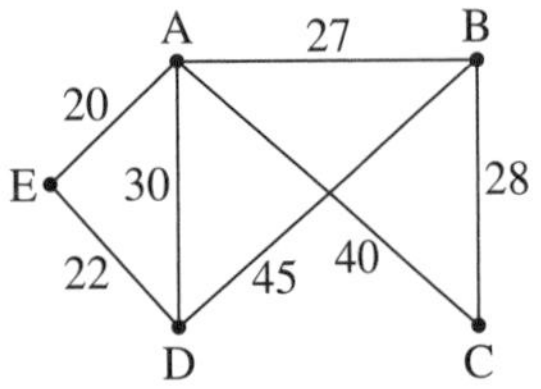

(b) Edge BD must be repeated.
ABCADBDEA of length
27 + 28 + 40 + 30 + 45 + 45 + 22 + 20 = 257 km

3 A pair of edges must be repeated. For a minimum route these edges are AD and BC.
A route is ABCDADBCA.
Length: 30 + 30 + 20 + 20 + 25 + 35 + 30 + 60 = 250 metres

4 A pair of edges must be repeated. For a minimum route these edges are AB and CD.
A route is ABCDEACDBA.
Length: 20 + 20 + 50 + 50 + 50 + 60 + 70 + 40 + 45 = 405 m

5 A cheapest route is
PSTUQRTSQPSVUTVP
(routes PSTU are repeated to allow for the fact that P and U are odd). cost = £9.62

6 (a) odd vertices: A, F, G, C
(b) pairings: AF 5 + GC 3 = 8; AG 9 + FC 5 = 14; AC 8 + FG 4 = 12

Therefore, repeat AF and GC.
length = 48
order (for example):
AFAEFBGFCGCBA

7 (a) odd nodes: A, C, F, I
pairings: AC 260 + FI 60 = 320; AF 44 + CI 240 = 680; AI 500 + CF 180 = 680

Therefore, minimum distance = 2300 + 320 = 2620 m

B	C	D	E	F	G	H	I
3	2	2	2	3	2	2	2

(b) All edges need to be repeated.
All vertices will now be even, therefore traversable.
distance = 2 × 2300 = 4600 m

(c) The street cleaner has to go in the correct direction along each side of the road. Needs 2 directed edges for each original edge, the 2 directions being opposite to each other.

8 (a) Queue time for B + time to ski from top of B to bottom of A = 10 + 8 = 18

(b)

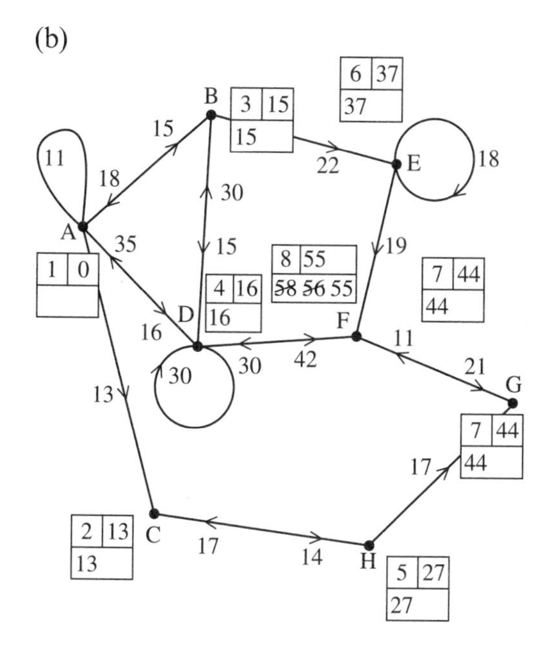

route: ACHGF time: 55 min

(c) *All* directions must be traversed (AB is not the same ski run as BA). Two runs into G and only one out, so the out run, GF, must be repeated.
Similarly, CH must be repeated.
e.g. ACHCHGFGF
Similarly, FD will have to be repeated twice (3 routes in and effectively only 1 out). Two repeats out of D will then be needed. But one repeat into B is needed and one into A, so repeat DA and DB.
e.g.
AACHCHGFGFDDBEEFDFDADABDBA

$$\begin{aligned}\text{time} &= 11+13+14+17+14\\&\quad+17+11+21+11+30\\&\quad+30+30+22+18+19\\&\quad+30+42+30+35+16\\&\quad+35+15+15+30+18\\&= 544\text{ min} \approx 9\text{ hours}\end{aligned}$$

Exercise 4A

1

Activity	*Depends on*
A	—
B	—
C	—
D	A
E	B
F	B, C
G	B, D
H	G
I	E, F, H

2

Activity	*Depends on*
A	—
B	—
C	—
D	—
E	—
F	B, D, E
G	C, F
H	A, G
I	H

3

Activity	*Depends on*
A	—
B	—
C	A
D	B
E	C, D

4

Activity	Depends on
A	—
B	—
C	—
D	A, B
E	D
F	D
G	E, F
H	F
I	C, G
J	H, I

Exercise 4B

1

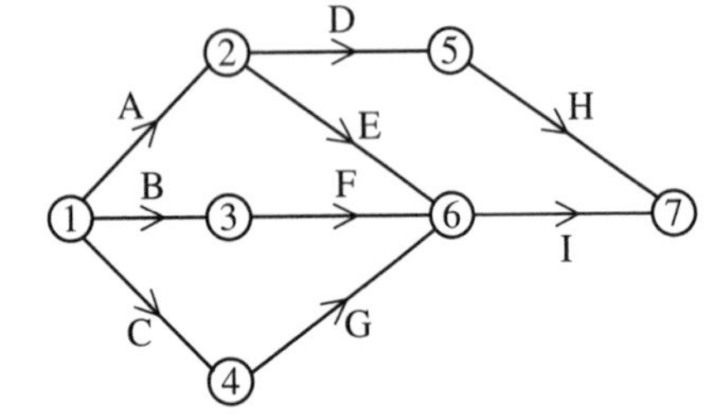

2

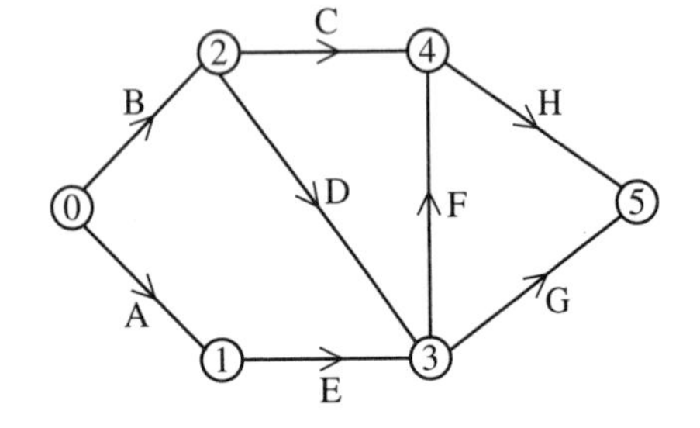

3

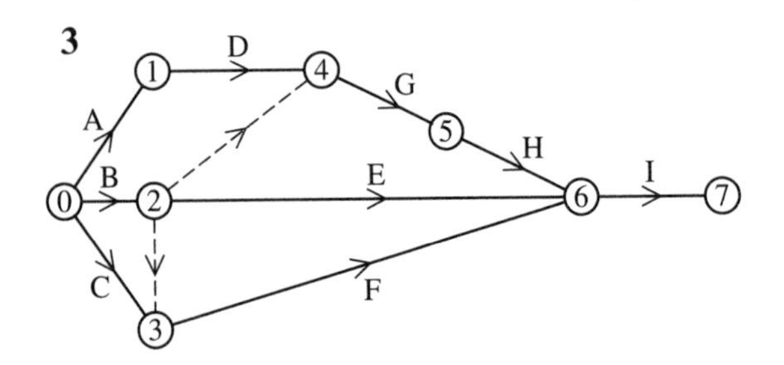

4

Activity	Depends on
A	—
B	A
C	A
D	A
E	A
F	A
G	E
H	D, G
I	D, G
J	B
K	C
L	H, J, K
M	F, I
N	L, M

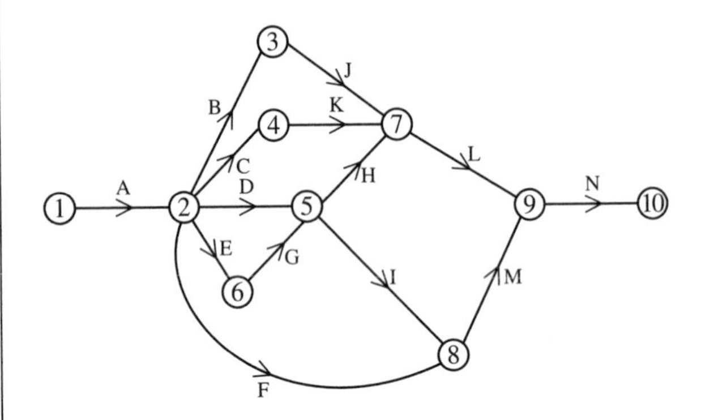

Exercise 4C

1 Completed table is

i	0	1	2	3	4
e_i	0	8	4	10	18
l_i	0	8	10	15	18

2 $e_2 = 7$, $e_4 = 16$, $e_5 = 20$
$l_4 = 16$, $l_3 = 14$, $l_2 = 7$

3

i	0	1	2	3	4	5	6
e_i	0	2	7	9	11	13	21
l_i	0	2	7	9	18	13	21

Exercise 4D

1 (a) $s_0 = 0, s_1 = 0, s_2 = 6, s_3 = 5, s_4 = 0$

So critical events are ⓪, ① and ④.

(b) Activity A — (0, 1) — 0

Activity B — (0, 2) — 6

Activity C — (2, 3) — 6

Activity D — (1, 3) — 5

Activity E — (1, 4) — 0

Activity F — (3, 4) — 5

So critical activities are A and E.

(c) Hence the critical path is

⓪ —A→ ① —E→ ④

which is of length 18.

2 (a) $s_0 = 0, s_1 = 0, s_2 = 0, s_3 = 8, s_4 = 0,$

$s_5 = 0, s_6 = 0.$

So critical events are ⓪, ①, ②, ④, ⑤ and ⑥.

(b) Activity A — (0, 1) — 0

Activity B — (1, 2) — 0

Activity C — (1, 3) — 8

Activity D — (2, 4) — 0

Activity E — (2, 5) — 12

Activity F — (3, 5) — 8

Activity G — (4, 5) — 0

Activity H — (5, 6) — 0

So critical activities are A, B, D, G and H.

(c) Hence the critical path is

⓪ —A→ ① —B→ ② —D→ ④ —G→ ⑤ —H→ ⑥

which is of length 29.

3 (a) $s_0 = 0, s_1 = 0, s_2 = 0, s_3 = 0, s_4 = 7,$

$s_5 = 0, s_6 = 0.$

So critical events are ⓪, ①, ②, ③, ⑤ and ⑥.

(b) Activity A — (0, 1) — 0

Activity B — (1, 2) — 0

Activity C — (1, 3) — 0

Activity D — (1, 4) — 7

Activity E — (2, 5) — 0

Activity F — (3, 5) — 0

Activity G — (5, 6) — 0

Activity H — (4, 6) — 7

So critical activities are A, B, C, E, F and G.

(c) There are two critical paths:

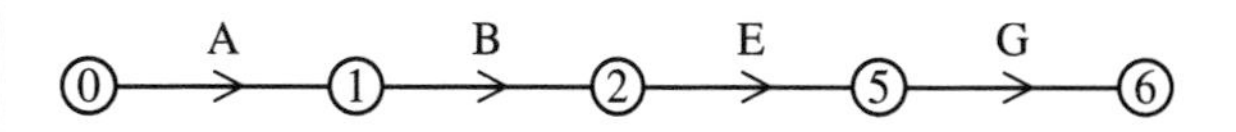

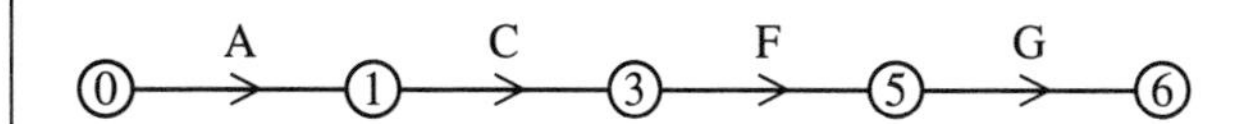

They are both of length 21.

Exercise 4E

		Start		Finish		
Activity	Duration	Earliest	Latest	Earliest	Latest	Float
A (1,2)	2	0	0	2	2	0
B (1,3)	7	0	0	7	7	0
C (1,4)	10	0	1	10	11	1
D (2,3)	5	2	2	7	7	0
E (3,6)	8	7	8	15	16	1
F (3,5)	6	7	7	13	13	0
G (4,5)	2	10	11	12	13	1
H (5,7)	4	13	13	17	17	0
I (4,7)	3	10	14	13	17	4
J (6,7)	1	15	16	16	17	1

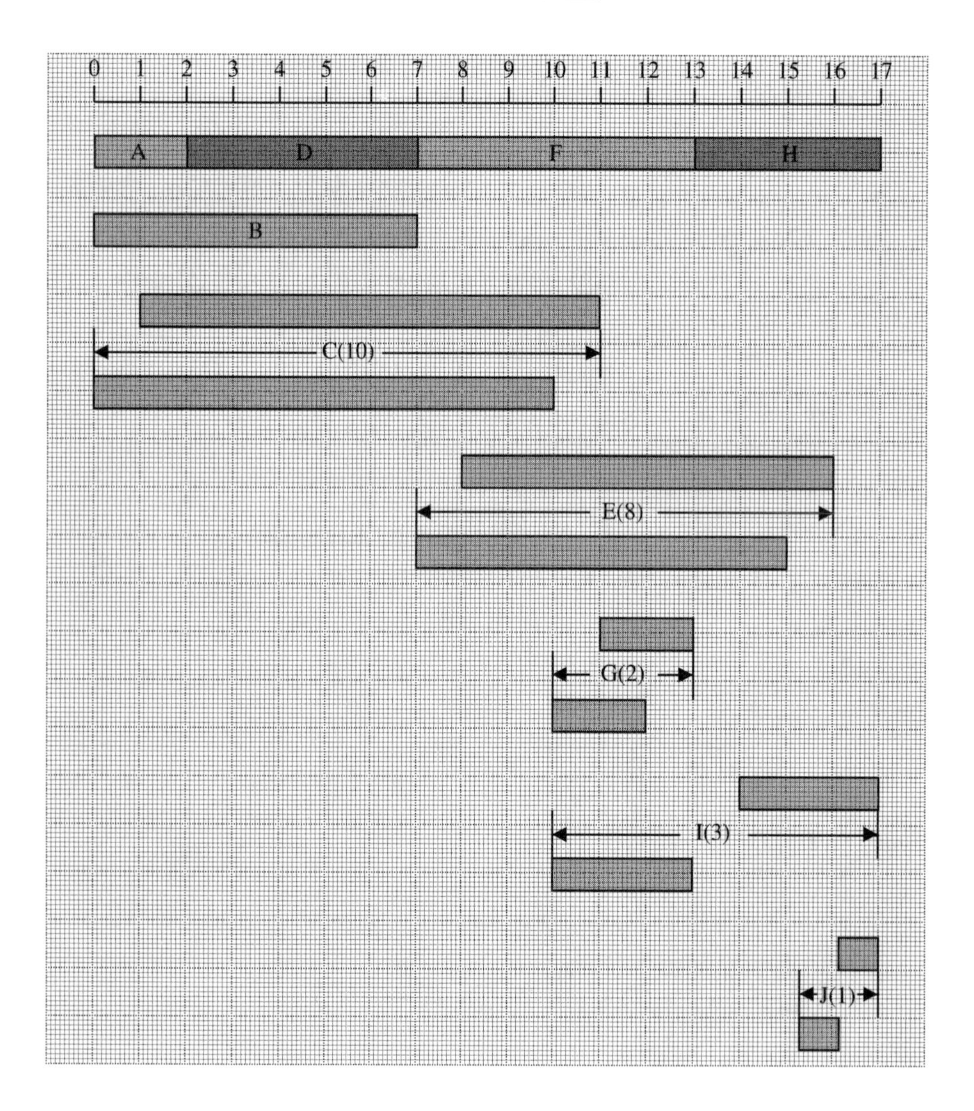

Exercise 4F

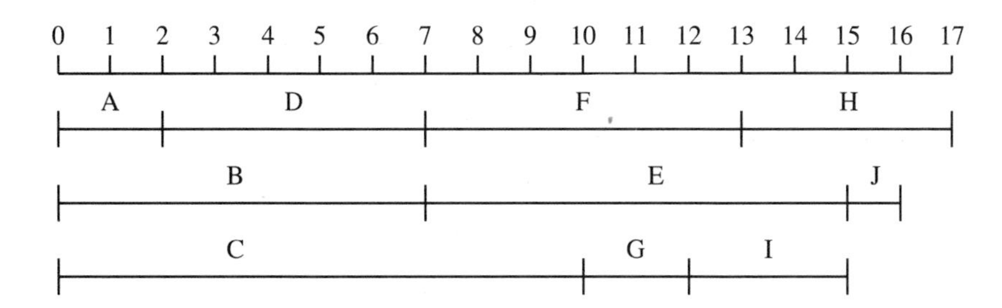

3 workers required

Exercise 4G

1

i	0	1	2	3	4	5
e_i	0	3	4	9	14	16
l_i	0	3	5	9	14	16

Floats A–0, B–1, C–3, D–1, E–0, F–0, G–2, H–0

Length of critical path 16

Critical events 0,1,3,4,5

Critical activities A,E,F,H

2

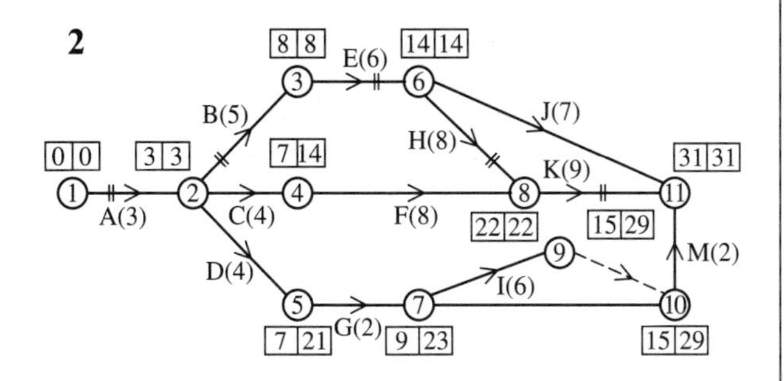

3 (a) The earliest time for event 8 is now increased to 23 days.

(b) This implies that the earliest time for the sink event is now 32. Thus the new project time is 32 days.
The critical path is then A, C, F, K.

4 (a)

i	1	2	3	4	5	6
e_i	0	3	5	9	8	14
l_i	0	3	11	9	12	14

Project time 14 days

Critical path ACF

(b)

Activity	Duration	Start Earliest	Start Latest	Finish Earliest	Finish Latest	Float
A (1,2)	3	0	0	3	3	0
B (1,3)	5	0	6	5	11	6
C (2,4)	6	3	3	9	9	0
D (2,5)	5	3	7	8	12	4
E (3,5)	1	5	11	6	12	6
F (4,6)	5	9	9	14	14	0
G (5,6)	2	8	12	10	14	4

(c) Schedule for 2 workers

Minimum number of workers required if the project is to be completed in the project time is 2.

5 (a)

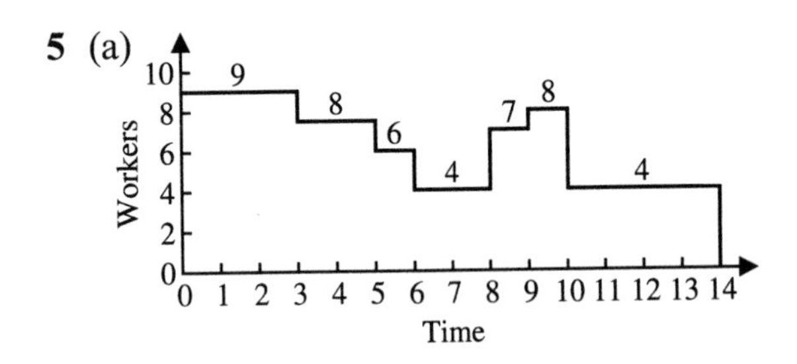

(b)

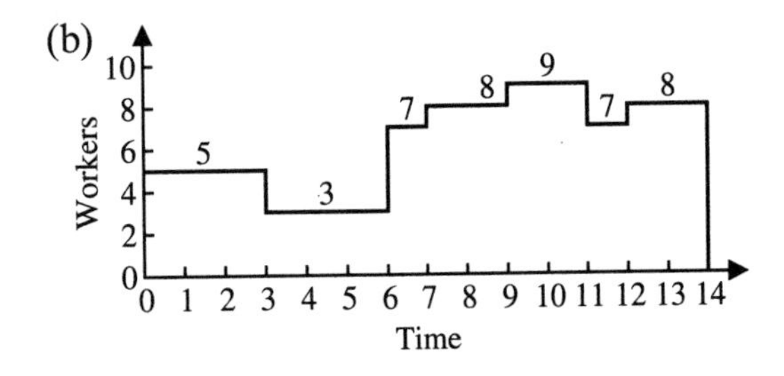

Exercise 5A

1

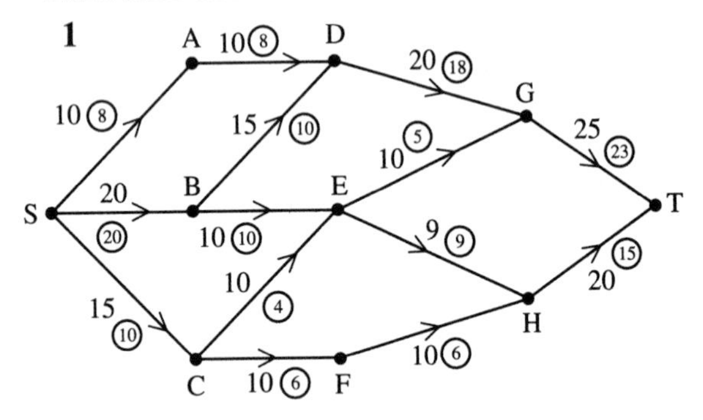

Capacity constraints:

AD	BD	BE	CE	CF	DG	EG
$8 \leqslant 10$	$8 \leqslant 10$	$10 \leqslant 10$	$4 \leqslant 10$	$6 \leqslant 10$	$18 \leqslant 20$	$5 \leqslant 10$

EH	FH	SA	SB	SC	GT	HT
$9 \leqslant 9$	$6 \leqslant 10$	$8 \leqslant 10$	$20 \leqslant 20$	$10 \leqslant 15$	$23 \leqslant 25$	$15 \leqslant 20$

Flow constraints:

A in = (8)
out = (8)

B in = (20)
out = (10) + (10) = (20)

C in = (10)
out = (4) + (6) = (10)

D in = (8) + (10) = (18)
out = (18)

E in = (10) + (4) = (14)
out = (5) + (9) = (14)

F in = (6)
out = (6)

G in = (18) + (5) = (23)
out = (23)

H in = (9) + (6) = (15)
out = (15)

Source:
Flow out of S = (8) + (20) + (10) = (38)

Sink:
Flow into T = (23) + (15) = (38)

2 (a) 3 (b) 6 (c) 6 (d) 2

3

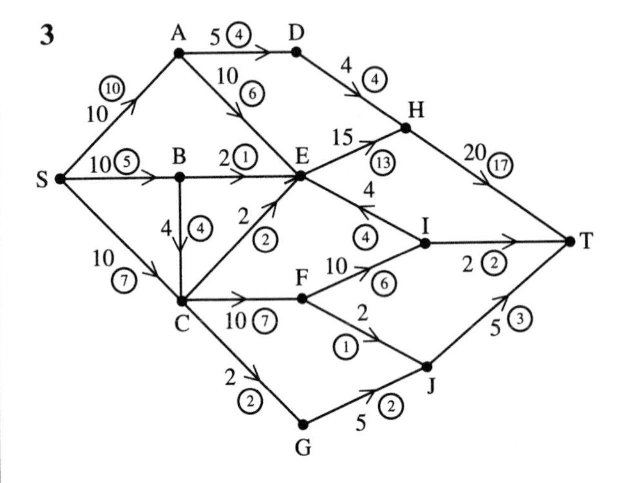

Exercise 5B

1

SA	SB	SC	SD	SE	SF	ST	
15	20	–	–	–	–	–	35

SC	SD	SE	SF	ST	AC	AD	AE	AF	AT	BC	BD	BE	BF	BT	
–	–	–	–	–	6	–	5	–	–	8	15	–	7	–	41

SA	SB	SC	SD	SE	ST	FA	FB	FC	FD	FG	FT	
15	20	–	–	–	–	–	7	12	6	–	20	80

SC	SE	SF	ST	AC	AE	AF	AT	BC	BE	BF	BT	DC	DE	DF	DT	
–	–	–	–	6	5	–	–	8	–	7	–	–	–	6	–	32

2

SA	SB	SC	SD	SE	SF	ST	
15	20	–	–	–	–	–	35

SC	SD	SE	SF	ST	AC	AD	AE	AF	AT	BC	BD	BE	BF	BT	
–	–	–	–	–	–	–	5	–	–	8	15	–	7	–	35

SA	SB	SC	SD	SE	ST	FA	FB	FC	FD	FE	FT	
15	20	–	–	–	–	–	–	12	–	–	20	67

SC	SE	SF	ST	AC	AE	AF	AT	BC	BE	BF	BT	DC	DE	DF	DT	
–	–	–	–	–	5	–	–	8	–	7	–	–	–	6	–	26

3 We have to place each of the letters A, B, J, K, L, M, X, Y, Z either with S or with T in $\{S,\ldots\} \mid \{\ldots,T\}$. That is 2 possibilities for A, 2 possibilities for B, etc. The total number of possibilities is then given by multiplying all of these nine 2s together, since for each one of the A possibilities there are 2 B possibilities, and for each of the B possibilities …
This gives $2^9 = 512$ possible cuts.

4 (a) A possible flow pattern is:

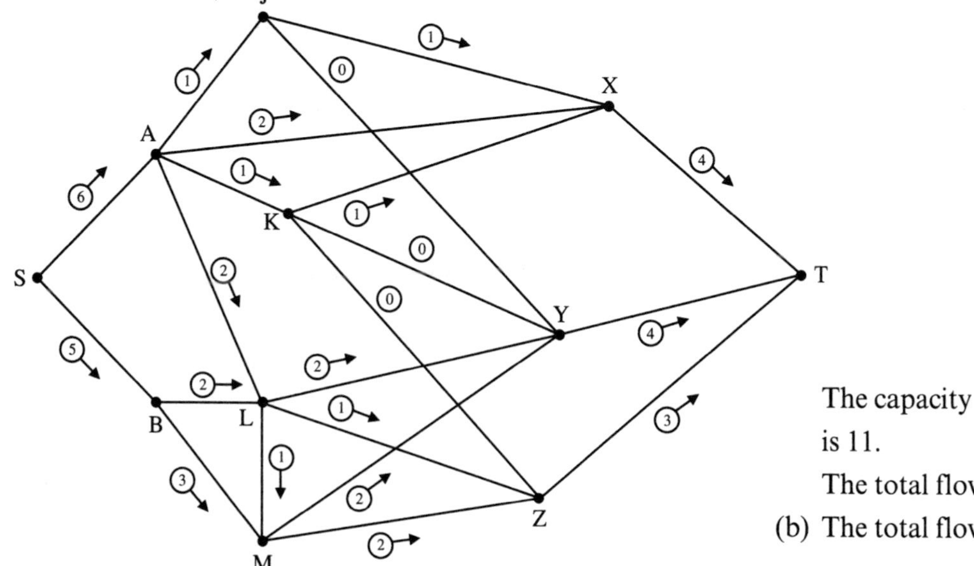

The capacity of the given cut is 11.

The total flow is also 11.

(b) The total flow is maximal.

Exercise 5C

1

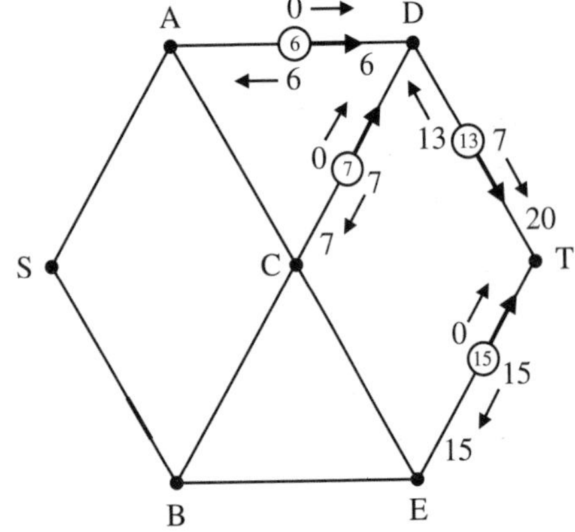

There are many possibilities for flows in edges other than those marked. Just one possibility is shown below:

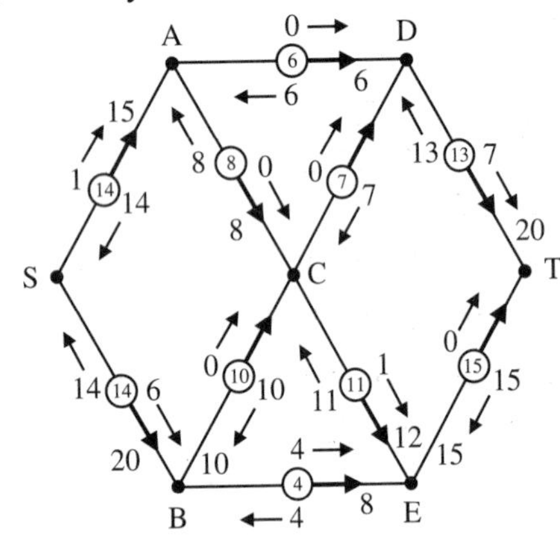

Total flow through network = 28

2 (a)

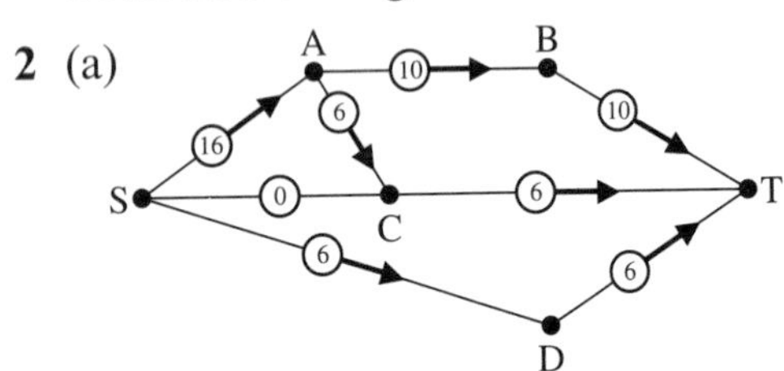

{S, A, C, D}|{B, T}

(b)

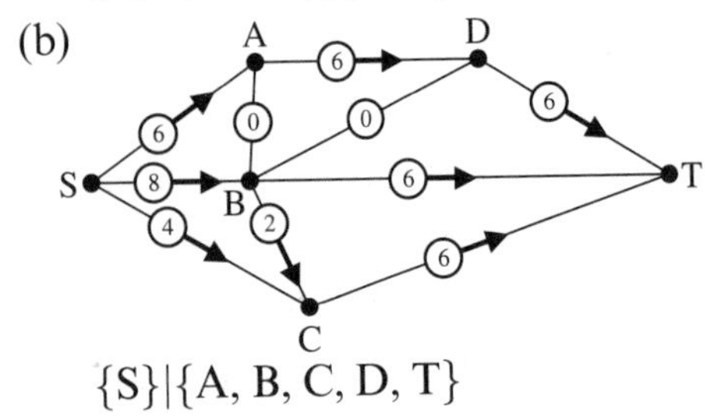

{S}|{A, B, C, D, T}

(c)

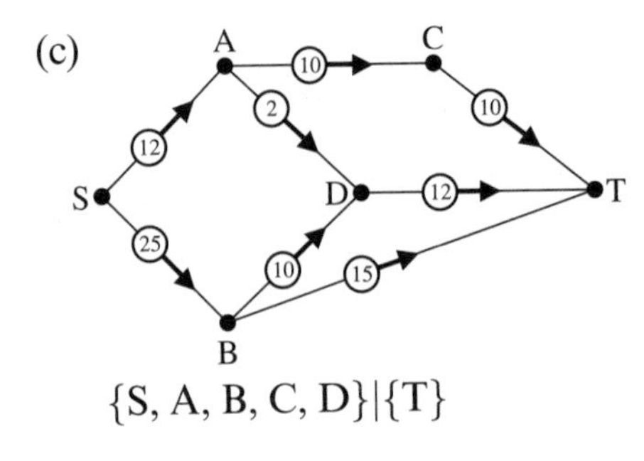

{S, A, B, C, D}|{T}

Exercise 5D

1

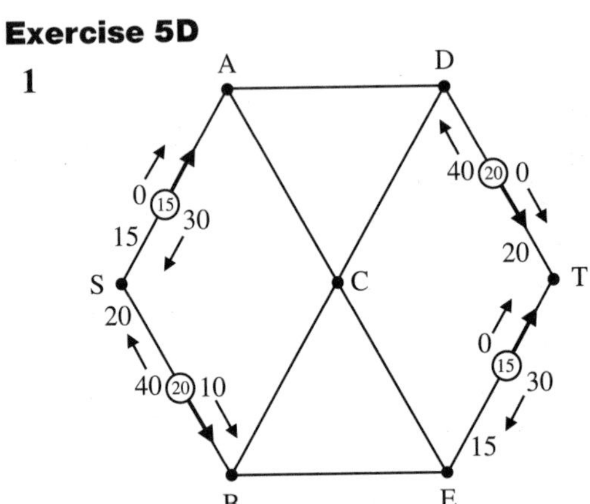

There are alternatives for the flow in edges other than those marked. All the alternatives involve some flow in the direction EC.

One feasible set of flows giving this maximum flow of 35 is:

Edge	SA	SB	AC	AD	BC	BE	CD	EC	DT	ET
Flow	15	20	9	6	4	16	14	1	20	15

2 Example of a final flow pattern:

Edge	SA	SC	SD	AB	AC	BT	CT	DT
Flow	16	0	6	10	6	10	6	6

3 Example of a final flow pattern:

Edge	SA	SB	SC	AD	BA	BD	BT	BC	DT	CT
Flow	6	8	4	6	0	0	6	2	6	6

4 Example of a final flow pattern:

Edge	SA	SB	AC	AD	BD	BT	CT	DT
Flow	12	25	10	2	10	15	10	12

Exercise 5E

1 (a) Min. cut: {S, B, D} | {A, C, T}

Capacity:

SA	SC	ST	BA	BC	BT	DA	DC	DT	
14	–	–	–	–	–	–	–	15	29

(b) Existing flow = 22 less than 29

(c)

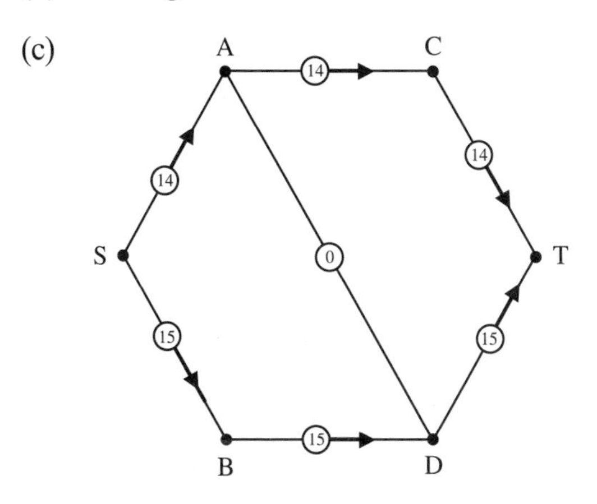

2 (a) Augmenting path SBDACT now exists.

(b) 30

(c)

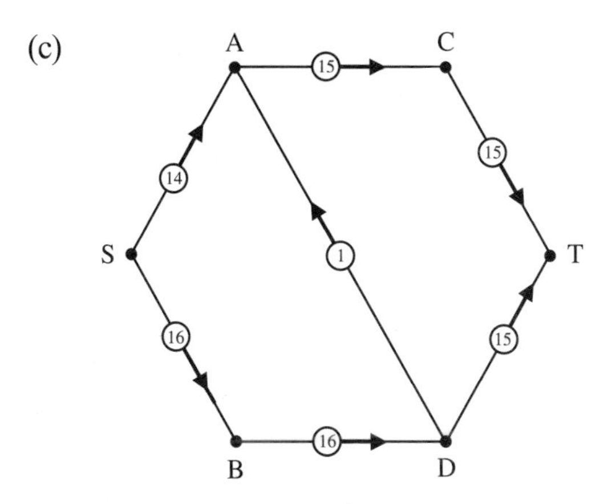

3 (a) Source is B; sink is F

(b) A partition of the vertices into 2 sets, one containing the source and the other containing the sink.

Capacity of cut = 12 + 13 + 18 = 43

(c) Flow augmenting path: BCHGF

Extra flow: 4

No. Flow augmenting path: BCHGEF

Extra flow: 1

Now maximal

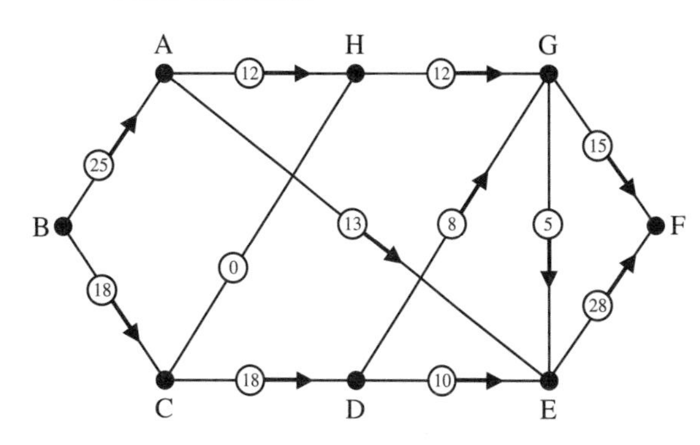

(e) Total flow = 43

Cut {B, A, C} | {D, E, G, H, F} has capacity 43.

4 (a), (b)

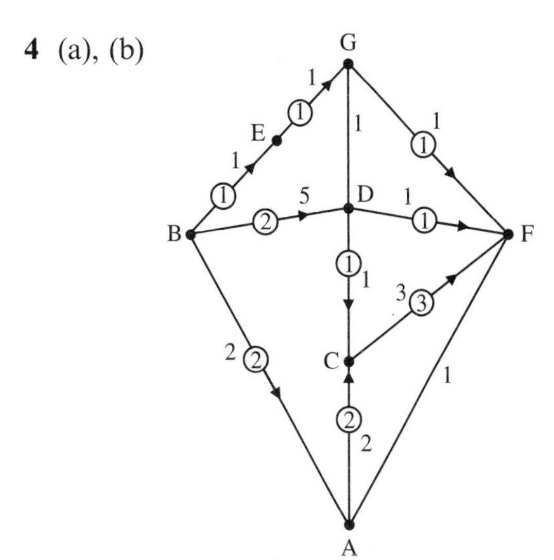

(There are other flow patterns which achieve the maximal flow.)

Cut {B, D, E, G} | {A, C, F} has capacity 5.

Therefore, max. hourly flow of cars from B to F is 5000.

5 (a)

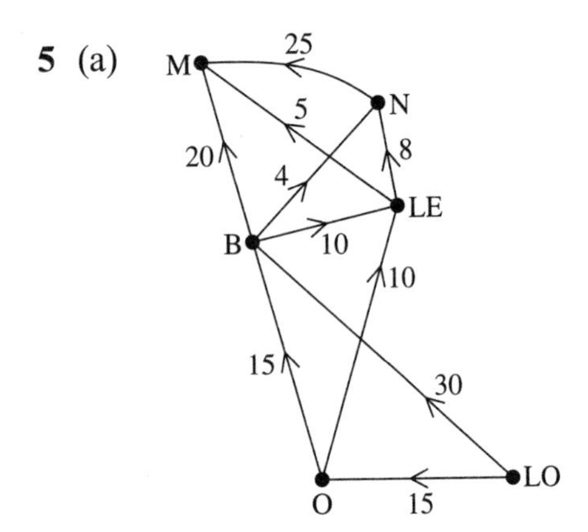

(b)

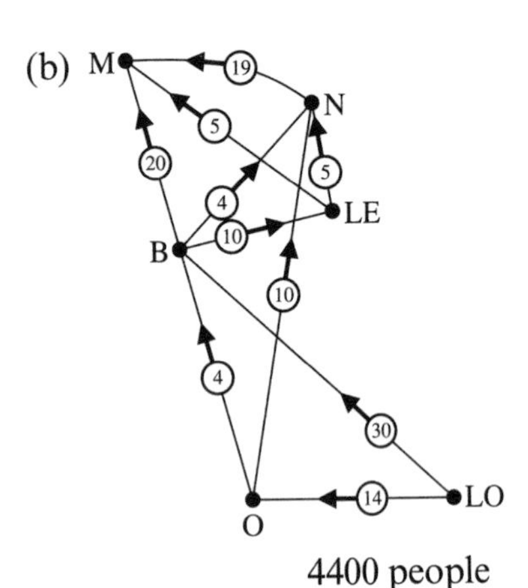

4400 people

(c)

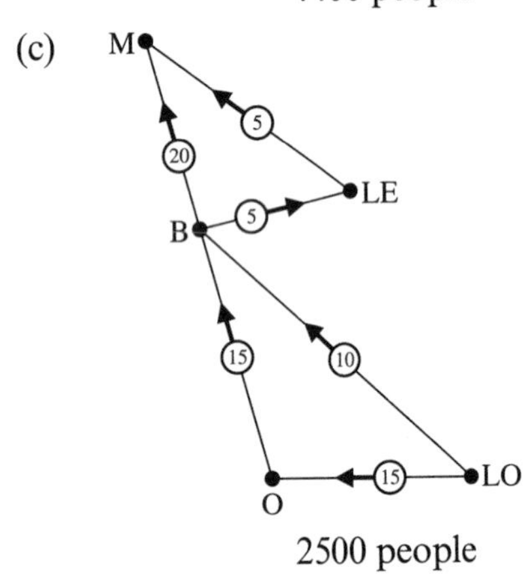

2500 people

6 (a)

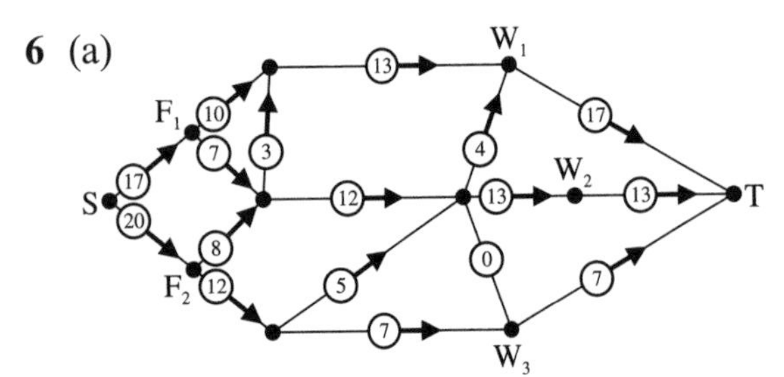

(b) This pattern shows 17 leaving F_1 and 20 leaving F_2. It has 17 reaching W_1, 13 reaching W_2 and 7 reaching W_3. (There are many other possible distributions, all with a total flow of 37).

7 (a), (b)

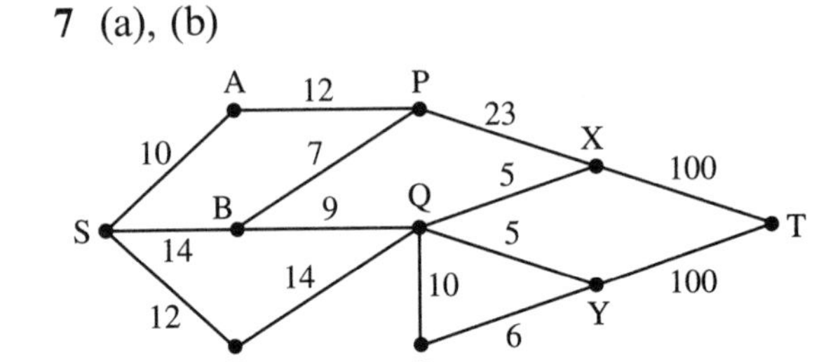

(c) For example:

Pipe/link	SA	SB	SC	AP	BP	BQ	CQ	PX	QX	QY	QR	RY	XT	YT
Flow	10	14	9	10	7	7	9	17	5	5	6	6	22	11

(d) Cut {S, B, C, Q, R} | {A, P, X, Y, T} has capacity 33.
Established flow = 33

(e) A is supplying 10 units (SA)
B is supplying 14 units (SB)
C is supplying 9 units (SC)
X is taking 22 units (XT)
Y is taking 11 units (YT)

(f)

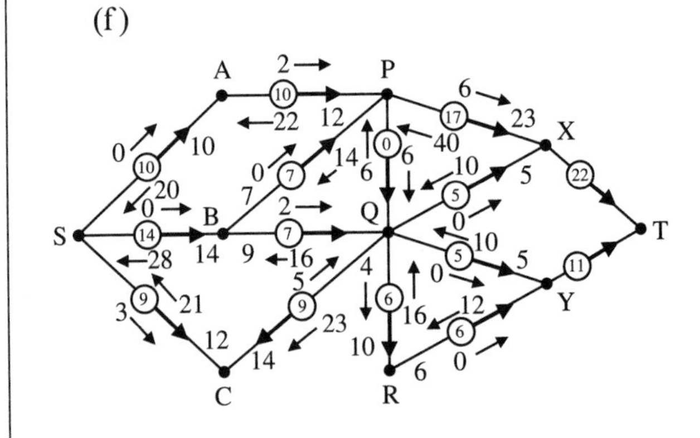

Flow augmenting path: SCQPXT
Extra capacity: 3

This will exhaust supplies, establishing a maximal flow of 36 units using the following flow pattern:

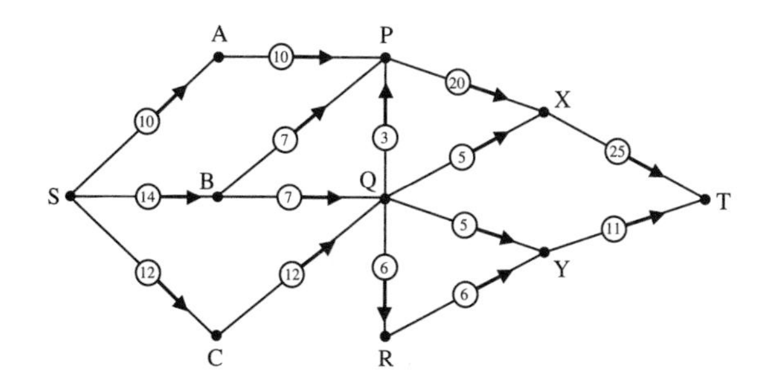

(g) Connect R to T.

8 There is one direct route.
There are $(n-2)$ routes which pass through 1 vertex only.
There are $(n-2)(n-3)$ routes which pass through 2 vertices, etc.
Total number of routes $= 1 + (n-2) + (n-2)(n-3) + (n-2)(n-3)(n-4) + \ldots + (n-2)!$, the last term counting the number of routes passing through all vertices.
e.g. $n = 5$, vertices, S, A, B, C, T
Direct route: ST
Passing through 1 vertex only: SAT, SBT, SCT; $(n-2) = 5 - 2 = 3$
Passing through 2 vertices: SABT, SACT, SBAT, SBCT, SCAT, SCBT;
$(n-2)(n-3) = 3 \times 2 = 6$
Passing through all vertices: SABCT, SACBT, SBACT, SBCAT, SCABT, SCBAT; $(n-2)! = 3! = 6$
The algorithm lists all of these, so it has factorial complexity.

Exercise 6A

1 $x =$ no. of type A tables
$y =$ no. of type B tables
Maximise $z = 15x + 17y$
Subject to $20x + 12y \leqslant 480$
$10x + 15y \leqslant 330$
$x \geqslant 0,\ y \geqslant 0$

2 Suppose she buys
x (hundred) grams of α and
y (hundred) grams of β
Minimise $C = 40x + 30y$
Subject to $30x + 10y \geqslant 25$
$20x + 25y \geqslant 30$
$10x + 40y \geqslant 15$
$x \geqslant 0,\ y \geqslant 0$

3 $x =$ no. of type A machines
$y =$ no. of type B machines
Maximise $z = 75x + 120y$
Subject to $x + 2y \leqslant 40$
$3x + 4y \leqslant 100$
$x \geqslant 0,\ y \geqslant 0$

4 $f =$ no. of full-fare passengers
$h =$ no. of half-fare passengers
Maximise $z = 10f + 5h$
Subject to $f + h \leqslant 14$
$f + h \geqslant 10$
$f \geqslant 0,\ h \geqslant f$

5 $x =$ no. of small bookshelves
$y =$ no. of medium bookshelves
$z =$ no. of large bookshelves
Maximise $P = 4x + 6y + 12z$
Subject to $4x + 8y + 16z \leqslant 500$
$2x + 4y + 6z \leqslant 400$
$x \geqslant 0,\ y \geqslant 0,\ z \geqslant 0$

6 $x =$ no. of type A made
$y =$ no. of type B made
$z =$ no. of type C made

$$\begin{aligned} \text{Maximise} \quad & P = 15x + 20y + 25z \\ \text{Subject to} \quad & 9x + 6y + 4z \leqslant 3600 \\ & 2x + 9y + 12z \leqslant 3600 \\ & 18x + 4y + 6z \leqslant 3600 \\ & 6x + 9y + 8z \leqslant 3600 \\ & x \geqslant 0, \ y \geqslant 0, \ z \geqslant 0 \end{aligned}$$

Exercise 6B

1

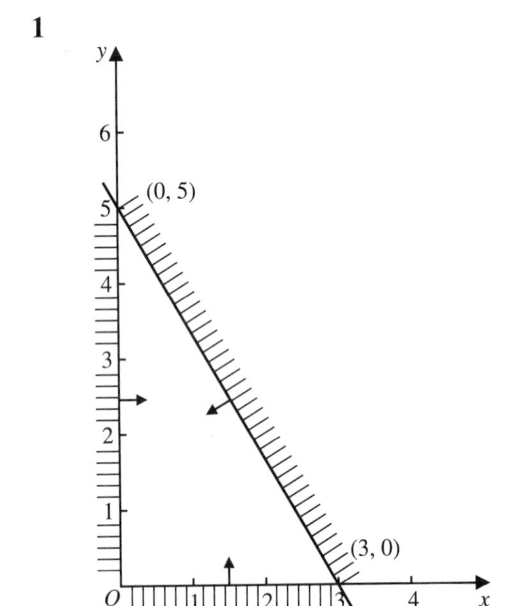

2

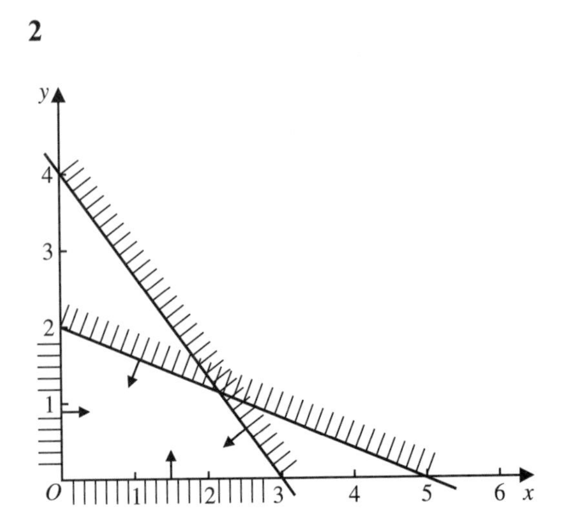

3

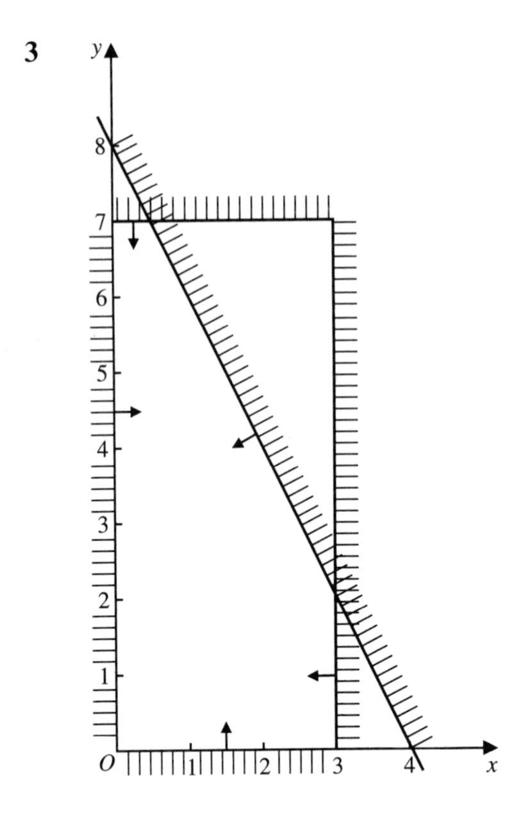

4

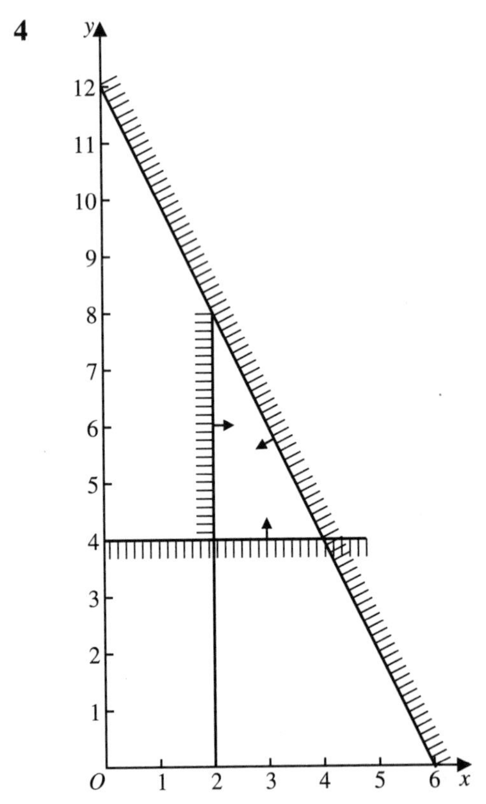

5

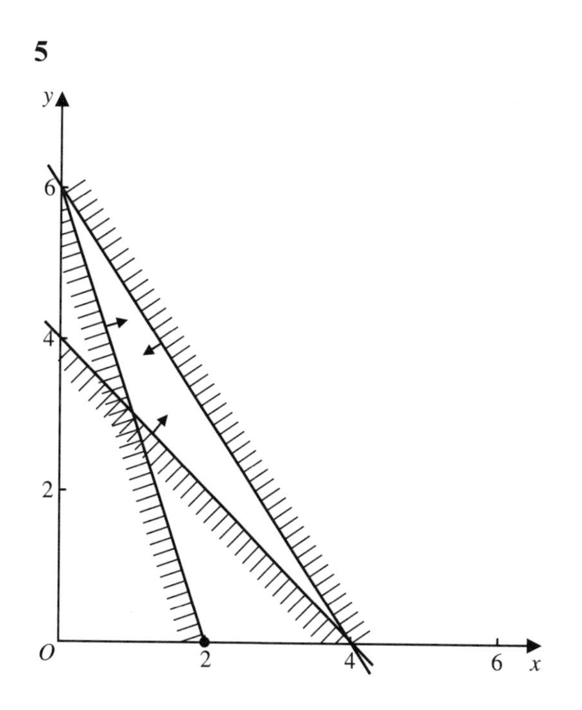

Exercise 6C

1

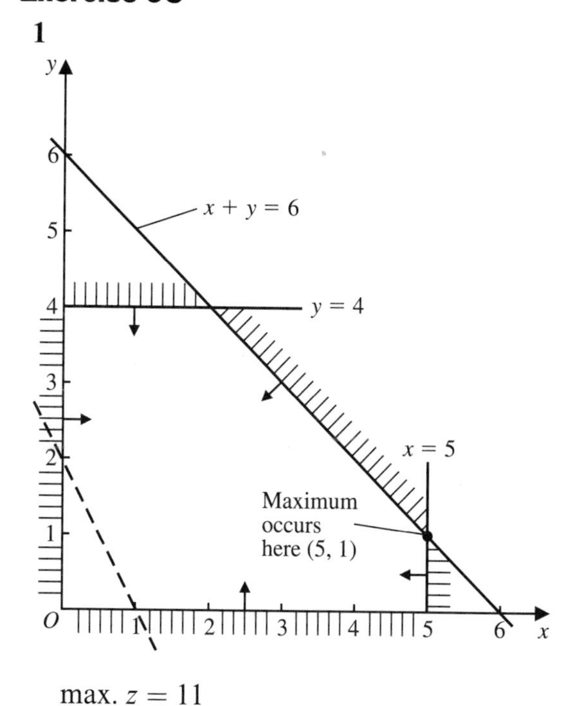

max. $z = 11$

2

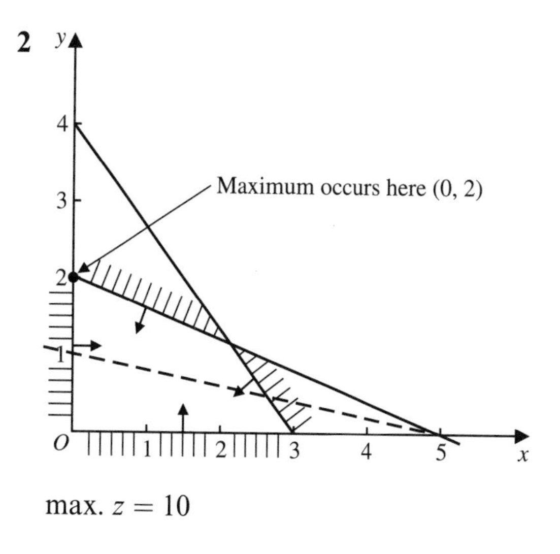

max. $z = 10$

3

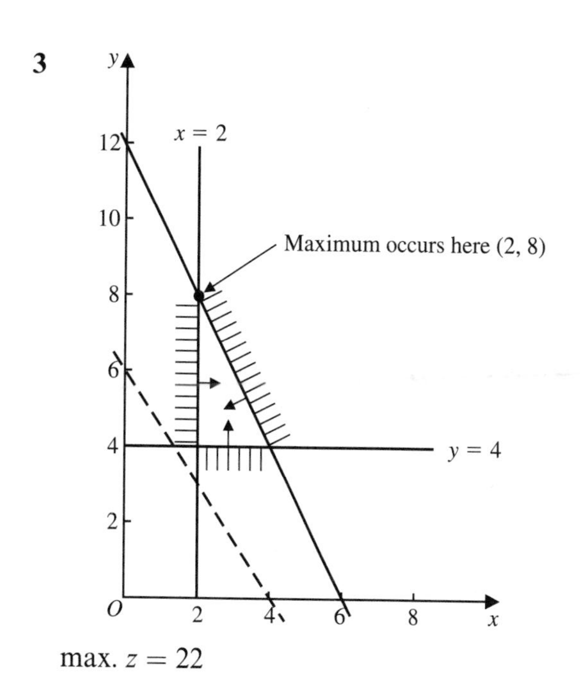

max. $z = 22$

4

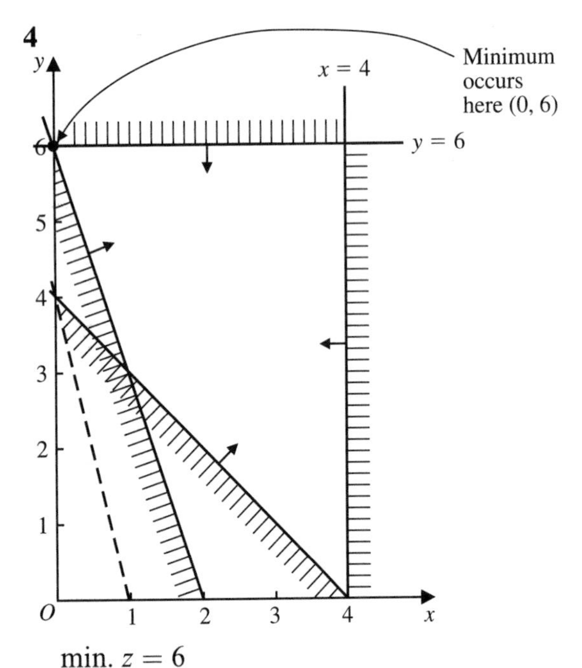

min. $z = 6$

5

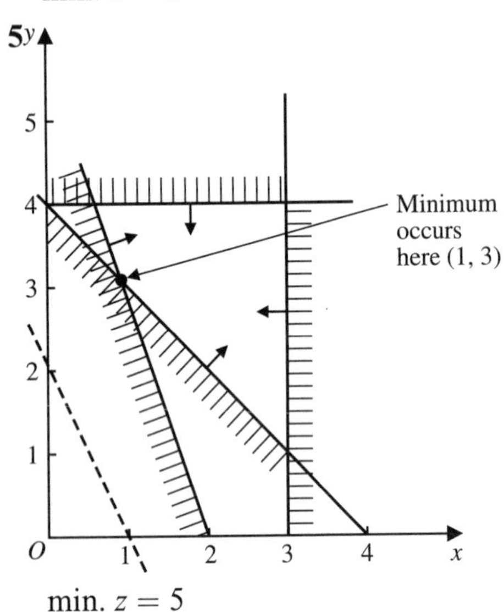

min. $z = 5$

Exercise 6D

1 (a) $(0, 0)$, $(5, 0)$, $(5, 1)$, $(2, 4)$, $(0, 4)$
(b) Corresponding values of z are 0, 10, 11, 8, 4.
(c) The maximum value 11 occurs at $(5, 1)$.

2 (a) $(0, 0)$, $(3, 0)$, $\left(2\frac{1}{7}, 1\frac{1}{7}\right)$, $(0, 2)$
(b) Corresponding values of z are 0, 3, $7\frac{6}{7}$, 10.
(c) The maximum value 10 occurs at $(0, 2)$.

3 (a) $(2, 4)$, $(4, 4)$, $(2, 8)$
(b) Corresponding values of z are 14, 20, 22.
(c) The maximum value 22 occurs at $(2, 8)$.

4 (a) $(4, 0)$, $(4, 6)$, $(0, 6)$, $(1, 3)$
(b) Corresponding values of z are 16, 22, 6, 7.
(c) The minimum value of 6 occurs at $(0, 6)$.

5 (a) $(3, 1)$, $(3, 4)$, $\left(\frac{2}{3}, 4\right)$, $(1, 3)$
(b) Corresponding values of z are 7, 10, $5\frac{1}{3}$, 5.
(c) The minimum value 5 occurs at $(1, 3)$.

Exercise 6E

1 $x = 6, \quad y = 12, \quad z = 48$
Vertices are $(0, 0)$, $(10, 0)$, $(0, 15)$ and $(6, 12)$.

2 $u = 5, \quad v = 2, \quad w = 80$
Vertices are $(0, 12)$, $(5, 2)$, $(8, 0)$.

3 All points on line segment joining $(12, 12)$ and $(15, 8)$ give maximum z. Maximum value of z is 336.
Vertices are $(0, 0)$, $(0, 21)$, $(12, 12)$, $(15, 8)$, $(15, 0)$.

4 (a) Max. $z = 240$ at $x = 12, y = 12$
(b) Max. $z = 570$ at $x = 15, y = 8$

5 (a) $x = \frac{4}{5}, \quad y = 2\frac{2}{5}, \quad z = 3\frac{1}{5}$
(b) $(0, 0)$, $(0, 1)$, $(0, 2)$, $(0, 3)$, $(1, 0)$, $(1, 1)$, $(1, 2)$, $(2, 0)$;
$(0, 3)$ and $(1, 2)$ gives max. z of 3
(c) $(0, 3)$ and $(1, 2)$ both give $z = 3$

6 (a) l = no. of luxury houses
s = no. of standard houses
Maximise $P = 12\,000\,l + 8000\,s$
Subject to $l \geqslant 5$, $s \geqslant 10$, $l + s \leqslant 30$
$300\,l + 150\,s \leqslant 6000$

(b) $l = 10$, $s = 20$, max. $P = £280\,000$.
That is, 10 luxury homes and 20 standard homes should be built.

7 $l = 8$, $s = 0$ and minimum cost £320

8 $x = £20\,000$, $y = £10\,000$
and maximum yield is £1900.

Exercise 6F

1 $x = 6$, $y = 12$, max. $z = 48$

2 $x = 15$, $y = 8$, max. $z = 570$

3 $x = 4$, $y = 2$, $z = 0$, max. $P = 24$

4 (a) x = no. of lengths of A made
y = no. of lengths of B made
Maximise $P = 12x + 15y$
Subject to $4x + y \leqslant 56$
$5x + 3y \leqslant 105$
$x + 2y \leqslant 56$
$x \geqslant 0, y \geqslant 0$

(b) $x = 6$, $y = 25$ and max. $P = £447$

Exercise 7A

1

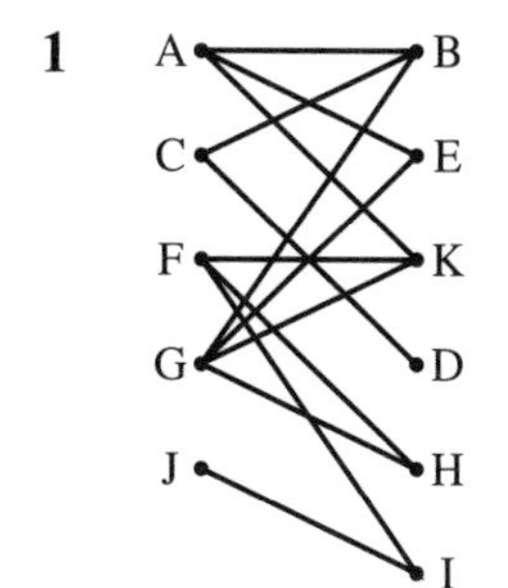

2 (a) A, C, D — B, E bipartite

(b) A, B — C, D not bipartite

(c) A, D, E — B, C bipartite

3 (a) Heathrow, Gatwick, Stanstead, Luton — Bristol, Birmingham, E. Midlands, Manchester, Newcastle

(b) Not possible to form a bipartite graph because (for instance) Heathrow would need to link to E. Midlands, E. Midlands to Gatwick, and Gatwick to Heathrow. So you cannot split the airports into two sets.

4

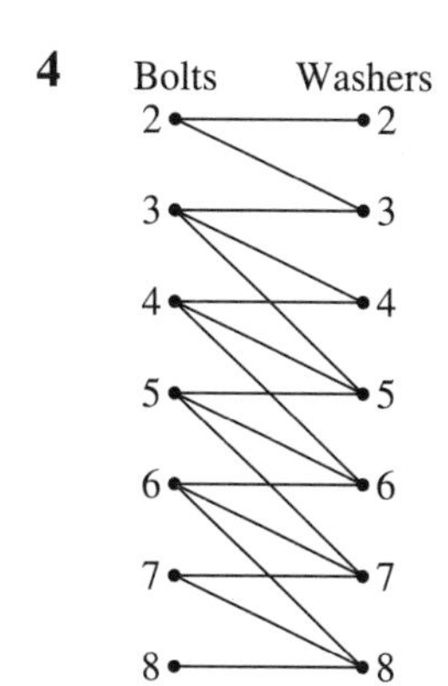

Exercise 7B

1 (a)

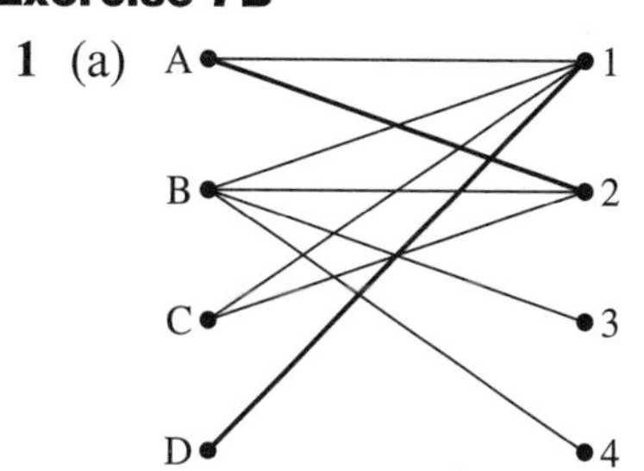

(b) D must teach 1 as only qualified for this.
As 1 has been allocated, A must teach 2 (only qualified to teach 1 and 2).

As C can only teach 1 and 2, both of which are allocated, there is no subject for C to teach.
B can now teach either 3 or 4.
Best possible matching only 3 languages: D $\leftrightarrow$ 1, A $\leftrightarrow$ 2, B $\leftrightarrow$ 3 or 4
A, C and D together are only qualified to teach 2 languages and so one will not be matched
or 3 and 4 can only be taught by B.

2 (a)

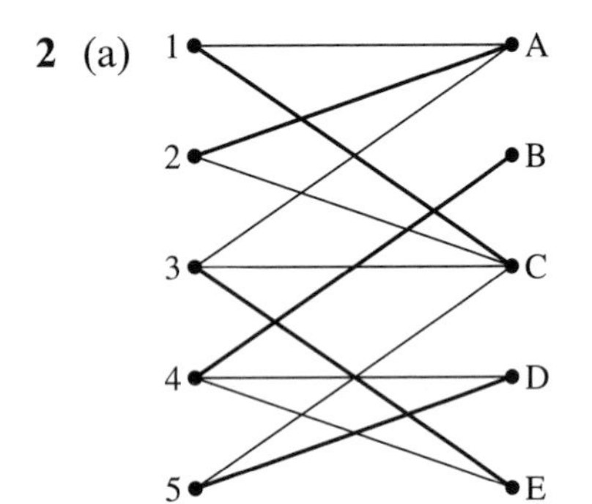

(b) Heavy lines show a complete matching so they can all be satisfied.
1 $\leftrightarrow$ C, 2 $\leftrightarrow$ A, 3 $\leftrightarrow$ E, 4 $\leftrightarrow$ B, 5 $\leftrightarrow$ D

3

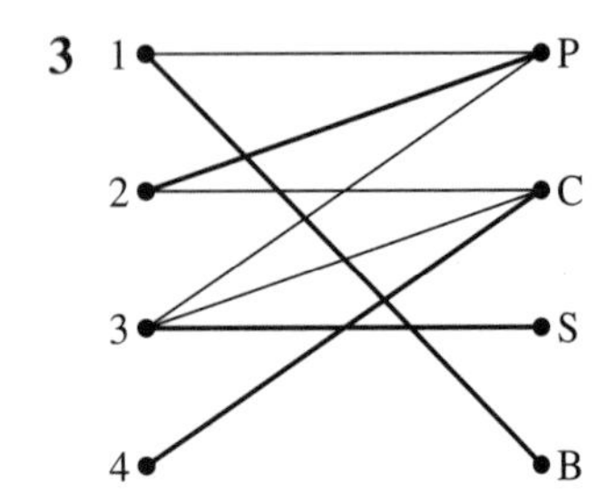

4 must grow C
1 must grow B
2 must grow P
and 3 must grow S
All crops can be grown.

4

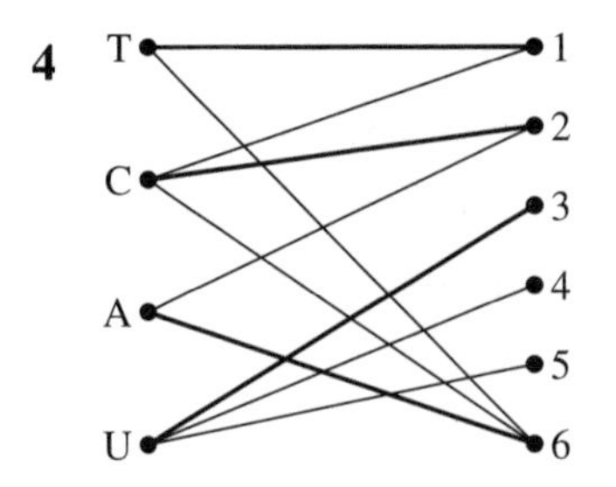

A possible committee is:
(1) represents T
(2) represents C
(6) represents A
(3) represents U

5

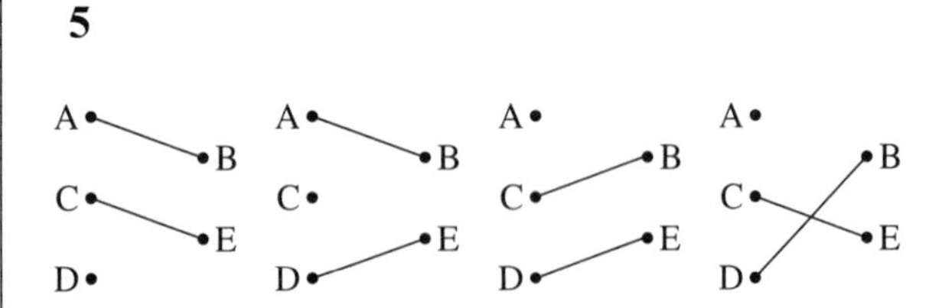

6

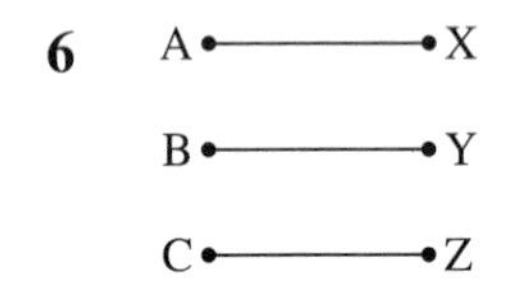

This is the only complete matching in the given graph.

7 Washer 2 has to match with bolt 2, which means that washer 3 has to match with bolt 3 and so on.

8 (a)

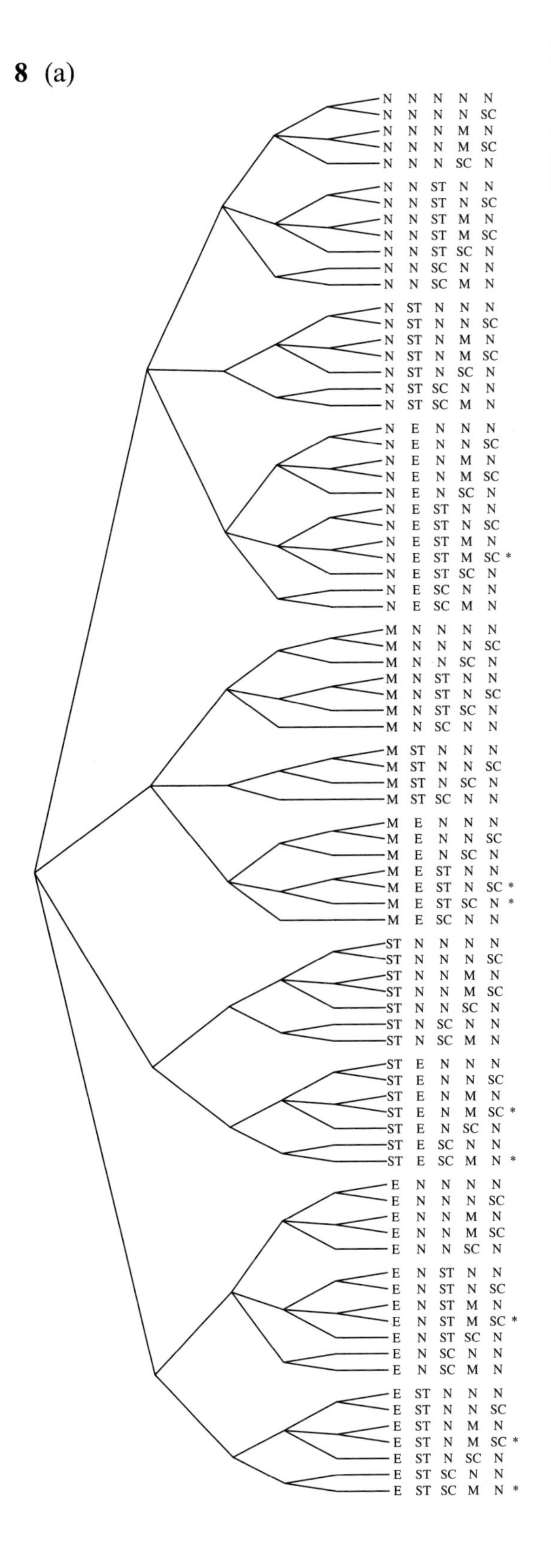

Key: N = not used
M = mathematics
ST = statistics
SC = science
E = economics

Branching is by each teacher in turn (alphabetically).

(b) Maximal matchings contain only one 'N'. They are marked with an asterisk. There are 8 of them.

Exercise 7C

1

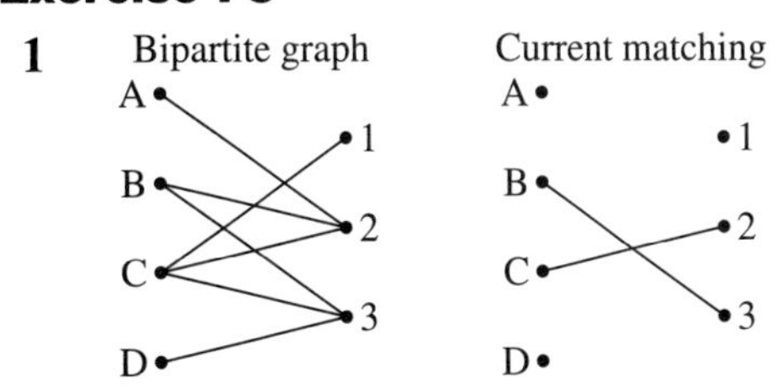

Starting with A (say):

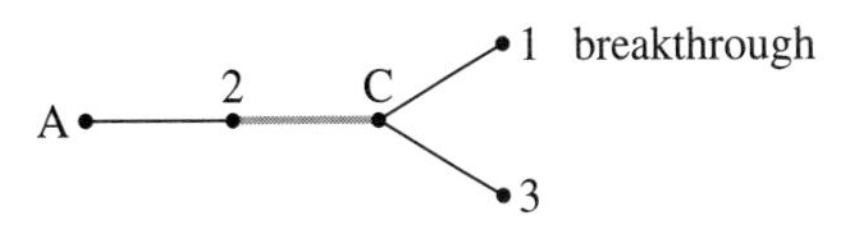

Changing status:

A 2 C 1

giving:

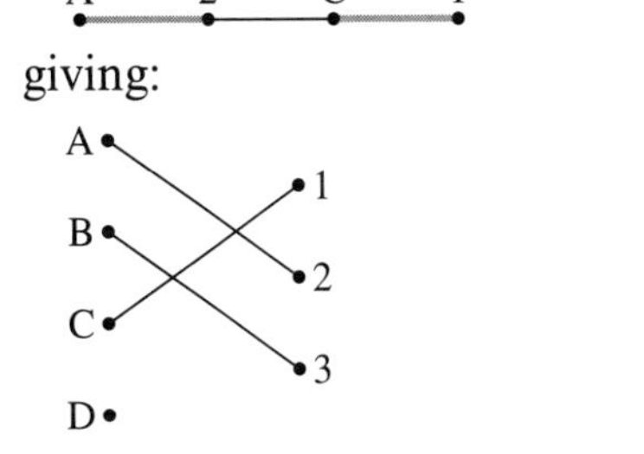

This is maximal.

2

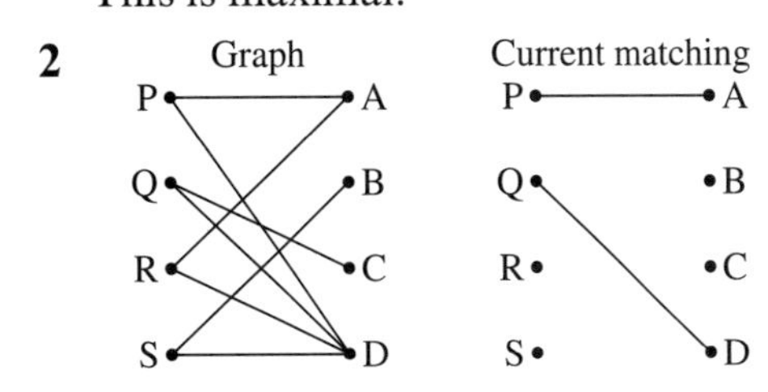

Starting with B (say):

B S breakthrough

Changing status:

B S

giving:

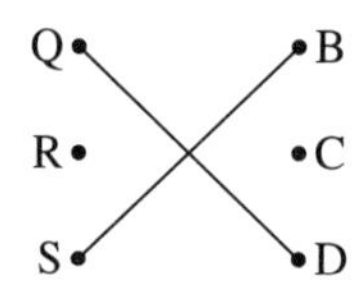

Starting again with R (say):

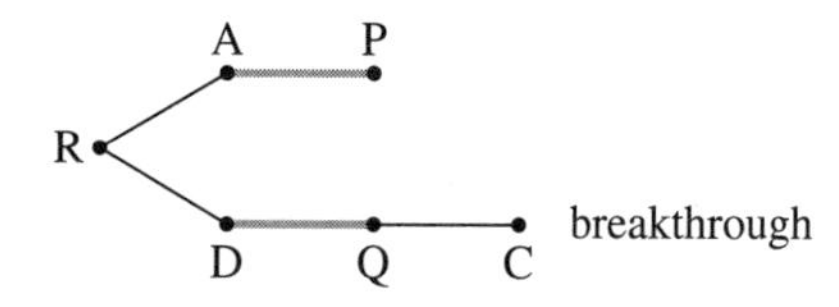

Changing status:

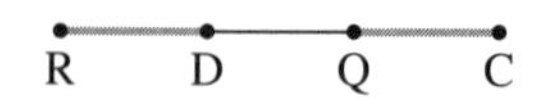

giving:

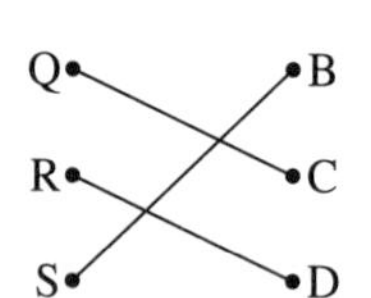

This is a complete matching.

3 Bipartite graph Current matching

Starting with vertex 1 (say):

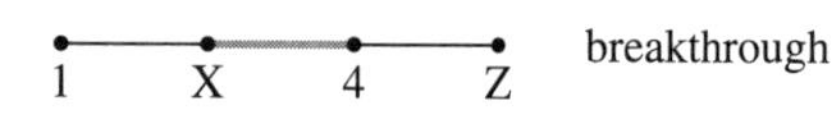

Changing status:

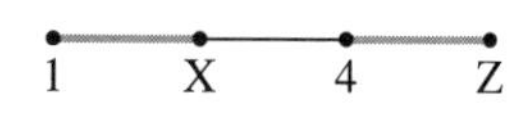

giving:

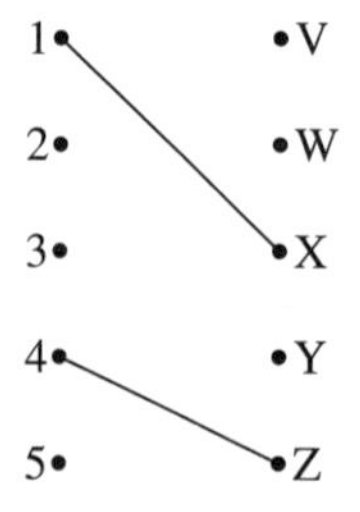

Restarting with V (say) → V–5 breakthrough

Restarting with 2 (say) → 2–W breakthrough

Restarting with 3 (say) → 3–Y breakthrough

giving:

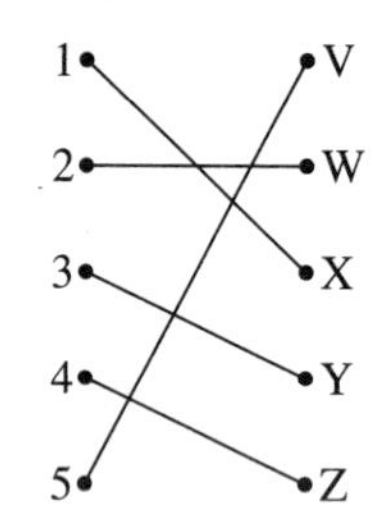

4 Starting with A (say) → A–1 breakthrough

Restarting with B (say) → B–2 breakthrough

Restarting with C (say) → C–3 breakthrough

Restarting with D:

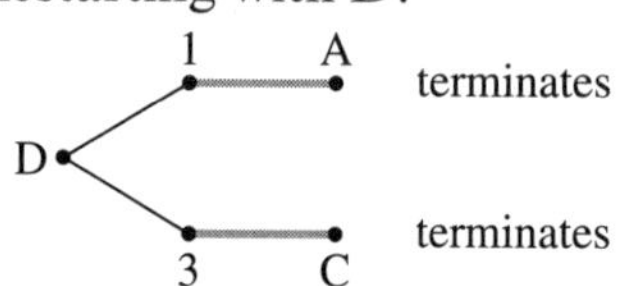

5

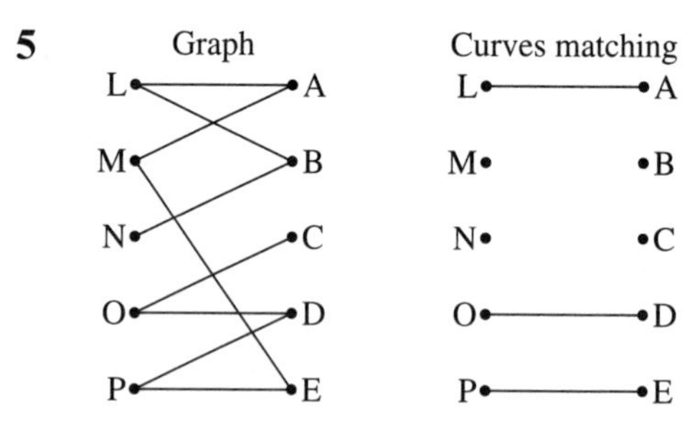

Starting from M (say):

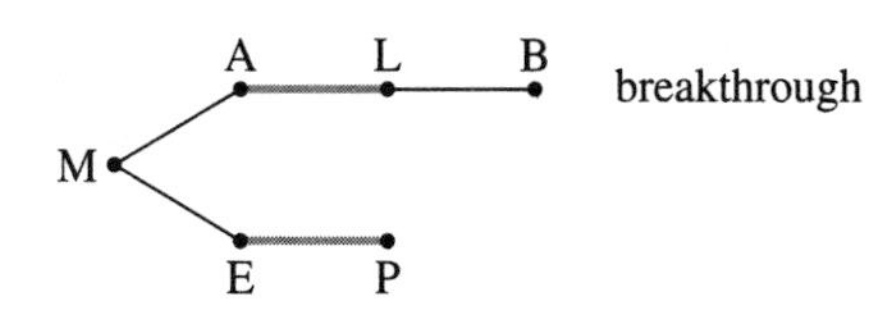

Changing status:

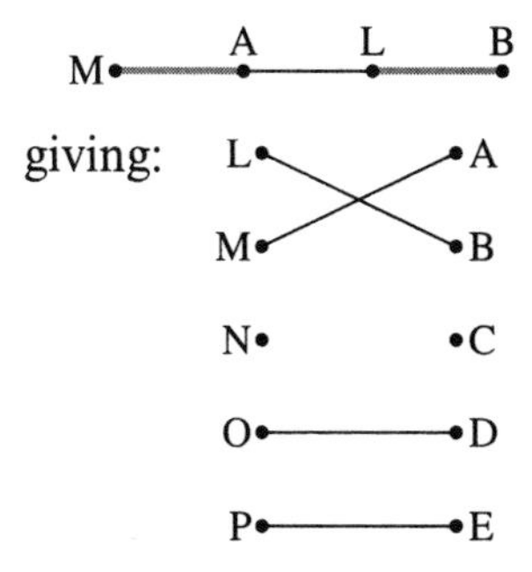

Restarting from N (say):

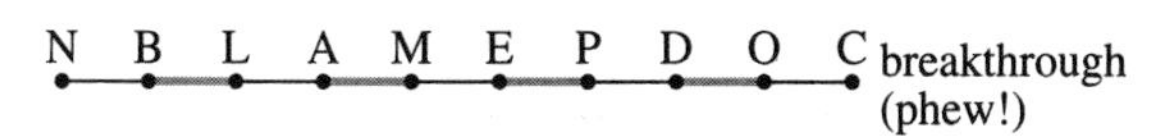

Changing status:

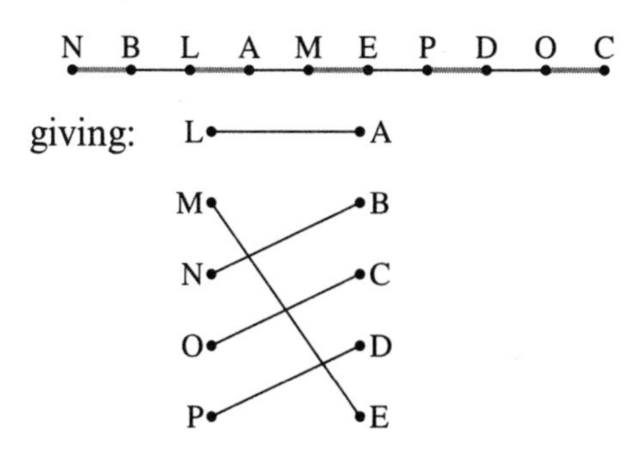

This is a complete matching.

6

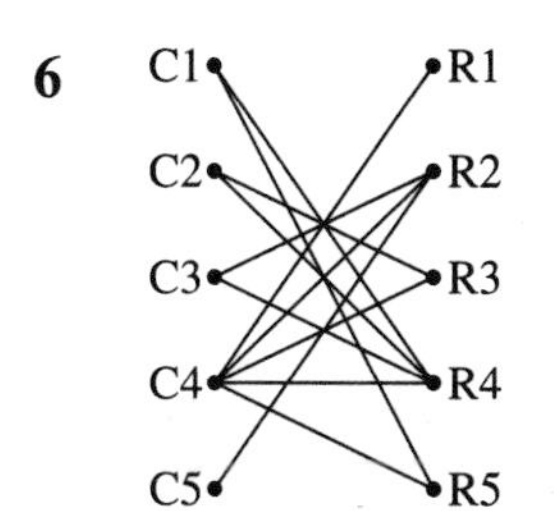

Starting with C1 (say) → C1–R4 breakthrough

Restarting with R1 (say) → R1–C4 breakthrough

Restarting with C2 (say) → C2–R3 breakthrough

Restarting with R2 (say) → R2–C3 breakthrough

Restarting from C5 (say):

Changing status:

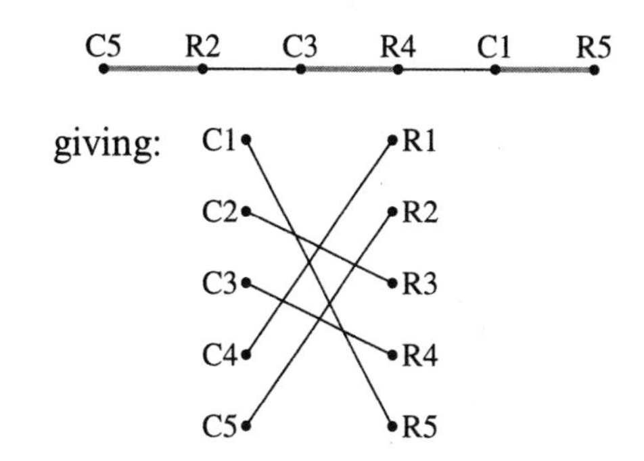

7 (a)

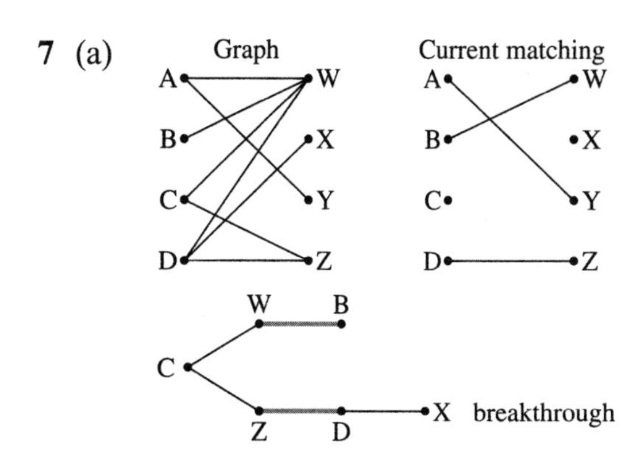

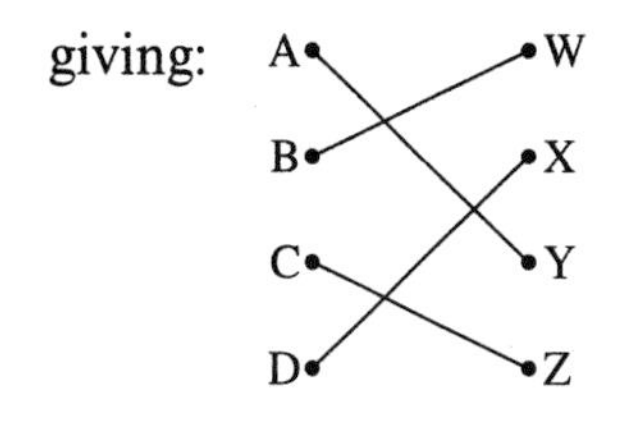

(b)

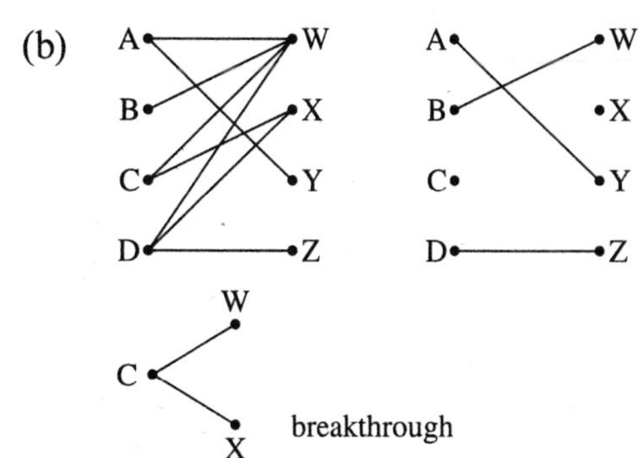

giving:

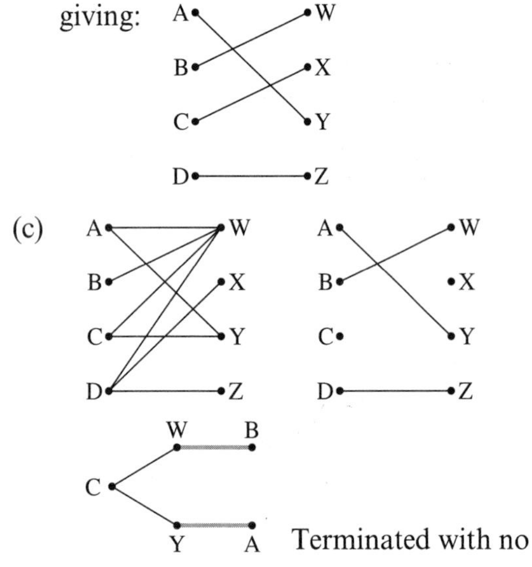

Terminated with no breakthrough, so current matching is maximal.

8 (a) e.g. SB = 1 Be = 1 mT = 1
SC = 1 Cst = 1 stT = 1
SD = 1 Dm = 1 scT = 1
SE = 1 Esc = 1 eT = 1
Max. flow = 4

(b) This solution to (a) implies a solution to the matching problem, i.e.
B $\leftrightarrow$ e
C $\leftrightarrow$ st
D $\leftrightarrow$ m
E $\leftrightarrow$ sc
There is a 1–1 correspondence between solutions to the transmission problem and solutions to the matching problem. The problems are equivalent.

Exercise A1

1 (a) A Hamilton cycle must visit every vertex, including the vertex of order 1. Any attempt to construct a cycle visiting a vertex of order 1 will involve revisiting the vertex to which it is connected. Revisiting is not allowed in a cycle.

(b) ABCADA; length = 17

(c)

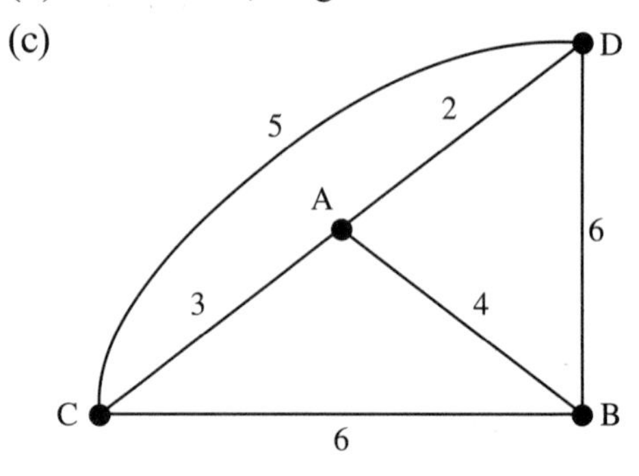

(d) ABCDA; length = 17
Corresponds to the tour given in (b) above.

2 (a) AEBCDA (or ADCBEA); cost = £350.

(b) e.g. ABCDBEA; £340

(c)

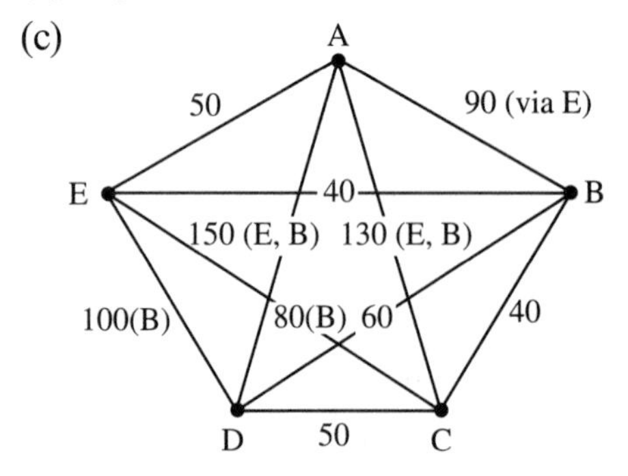

(d) e.g. AEBCDA; £330
corresponds to AEBCDBEA

Exercise A2

1 (a) ABCFED – algorithm terminates (C already visited)

(b) no

(c) yes, e.g. ABDEFCA

(d) ABCFED – algorithm terminates (B and C already visited)

(e) CFEDBAC – same as that given in (c) above

2 ADCBA; weight = 15
BCDAB; weight = 15
CDABC; weight = 15
DACBA; weight = 16
but ACBDA has weight 13

3 starting at A: AFBCDEA with length 99

starting at B:	98
starting at C:	96
starting at D:	97
starting at E:	94
starting at F:	97

So best Hamilton cycle found by this method is EDABFCE, with length 94.

4 (a) TADEBCFT;
length $= 4 + 3 + 4 + 3 + 4 + 3 + \sqrt{153}$
$= 21 + \sqrt{153} \approx 33.4$

(b) TABCFEDT;
length $= 4 + 4 + 4 + 3 + 4 + 4 + 5 = 28$

(c) anywhere on the cycle ABCFEDA

5 ABDECFA; cost = 610
cf. ABDEFCA; cost = 507

6 (a) Any attempt to visit all sites must involve revisiting Aptera. Revisiting is not allowed in a cycle.

(b) KMNGIK$_Z$GPFAK; length = 699 km
Insert after A: Aptera to Kandanos to Falasarna to Aptera. This is an extra 231 km, giving 930 km.

Examination style paper D1a

1 J E C B S A **T**
E C B J A **S**
C B E A **J**
B C A **E**
B A **C**
A **B**

Adams, Basra, Chung, Evans, Jones, Smith, Takashima

2

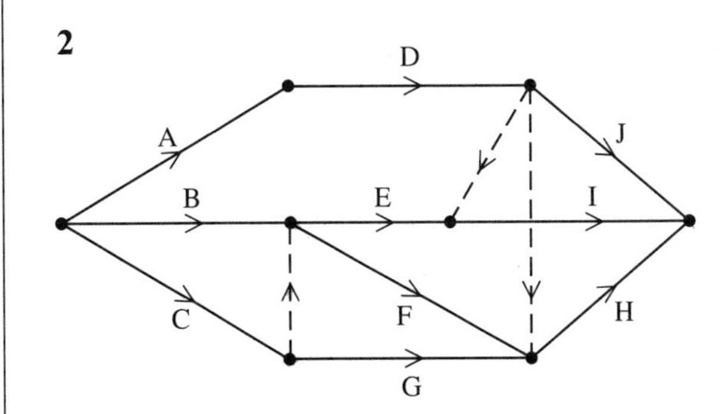

3 (a) Max flow 13

(b)

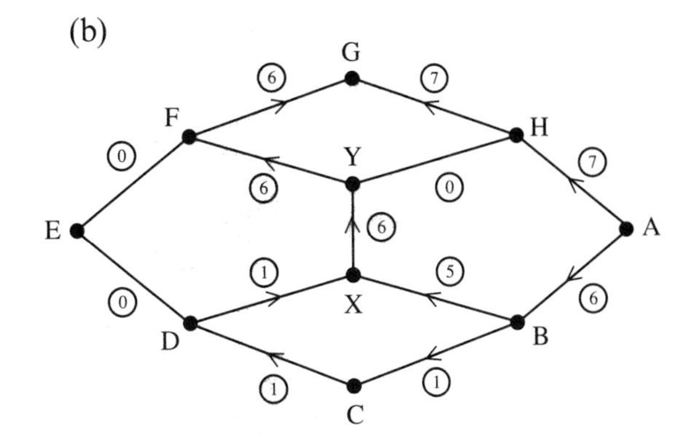

(c) Cut HG, YF, $\left\{\begin{array}{c}\text{DE}\\ \text{or}\\ \text{EF}\end{array}\right\}$

or AH, XY, $\left\{\begin{array}{c}\text{DE}\\ \text{or}\\ \text{EF}\end{array}\right\}$

4 (a) ABECDA (27), ABCEDA (30), ABCDEA (32), AEBCDA (26)

(b)

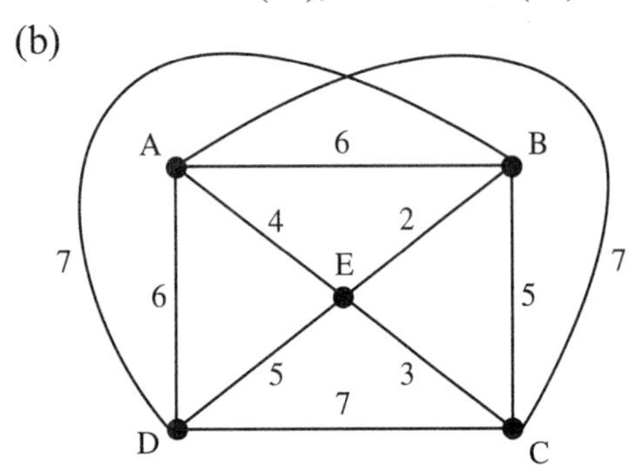

(c) AEBCDA length 24

(d) Corresponds to AEBECDA.

5 (a)

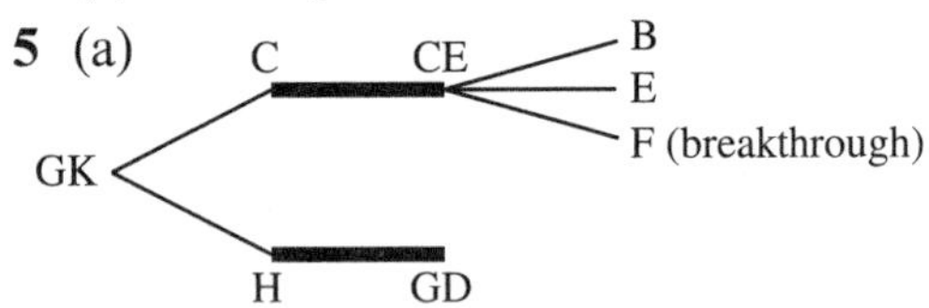

Implies Cathy → GK and Freda → CE, giving:
Aretha → GS, Bhavni → WA, Cathy → GK, Eve → WD, Freda → CE, Gaynor → GA, Helen → GD

(b)

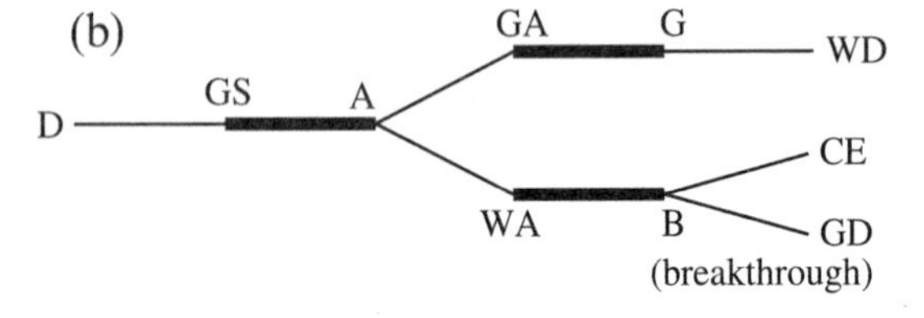

Implies Deepna → GS, Aretha → WA, Bhavni → GD, giving:
Aretha → WA, Bhavni → GD, Cathy → GK, Deepna → GS, Eve → WD, Freda → CE, Gaynor → GA

6 (a) $x \geqslant 24$, $y \geqslant 6$, $x \geqslant 2y$, $x + 3y \leqslant 80$

(b) $P = 1.5x + 7.5y$

$x = 32$, $y = 16$ Max. profit = £168

7 (a)

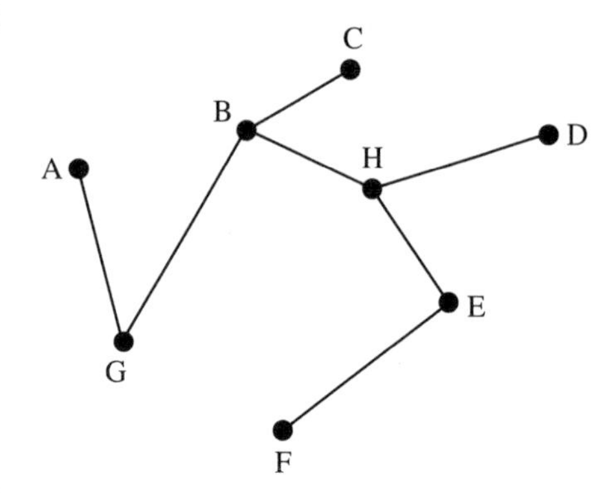

Length 89 (890 m)

(b), (c)

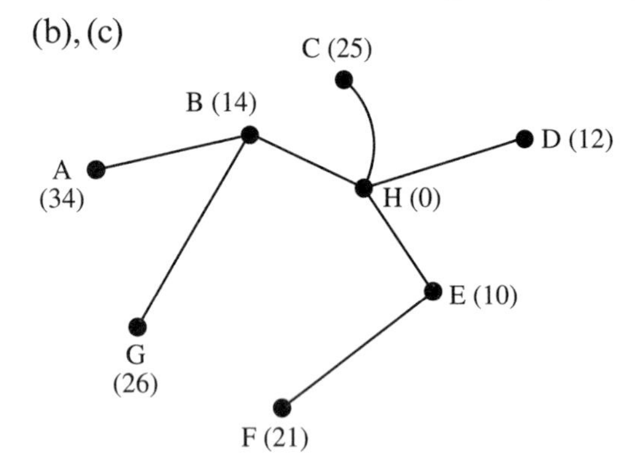

(d) Length 104, extra length 15 (1040 m, 150 m)

8 (a)

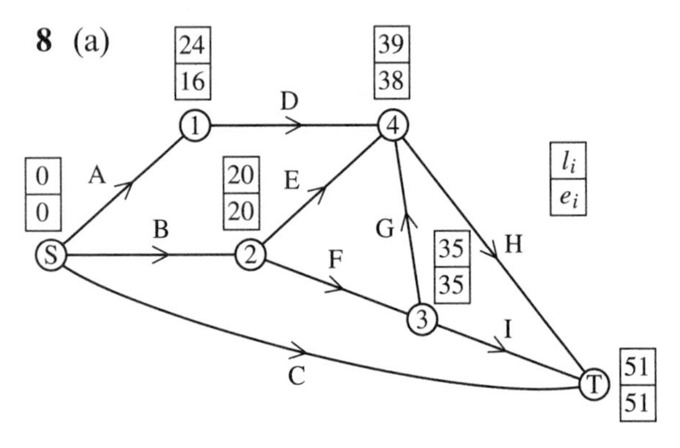

Length of critical path 51

(b) Critical events S, 2, 3, T
Critical activities B, F, I
Critical path S → 2 → 3 → T

(c) (i)

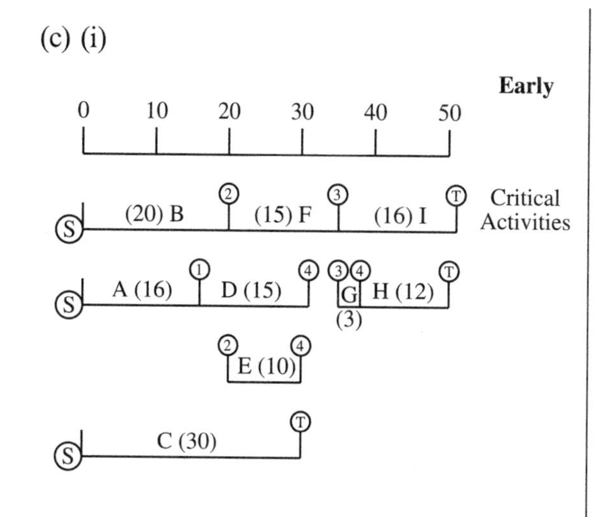

(ii)

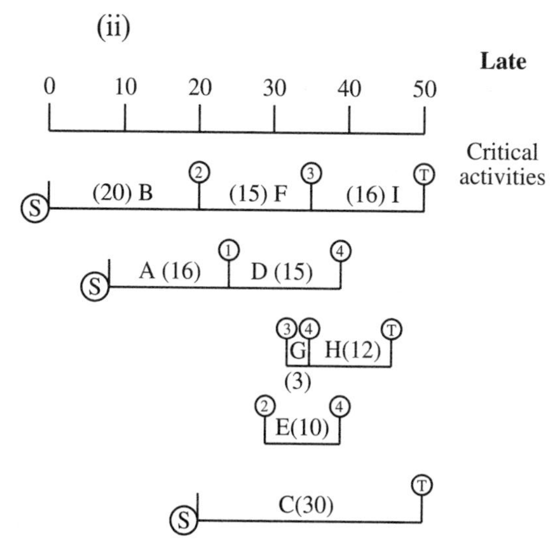

(d) 4 workers required

Examination style paper D1b

1

N	i	N divisible by i?
119	2	✗
	3	✗
	4	✗
	5	✗
	6	✗
	7	✓

119 is not prime. It is divisible by 7.

2 (a) Order size of packages in *decreasing* size.
Fit package into first available bin which has sufficient spare capacity.

(b) Total length of vehicles 59 m. Hence a minimum of 3 lanes, 60 m, are required.

(c) At least the vehicle of length 3 m will have to go into a 4th lane.

3 (a)

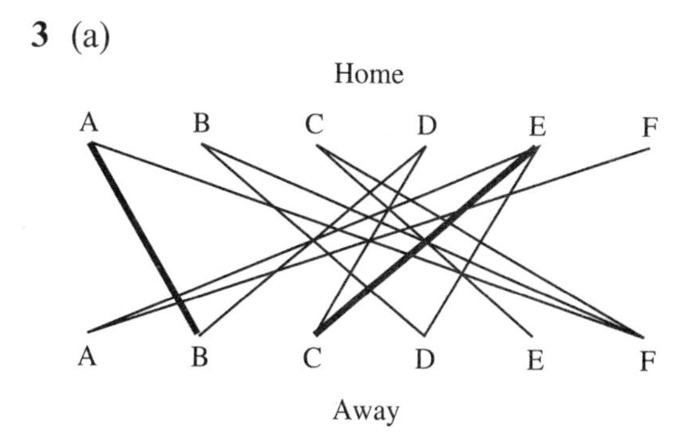

(b) Would imply D playing in two matches, one home and the other away at the same time.

(c) Home D — Away B — Home A — Away F

D_H —— B_A —— A_H —— F_A (breakthrough)

D at home to B
A at home to F
E at home to C

4 (a) Old grinding capacity can be represented by a single edge AB of capacity 40 (= 20 + 20).
New polishing capacity can be represented by a single edge CD of capacity 80 (= 40 + 40).

(b)

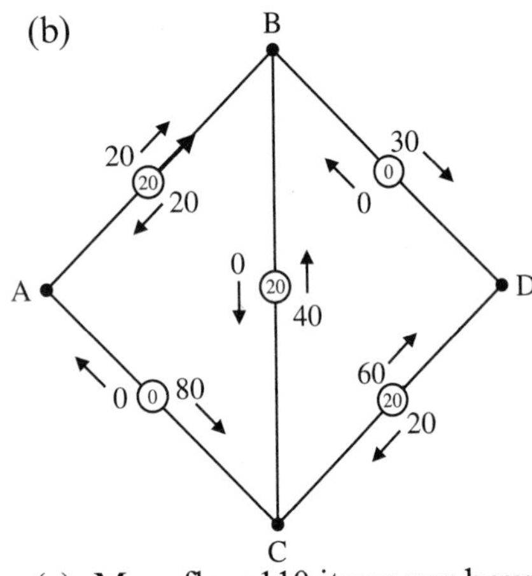

(c) Max. flow 110 items per hour

Flow pattern

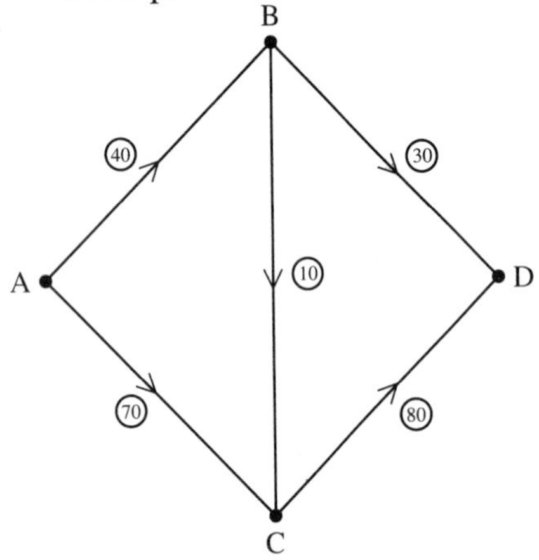

(d) Min. cut ABC/D with capacity 110

5 (a) $x \leqslant 3$, $y \leqslant 4$, $x + y \leqslant 5$, $3x + 2y \geqslant 10$

(b)

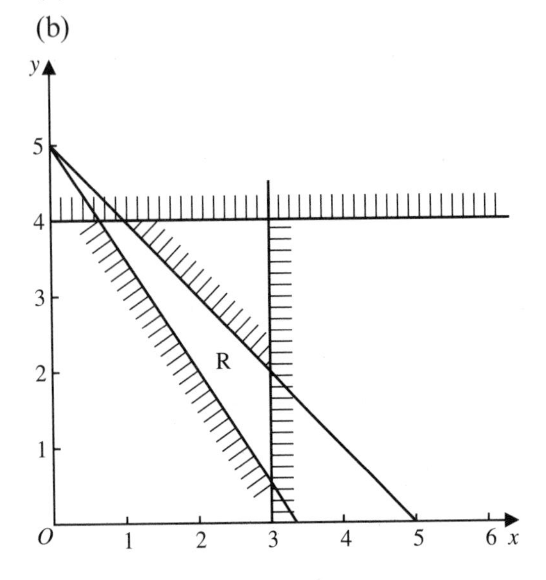

(c) $(1, 4)$ $(2, 2)$ $(2, 3)$ $(3, 1)$ $(3, 2)$

(d) Cost £C: $C = 50x + 40y$

$(1, 4) = 50 + 160 = 210$

$(2, 2) = 100 + 80 = 180$ *

$(2, 3) = 100 + 120 = 220$

$(3, 1) = 150 + 40 = 190$

$(3, 2) = 150 + 80 = 230$

Minimum cost £180; $x = 2$, $y = 2$

6

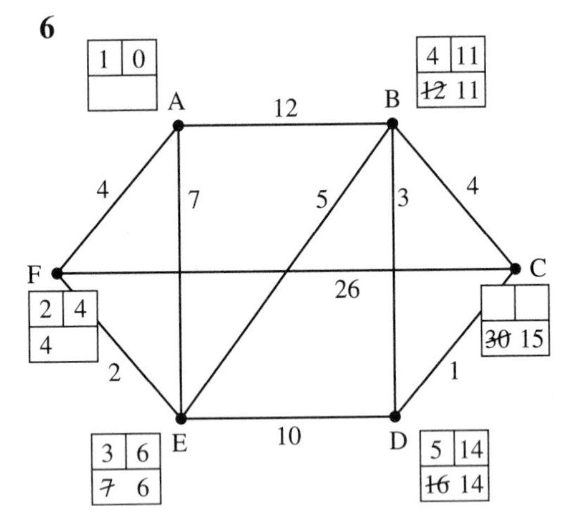

Shortest route is AFEBD, of length 14

(b) CD, EF, BD, AF, (not BC), EB

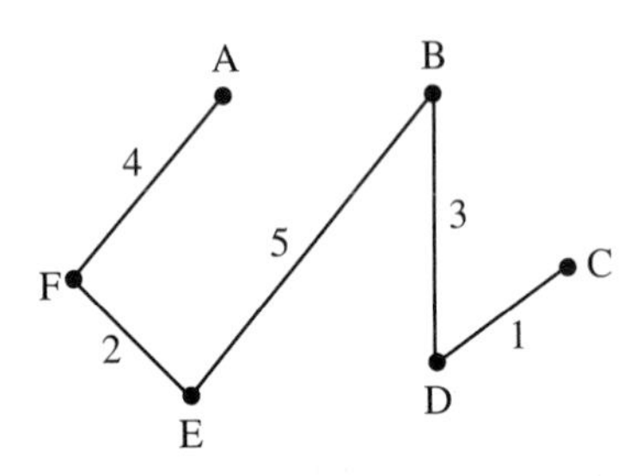

Total length of tree is 15.

7 Optimal tableau

	x	y	z	r	s	
y	$\frac{1}{8}$	1	0	$\frac{1}{4}$	$-\frac{1}{8}$	9
z	$\frac{1}{4}$	0	1	0	$\frac{1}{4}$	2
P	2	0	0	1	2	56

Maximum value of $P = 56$ occurs when $x = 0$, $y = 9$ and $z = 2$.

8 (a) Odd nodes B, C, E and F
ABXAYCAEDB*DC*DFE*FA
* not sewing
Length 300

(b) BC direct instead of BDC repeated, saving 10.

(c) Choose a set of pairings such that the sum of all the extra links, except the longest, is as small as possible. Start at one of the end vertices of the longest link and finish at the other.

(d) BXABDEAYCAFE*FDC with length 280

List of symbols and notation

The following symbols and notation are used in the London modular mathematics examinations:

$\{\quad\}$	the set of
$\mathrm{n}(A)$	the number of elements in the set A
$\{x: \quad\}$	the set of all x such that
$\in$	is an element of
$\notin$	is not an element of
$\varnothing$	the empty (null) set
$\mathscr{E}$	the universal set
$\cup$	union
$\cap$	intersection
$\subset$	is a subset of
A'	the complement of the set A
PQ	operation Q followed by operation P
$\mathrm{f}: A \rightarrow B$	f is a function under which each element of set A has an image set in B
$\mathrm{f}: x \mapsto y$	f is a function under which x is mapped to y
$\mathrm{f}(x)$	the image of x under the function f
f^{-1}	the inverse relation of the function f
fg	the function f of the function g
o—o	open interval on the number line
•—•	closed interval on the number line
$\mathbb{N}$	the set of positive integers and zero, $\{0, 1, 2, 3, \ldots\}$
$\mathbb{Z}$	the set of integers, $\{0, \pm1, \pm2, \pm3, \ldots\}$
$\mathbb{Z}^+$	the set of positive integers, $\{1, 2, 3, \ldots\}$
$\mathbb{Q}$	the set of rational numbers
$\mathbb{Q}^+$	the set of positive rational numbers, $\{x : x \in \mathbb{Q}, x > 0\}$
$\mathbb{R}$	the set of real numbers
$\mathbb{R}_0^+$	the set of positive real numbers, $\{x : x \in \mathbb{R}, x > 0\}$
$\mathbb{C}$	the set of complex numbers
$\sqrt{\ }$	the positive square root
$[a, b]$	the interval $\{x : a \leqslant x \leqslant b\}$
$(a, b]$	the interval $\{x : a < x \leqslant b\}$
(a, b)	the interval $\{x : a < x < b\}$

$\lvert x \rvert$	the modulus of $x = \begin{cases} x \text{ for } x \geqslant 0 \\ -x \text{ for } x < 0 \end{cases}, x \in \mathbb{R}$
$\approx$	is approximately equal to
A^{-1}	the inverse of the non-singular matrix A
A^T	the transpose of the matrix A
det A	the determinant of the square matrix A
$\sum_{r=1}^{n} f(r)$	$f(1) + f(2) + \ldots + f(n)$
$\prod_{r=1}^{n} f(r)$	$f(1)f(2)\ldots f(n)$
$\binom{n}{r}$	the binomial coefficient $\dfrac{n!}{r!(n-r)!}$ for $n \in \mathbb{Z}^+$; $\dfrac{n(n-1)\ldots(n-r+1)}{r!}$ for $n \in \mathbb{Q}$
$\exp x$	e^x
$\ln x$	the natural logarithm of x, $\log_e x$
$\lg x$	the common logarithm of x, $\log_{10} x$
arcsin	the inverse function of sin with range $[-\pi/2, \pi/2]$
arccos	the inverse function of cos with range $[0, \pi]$
arctan	the inverse function of tan with range $(-\pi/2, \pi/2)$
arsinh	the inverse function of sinh with range $\mathbb{R}$
arcosh	the inverse function of cosh with range $\mathbb{R}_0^+$
artanh	the inverse function of tanh with range $\mathbb{R}$
$f'(x), f''(x), f'''(x)$	the first, second and third derivatives of $f(x)$ with respect to x
$f^{(r)}(x)$	the rth derivative of $f(x)$ with respect to x
$\dot{x}, \ddot{x}, \ldots$	the first, second, ... derivatives of x with respect to t
z	a complex number, $z = x + iy = r(\cos\theta + i\sin\theta) = re^{i\theta}$
Re z	the real part of z, $\text{Re}\, z = x = r\cos\theta$
Im z	the imaginary part of z, $\text{Im}\, z = y = r\sin\theta$
z^*	the conjugate of z, $z^* = x - iy = r(\cos\theta + i\sin\theta) = re^{-i\theta}$
$\lvert z \rvert$	the modulus of z, $\lvert z \rvert = \sqrt{(x^2 + y^2)} = r$
arg z	the principal value of the argument of z, $\arg z = \theta$, where $\left.\begin{matrix} \sin\theta = y/r \\ \cos\theta = x/r \end{matrix}\right\} -\pi < \theta \leqslant \pi$
a	the vector **a**
$\overrightarrow{AB}$	the vector represented in magnitude and direction by the directed line segment AB
$\hat{\mathbf{a}}$	a unit vector in the direction of **a**
i, **j**, **k**	unit vectors in the directions of the cartesian coordinate axes
$\lvert\mathbf{a}\rvert$	the magnitude of **a**
$\lvert\overrightarrow{AB}\rvert$	the magnitude of $\overrightarrow{AB}$
a.b	the scalar product of **a** and **b**
$\mathbf{a} \times \mathbf{b}$	the vector product **a** and **b**

A'	the complement of the event A
$\mathrm{P}(A)$	probability of the event A
$\mathrm{P}(A\|B)$	probability of the event A conditional on the event B
$\mathrm{E}(X)$	the mean (expectation, expected value) of the random variable X
X, Y, R, etc.	random variables
x, y, r, etc.	values of the random variables X, Y, R, etc.
$x_1, x_2 \ldots$	observations
$f_1, f_2, \ldots$	frequencies with which the observations $x_1, x_2 \ldots$ occur
$\mathrm{p}(x)$	probability function $\mathrm{P}(X = x)$ of the discrete random variable X
$p_1, p_2, \ldots$	probabilities of the values $x_1, x_2 \ldots$ of the discrete random variable X
$\mathrm{f}(x), \mathrm{g}(x), \ldots$	the value of the probability density function of a continuous random variable X
$\mathrm{F}(x), \mathrm{G}(x), \ldots$	the value of the (cumulative) distribution function $\mathrm{P}(X \leqslant x)$ of a continuous random variable X
$\mathrm{Var}(X)$	variance of the random variable X
$\mathrm{B}(n, p)$	binomial distribution with parameters n and p
$\mathrm{N}(\mu, \sigma^2)$	normal distribution with mean μ and variance σ^2
μ	population mean
σ^2	population variance
σ	population standard deviation
$\bar{x}$	sample mean
s^2	unbiased estimate of population variance from a sample, $s^2 = \frac{1}{n-1}\sum (x - \bar{x})^2$
ϕ	probability density function of the standardised normal variable with distribution $\mathrm{N}(0, 1)$
Φ	corresponding cumulative distribution formation
α, β	regression coefficients
ρ	product-moment correlation coefficient for a population
r	product-moment correlation coefficient for a sample
$\sim p$	not p
$p \Rightarrow q$	p implies q (if p then q)
$p \Leftrightarrow q$	p implies and is implied by q (p is equivalent to q)

Index

activities 99
  critical 116
  total float of 116
activity networks 103–110
  drawing 104–5
  dummies in 107–110
  example 126–8
admissible sets 169
algebraic methods, for linear programming problems 189–91
algorithms 1–20
  bin packing 13–18
  bubble sort 7–10
  complexity 62
  critical path analysis 111–14
  definition 1
  Dijkstra's 63–6, 233
  enumeration 62
  Euclidean 5–6
  examples 2–6
  first fit 14–15
  first fit decreasing 15–18
  flow augmentation 148
  flow charts for 4–7
  on graphs 21–71
  greedy 55, 235–6
  heuristic 15, 234
  Hungarian A5
  Kruskal's 52–5
  labelling 63, 141
  matching improvement 212–20
  nearest neighbour 234–6
  Prim's 46–52
  quick sort 10–12
  route inspection 92–3
  Russian peasant's 3
  sorting 7–12
allocation problems 227
alternating path 213
arcs 21
  in activity networks 103

basic feasible solution 191
basic solutions 191–2
basic variables 191, 193
bin packing algorithms 13–18
bipartite graphs 203–211
bounds 76–89, 234–6
breakthrough 213
bubble sort algorithm 7–10

capacities, and flows 134–6
changing status 214–17
Chinese postman problem 90–3
circuit 26
complete matching 209
complexity 62, 233
connections, in graphs 26
constraints 160
  integer 183–7
  non-negativity 161–2
  in tableaux 193–9
CPM *see* critical path analysis
critical activities 116
critical events 116
critical path analysis 99–131
  algorithm 111–15
  applications 99
cut capacity 137
cycle, definition 26

decision making
  applications 159
  in graphs 73–98, 229–38
  historical background 159
decision variables 160
  integers 183
degree, of vertex 27
dependence table 100
digraph 27
Dijkstra's algorithm 63–6, 233
directed edges 27

dummies, in activity networks 107–10
duration, activity networks 103

earliest finish time 117
earliest start time 117
earliest time 112
edge set 26
edges 21
  directed 27
  undirected 139–41
  weighted 37–8
electricity distribution
  network analysis 133
  see also transmission systems
entering variable 194
enumeration algorithm 62
Euler, Leonard (1707–83), graph theories 34
even vertex 27
events
  activity networks 103–4
  critical 116
extreme points, and optimal solutions 181–3

feasible region 172
  and optimal solutions 181
feasible solutions, linear programming 172
first fit algorithm 14–15
first fit decreasing algorithm 15–18
float 116, 119–28
flow augmentation algorithm 148
flow augmenting paths 141
  identification 148–9
flow charts
  for algorithms 4–8
  symbols 4
flows
  and capacities 134–6
  in networks 133-57
  in undirected edges 147–9
full-bin combinations 14

generator maintenance, critical path analysis 99
graphical solutions, of linear programming problems
  157–62
graphs
  algorithms on 21–71
  bipartite 203–10
  decision making in 73–98
  in linear programming problems 167–72
  in problem solving 21–5
  representations 25–8
  straight line 167–72
greedy algorithms 47, 74, 235–6

Hamilton cycle 75–6, 230–3
  definition 75
  minimum 82–3
head event, definition 104
heuristic algorithms 15, 73, 234
Huffman code 40
Hungarian algorithm 227
incidence matrix, definition 32
inequalities, in linear programming problems 170–2
initial tableau 193
integer constraints 183
integer-valued solutions, linear programming problems
  183–9
iteration 8

Kruskal's algorithm 52–5

labelling algorithms 63, 141
latest finish time 117
latest start time 117
latest time 111
leaving variable 194
linear equations 167
linear inequality 167
linear programming 159–202, 223–9
  algebraic solutions 189–92
  basic solutions 191–2
  constraints 160–5
  feasible solutions 172
  graphical solutions 167–72
  integer-valued solutions 183–9
  in matching 228–9
  maximum flow problems 225–6
  and networks 223–7
  optimal solutions 172–82
  problem formulation 160–7
  simplex method 192–202
  in transmission systems 225–6
longest paths 111
loops, definition 4
lower bounds 72–85, 236

matching improvement algorithm 212–20
matchings 203–221
  alternating path 213–20
  breakthrough 213
  complete 209
  definition 208
  linear programming in A6–7
  maximal 208
  non-maximal 212
  status change 214–17
mathematical modelling 41–3
matrices, Prim's algorithm 48–51

maximal matchings 208
maximum flow augmentation 141–5
maximum flow problems, linear programming 225–6
maximum flow–minimum cut theorem 137–9
  applications 139–40
minimum connector *see* minimum spanning tree
minimum spanning tree 43–55
modelling
  bipartite graphs 203–6
  complex projects 99
  mathematical 41–3
MST *see* minimum spanning tree
multiplication, algorithms for 3–4

nearest neighbour algorithm 234–6
network analysis *see* critical path analysis
networks 37–8
  flows in 133–57
  Hamilton cycles 82–4, 230–1
  and linear programming 223–6
  shortest paths 59–63, 223–5, 233
  time analysis 117–20
  see also activity networks
nodes 21
  activity networks 103
non-basic variables 191
non-maximal matchings 212
non-negativity constraints 161
NP complete problems 234

objective function 160
objective line 180
odd vertex 27
optimal solution 159
  and extreme points 181–3
  linear programming problems 172–82
optimal tableaux 197

partition, of sets 203
path
  alternating 213–20
  definition 25
  flow augmenting 141, 148–9
  longest 111
  shortest 59–66, 223–5, 233
  see also critical path analysis
pivot 195
pivotal columns 194
pivotal row 195
precedence tables 99–103
Prim's algorithm 46–52
problem solving, graphs in 21–5
programming, linear 159–202
projects
  critical path analysis 111–14
  modelling 99

quick sort algorithm 10–12

resource levelling 123–8
route inspection problem 90–3
Russian peasant's algorithm 3

scheduling 120–3
sets
  admissible 169
  partition of 203
  of points 170
shortcuts 72
shortest paths, determination 59–66, 223–5, A11
simplex method 192–201
sink node 103
slack 116
slack variables 189–91
sorting, algorithms 7–12
source node 103
straight line graphs, linear programming 167–72
subgraph 26

tableaux 192–201
  formation 195–201
  initial 193
  optimal 197
tail event 104
time
  direction of flow 104
  earliest 112–13
  latest 113–14
time analysis, networks 117–28
total float
  of activities 116
  meaning of 119–20
  using 120–8
tour 74
  length 75–89
trail 25
transmission systems
  linear programming A3–4
  network analysis 133–4
transportation problems 227
travelling salesman problem (TSP) 73–89
  and complexity 75–6, 233
trees 36–7
  minimum spanning 43–59

undirected edges, flows in 147–9
upper bounds 77–9, 234

valency, of vertices 27
vertex
  activity networks 103
  definition 21
  degrees of 27
  even 27
  odd 27
  order 27
  pairing 85–6
  weighting 227
  vertex set 26
viability, definition 73, 233

walk 25
weighted edges, concept of 37–8
weights, activity networks 103